# METHODS IN MOLECULAR BIOLOGY

*Series Editor*
**John M. Walker**
**School of Life and Medical Sciences**
**University of Hertfordshire**
**Hatfield, Hertfordshire, AL10 9AB, UK**

For further volumes:
http://www.springer.com/series/7651

# CRISPR

## Methods and Protocols

Edited by

## Magnus Lundgren

*Department of Cell and Molecular Biology, Uppsala University, Uppsala, Sweden*

## Emmanuelle Charpentier

*Helmholtz Centre for Infection Research, Department of Regulation in Infection Biology, Braunschweig, Germany; The Laboratory for Molecular Infection Medicine Sweden (MIMS), Umeå Centre for Microbial Research (UCMR), Department of Molecular Biology, Umeå University, Umeå, Sweden; Hannover Medical School, Hannover, Germany*

## Peter C. Fineran

*Department of Microbiology and Immunology, University of Otago, Dunedin, New Zealand*

*Editors*
Magnus Lundgren
Department of Cell and Molecular Biology
Uppsala University
Uppsala, Sweden

Peter C. Fineran
Department of Microbiology and Immunology
University of Otago
Dunedin, New Zealand

Emmanuelle Charpentier
Helmholtz Centre for Infection Research
Department of Regulation in Infection Biology
Braunschweig, Germany

The Laboratory for Molecular Infection
Medicine Sweden (MIMS)
Umeå Centre for Microbial Research (UCMR)
Department of Molecular Biology
Umeå University
Umeå, Sweden

Hannover Medical School
Hannover, Germany

ISSN 1064-3745     ISSN 1940-6029  (electronic)
Methods in Molecular Biology
ISBN 978-1-4939-2686-2     ISBN 978-1-4939-2687-9  (eBook)
DOI 10.1007/978-1-4939-2687-9

Library of Congress Control Number: 2015937760

Springer New York Heidelberg Dordrecht London
© Springer Science+Business Media New York 2015
This work is subject to copyright. All rights are reserved by the Publisher, whether the whole or part of the material is concerned, specifically the rights of translation, reprinting, reuse of illustrations, recitation, broadcasting, reproduction on microfilms or in any other physical way, and transmission or information storage and retrieval, electronic adaptation, computer software, or by similar or dissimilar methodology now known or hereafter developed.
The use of general descriptive names, registered names, trademarks, service marks, etc. in this publication does not imply, even in the absence of a specific statement, that such names are exempt from the relevant protective laws and regulations and therefore free for general use.
The publisher, the authors and the editors are safe to assume that the advice and information in this book are believed to be true and accurate at the date of publication. Neither the publisher nor the authors or the editors give a warranty, express or implied, with respect to the material contained herein or for any errors or omissions that may have been made.

Photo Credit: Magnus Lundgren and Stan Brouns

Printed on acid-free paper

Humana Press is a brand of Springer
Springer Science+Business Media LLC New York is part of Springer Science+Business Media (www.springer.com)

# Preface

All cellular organisms have developed strategies to combat parasitic genome invaders. In many bacteria and most archaea, CRISPR (*C*lustered *R*egularly *I*nterspaced *S*hort *P*alindromic *R*epeats)-Cas (*CR*ISPR-*as*sociated proteins) has recently emerged as an adaptive RNA-mediated defense mechanism that provides cells with acquired resistance to invading foreign genetic elements such as bacteriophages and plasmids. These defense systems are composed of Cas proteins, and one or more CRISPR arrays containing a collection of short repeat-spacer sequences that provide the CRISPR RNAs (crRNAs) that target the foreign genome. Immunity to invading genomes is mediated via a three-stage process: (1) adaptation via acquisition of viral or plasmid DNA-derived spacers into the CRISPR array, (2) expression and maturation of the short crRNAs, and (3) interference with the invading cognate foreign genomes.

The repeat-spacer CRISPR array contents and *cas* genes have undergone an extraordinarily dynamic rate of evolution leading to a complex diversity in the genetic architectures of the CRISPR-Cas systems. Indeed, currently there are three major types (I–III) consisting of more than 11 subtypes. Although the general principles of the three-stage immunity process are conserved, CRISPR-Cas has evolved a wide variety of precise and coordinated mechanisms that are mainly based on the Cas proteins involved. Interestingly, recent structural studies of Cas protein interference complexes in type I and III systems have demonstrated an overall similarity in architecture, despite a large difference in subunit sequence and composition.

CRISPR-Cas adaptive immune systems are currently one of the most flourishing and fascinating research themes in microbiology and much work has been performed in the last 10 years. Important findings have recently been made towards the understanding of the molecular mechanisms of CRISPR-Cas systems mediating crRNA expression and interference in many select CRISPR-Cas systems. In the last couple of years, we have also witnessed a rapid growth in our understanding of the process of spacer acquisition, although the mechanistic details are largely unknown. Other details around, for example, CRISPR-Cas regulation and the role of the systems in other processes remain obscure. By their remarkable mechanistic and functional diversity, CRISPR-Cas systems constitute an extraordinary source of proteins for RNA-DNA, DNA-protein, RNA-protein, and protein-protein interactions that can be developed for applied use. The applications are diverse and include the ability to repress or activate the transcription of genes of interest. The best known application to date is genome editing based on the type II CRISPR-Cas9 system. It is not an overstatement that the ability to use a single Cas9 protein with a guide RNA to direct site-specific cleavage of target DNA has revolutionized genome engineering in organisms ranging from bacteria and yeast to plants and animals. The therapeutic potential of CRISPR-Cas9 has been demonstrated by, e.g., curing genetic defects in adult mice, and there is a large interest in developing CRISPR-Cas9 for gene therapy in humans. These advances have resulted in the CRISPR-Cas9 technology being adopted as a common genetic tool in the scientific community in a very short time.

The purpose of this Methods in Molecular Biology volume is to provide a list of cutting-edge protocols for the study of CRISPR-Cas defense systems at the genomic, genetic, biochemical, and structural levels. A number of the tools and techniques described in this book have been developed specifically for the analysis of CRISPR-Cas; others have been adapted from more standard protocols of DNA, RNA, and protein biology. Here, we wish to offer a list of protocols to any scientist who would like to analyze CRISPR-Cas in their favorite bacterial or archaeal species or use CRISPR-based genetic tools. The large panel of techniques included in this volume should also be of high interest for bacterial or archaeal geneticists who study mobile genetic elements and horizontal gene transfer in general as well as researchers engaged in the analysis of general mechanisms of DNA/protein, RNA/protein, and protein/protein interactions. This list of methods is complemented by the description of genome editing protocols using the CRISPR-Cas9 technology.

Chapter 1 describes and compares methodologies of RNA sequencing to investigate crRNA biogenesis and function. Chapter 2 discusses methods to reconstitute Cas protein complexes in vitro. Chapter 3 details a method to analyze cleavage of precursor crRNA by Cas endonucleases. In Chapter 4, a methodology to annotate and classify CRISPR-Cas systems is detailed. In Chapter 5, instructions for use of a web-based bioinformatic tool for the identification of targets of the crRNAs from CRISPR arrays are illustrated. Chapters 6 and 7 describe the use of CRISPRs for quick typing of bacteria. Chapter 8 describes the use of mass spectrometry to analyze crRNA. Chapter 9 describes the ability to use primed adaptation plasmids to rapidly generate CRISPR arrays with spacers targeting the cloned region of interest. Chapter 10 provides a CRISPR-Cas plasmid transformation assay to quantitatively assess interference. Two different assays are outlined in Chapter 11 that enables the accurate measurement of Cas complex interactions with DNA in electrophoretic mobility shift assays. Chapter 12 describes the expression and purification of CMR complexes in *Sulfolobus*. Chapters 13 and 14 detail methods to generate and detect spacer acquisition into CRISPR arrays in laboratory phage-host challenge experiments in bacteria and archaea. A positive selection method to identify genes that are required for CRISPR-Cas activity is outlined in Chapter 15. Chapters 16 and 17 detail how to assay nuclease activity of Cas proteins while Chapter 18 describes how to perform helicase assays on Cas3. Chapter 19 provides methods for characterizing details of Cas complexes binding to target DNA. In Chapter 20 the bioinformatic workflow to use CRISPR spacers to create and analyze viromes from metagenomic data is described. Chapters 21 and 22 detail protocols for targeted mutagenesis in zebrafish and *Drosophila*, respectively, using CRISPR-Cas9. Finally, Chapter 23 describes how to use dCas9 for transcriptional repression.

In summary, this timely volume provides a broad list of tools and techniques to study the interdisciplinary aspects of the prokaryotic CRISPR-Cas defense systems. The detailed protocols should facilitate the experimental work of the CRISPR biologists who would like to investigate the activation, expression, and regulation of the CRISPR-Cas system, and the mechanisms of adaptation and interference with alien genomes. Recently, a number of bacterial and archaeal geneticists, bioinformaticians, biochemists, and structural biologists specialized in DNA, RNA, or/and protein biology have converged to the field of CRISPR-Cas research. It is our hope that the technical guidelines described in this volume will stimulate additional scientists to join the field and contribute to the understanding of the CRISPR-Cas immune system.

We would like to thank all the authors, who have shared their expertise and contributed the detailed protocols for this book, as well as their patience during the editing process. We are also grateful to the series editor John M. Walker without whom this book would not have been brought to completion.

*Uppsala, Sweden*   *Magnus Lundgren*
*Braunschweig, Germany*   *Emmanuelle Charpentier*
*Umeå, Sweden*
*Hannover, Germany*
*Dunedin, New Zealand*   *Peter C. Fineran*

# Contents

*Preface* .................................................................. *v*
*Contributors* ........................................................... *xi*

1  Investigating CRISPR RNA Biogenesis and Function Using RNA-seq ...... 1
   **Nadja Heidrich, Gaurav Dugar, Jörg Vogel, and Cynthia M. Sharma**

2  In Vitro Co-reconstitution of Cas Protein Complexes ................. 23
   **André Plagens and Lennart Randau**

3  Analysis of CRISPR Pre-crRNA Cleavage .............................. 35
   **Erin L. Garside and Andrew M. MacMillan**

4  Annotation and Classification of CRISPR-Cas Systems ................ 47
   **Kira S. Makarova and Eugene V. Koonin**

5  Computational Detection of CRISPR/crRNA Targets .................... 77
   **Ambarish Biswas, Peter C. Fineran, and Chris M. Brown**

6  High-Throughput CRISPR Typing of *Mycobacterium tuberculosis*
   Complex and *Salmonella enterica* Serotype Typhimurium ............. 91
   **Christophe Sola, Edgar Abadia, Simon Le Hello,
   and François-Xavier Weill**

7  Spacer-Based Macroarrays for CRISPR Genotyping .................... 111
   **Igor Mokrousov and Nalin Rastogi**

8  Analysis of crRNA Using Liquid Chromatography Electrospray
   Ionization Mass Spectrometry (LC ESI MS) .......................... 133
   **Sakharam P. Waghmare, Alison O. Nwokeoji, and Mark J. Dickman**

9  Rapid Multiplex Creation of *Escherichia coli* Strains Capable
   of Interfering with Phage Infection Through CRISPR ................ 147
   **Alexandra Strotksaya, Ekaterina Semenova, Ekaterina Savitskaya,
   and Konstantin Severinov**

10 Exploring CRISPR Interference by Transformation with Plasmid Mixtures:
   Identification of Target Interference Motifs in *Escherichia coli* ..... 161
   **Cristóbal Almendros and Francisco J.M. Mojica**

11 Electrophoretic Mobility Shift Assay of DNA and CRISPR-Cas
   Ribonucleoprotein Complexes ....................................... 171
   **Tim Künne, Edze R. Westra, and Stan J.J. Brouns**

12 Expression and Purification of the CMR (Type III-B) Complex
   in *Sulfolobus solfataricus* ...................................... 185
   **Jing Zhang and Malcolm F. White**

13 Procedures for Generating CRISPR Mutants with Novel Spacers
   Acquired from Viruses or Plasmids ................................. 195
   **Marie-Ève Dupuis, Rodolphe Barrangou, and Sylvain Moineau**

14  Archaeal Viruses of the Sulfolobales: Isolation, Infection,
    and CRISPR Spacer Acquisition . . . . . . . . . . . . . . . . . . . . . . . . . . . . .   223
    *Susanne Erdmann and Roger A. Garrett*

15  Using the CRISPR-Cas System to Positively Select Mutants
    in Genes Essential for Its Function . . . . . . . . . . . . . . . . . . . . . . . . . .   233
    *Ido Yosef, Moran G. Goren, Rotem Edgar, and Udi Qimron*

16  Analysis of Nuclease Activity of Cas1 Proteins Against Complex
    DNA Substrates . . . . . . . . . . . . . . . . . . . . . . . . . . . . . . . . . . . . . . . . .   251
    *Natalia Beloglazova, Sofia Lemak, Robert Flick,
    and Alexander F. Yakunin*

17  Characterizing Metal-Dependent Nucleases of CRISPR-Cas Prokaryotic
    Adaptive Immunity Systems . . . . . . . . . . . . . . . . . . . . . . . . . . . . . . .   265
    *Ki H. Nam, Matthew P. DeLisa, and Ailong Ke*

18  Cas3 Nuclease–Helicase Activity Assays . . . . . . . . . . . . . . . . . . . . . . .   277
    *Tomas Sinkunas, Giedrius Gasiunas, and Virginijus Siksnys*

19  Chemical and Enzymatic Footprint Analyses of R-Loop
    Formation by Cascade-crRNA Complex . . . . . . . . . . . . . . . . . . . . . .   293
    *Ümit Pul*

20  Creation and Analysis of a Virome: Using CRISPR Spacers . . . . . . . . . . . . .   307
    *Michelle Davison and Devaki Bhaya*

21  Targeted Mutagenesis in Zebrafish Using CRISPR
    RNA-Guided Nucleases . . . . . . . . . . . . . . . . . . . . . . . . . . . . . . . . . .   317
    *Woong Y. Hwang, Yanfang Fu, Deepak Reyon, Andrew P.W. Gonzales,
    J. Keith Joung, and Jing-Ruey Joanna Yeh*

22  Precise Genome Editing of Drosophila with CRISPR RNA-Guided Cas9 . . . .   335
    *Scott J. Gratz, Melissa M. Harrison, Jill Wildonger,
    and Kate M. O'Connor-Giles*

23  Targeted Transcriptional Repression in Bacteria Using CRISPR
    Interference (CRISPRi) . . . . . . . . . . . . . . . . . . . . . . . . . . . . . . . . . . .   349
    *John S. Hawkins, Spencer Wong, Jason M. Peters, Ricardo Almeida,
    and Lei S. Qi*

Index . . . . . . . . . . . . . . . . . . . . . . . . . . . . . . . . . . . . . . . . . . . . . . . . . . . . . . .   363

# Contributors

EDGAR ABADIA • *Instituto Venezolano de Investigaciones Científicas (IVIC), Caracas, Venezuela*

RICARDO ALMEIDA • *Department of Cellular & Molecular Pharmacology, University of California, San Francisco, San Francisco, CA, USA; UCSF Center for Systems and Synthetic Biology, University of California, San Francisco, San Francisco, CA, USA; California Institute for Quantitative Biomedical Research, San Francisco, CA, USA*

CRISTÓBAL ALMENDROS • *Departamento de Fisiología, Genética y Microbiología, Universidad de Alicante, Alicante, Spain*

RODOLPHE BARRANGOU • *Food, Bioprocessing and Nutrition Sciences Department, North Carolina State University, Raleigh, NC, USA*

NATALIA BELOGLAZOVA • *Department of Chemical Engineering and Applied Chemistry, University of Toronto, Toronto, ON, Canada*

DEVAKI BHAYA • *Department of Plant Biology, Carnegie Institution for Science, Stanford University, Stanford, CA, USA*

AMBARISH BISWAS • *Department of Biochemistry, University of Otago, Dunedin, New Zealand*

STAN J.J. BROUNS • *Laboratory of Microbiology, Department of Agrotechnology and Food Sciences, Wageningen University, Wageningen, The Netherlands*

CHRIS M. BROWN • *Department of Biochemistry, University of Otago, Dunedin, New Zealand*

MICHELLE DAVISON • *Department of Plant Biology, Carnegie Institution for Science, Stanford University, Stanford, CA, USA*

MATTHEW P. DELISA • *School of Chemical and Biomolecular Engineering, Cornell University, Ithaca, NY, USA*

MARK J. DICKMAN • *Department of Chemical and Biological Engineering, ChELSI Institute, University of Sheffield, Sheffield, UK*

GAURAV DUGAR • *Research Centre for Infectious Diseases (ZINF), University of Würzburg, Würzburg, Germany*

MARIE-ÈVE DUPUIS • *Département de Biochimie, de Microbiologie et de Bio-informatique, Faculté des Sciences et de Génie, Université Laval, Quebec City, QC, Canada; Groupe de Recherche en Écologie Buccale, Faculté de Médecine Dentaire, Félix d'Hérelle Reference Center for Bacterial Viruses, Université Laval, Quebec City, QC, Canada*

ROTEM EDGAR • *Department of Clinical Microbiology and Immunology, Sackler Faculty of Medicine, Tel Aviv University, Tel Aviv, Israel*

SUSANNE ERDMANN • *Archaea Centre, Department of Biology, University of Copenhagen, Copenhagen, Denmark; School of Biotechnology and Biomolecular Sciences, University of New South Wales, Sydney, Australia*

PETER C. FINERAN • *Department of Microbiology and Immunology, University of Otago, Dunedin, New Zealand*

ROBERT FLICK • *Department of Chemical Engineering and Applied Chemistry, University of Toronto, Toronto, ON, Canada*

YANFANG FU • *Molecular Pathology Unit, Center for Cancer Research, Massachusetts General Hospital, Charlestown, MA, USA; Center for Computational and Integrative Biology, Massachusetts General Hospital, Charlestown, MA, USA; Department of Pathology, Harvard Medical School, Boston, MA, USA*

ROGER A. GARRETT • *Archaea Centre, Department of Biology, University of Copenhagen, Copenhagen, Denmark*

ERIN L. GARSIDE • *Department of Biochemistry, University of Alberta, Edmonton, AB, Canada*

GIEDRIUS GASIUNAS • *Institute of Biotechnology, Vilnius University, Vilnius, Lithuania*

ANDREW P.W. GONZALES • *Cardiovascular Research Center, Massachusetts General Hospital, Charlestown, MA, USA; Department of Medicine, Harvard Medical School, Boston, MA, USA*

MORAN G. GOREN • *Department of Clinical Microbiology and Immunology, Sackler Faculty of Medicine, Tel Aviv University, Tel Aviv, Israel*

SCOTT J. GRATZ • *Genetics Training Program, University of Wisconsin-Madison, Madison, WI, USA*

MELISSA M. HARRISON • *Department of Biomolecular Chemistry, University of Wisconsin School of Medicine and Public Health, Madison, WI, USA*

JOHN S. HAWKINS • *Biophysics Graduate Program, San Francisco, CA, USA; Department of Cellular & Molecular Pharmacology, University of California, San Francisco, San Francisco, CA, USA; UCSF Center for Systems and Synthetic Biology, University of California, San Francisco, San Francisco, CA, USA*

NADJA HEIDRICH • *Institute for Molecular Infection Biology (IMIB), University of Würzburg, Würzburg, Germany*

SIMON LE HELLO • *Unité des Bactéries Pathogènes Entériques, Institut Pasteur, Paris, France*

WOONG Y. HWANG • *Cardiovascular Research Center, Massachusetts General Hospital, Charlestown, MA, USA; Department of Medicine, Harvard Medical School, Boston, MA, USA*

J. KEITH JOUNG • *Molecular Pathology Unit, Center for Cancer Research, Massachusetts General Hospital, Charlestown, MA, USA; Center for Computational and Integrative Biology, Massachusetts General Hospital, Charlestown, MA, USA; Department of Pathology, Harvard Medical School, Boston, MA, USA*

AILONG KE • *Department of Molecular Biology and Genetics, Cornell University, Ithaca, NY, USA*

EUGENE V. KOONIN • *National Center for Biotechnology Information, National Library of Medicine, National Institutes of Health, Bethesda, MD, USA*

TIM KÜNNE • *Laboratory of Microbiology, Department of Agrotechnology and Food Sciences, Wageningen University, Wageningen, The Netherlands*

SOFIA LEMAK • *Department of Chemical Engineering and Applied Chemistry, University of Toronto, Toronto, ON, Canada*

ANDREW M. MACMILLAN • *Department of Biochemistry, University of Alberta, Edmonton, AB, Canada*

KIRA S. MAKAROVA • *National Center for Biotechnology Information, National Library of Medicine, National Institutes of Health, Bethesda, MD, USA*

SYLVAIN MOINEAU • *Département de Biochimie, de Microbiologie et de Bio-informatique, Faculté des Sciences et de Génie, Université Laval, Quebec City, QC, Canada; Groupe de Recherche en Écologie Buccale, Faculté de Médecine Dentaire, Félix d'Hérelle Reference Center for Bacterial Viruses, Université Laval, Quebec City, QC, Canada*

Francisco J.M. Mojica • *Departamento de Fisiología, Genética y Microbiología, Universidad de Alicante, Alicante, Spain*

Igor Mokrousov • *Laboratory of Molecular Microbiology, St. Petersburg Pasteur Institute, St. Petersburg, Russia*

Ki H. Nam • *Department of Molecular Biology and Genetics, Cornell University, Ithaca, NY, USA*

Alison O. Nwokeoji • *Department of Chemical and Biological Engineering, ChELSI Institute, University of Sheffield, Sheffield, UK*

Kate M. O'Connor-Giles • *Genetics Training Program, University of Wisconsin-Madison, Madison, WI, USA; Laboratory of Genetics, University of Wisconsin-Madison, Madison, WI, USA; Laboratory of Cell and Molecular Biology, University of Wisconsin-Madison, Madison, WI, USA*

Jason M. Peters • *Department of Microbiology and Immunology, University of California, San Francisco, San Francisco, CA, USA*

André Plagens • *Prokaryotic Small RNA Biology, Max-Planck Institute for Terrestrial Microbiology, Marburg, Germany*

Ümit Pul • *B.R.A.I.N AG, Zwingenberg, Germany*

Lei S. Qi • *Department of Cellular & Molecular Pharmacology, University of California, San Francisco, San Francisco, CA, USA; UCSF Center for Systems and Synthetic Biology, University of California, San Francisco, San Francisco, CA, USA; California Institute for Quantitative Biomedical Research, San Francisco, CA, USA; Center for Systems & Synthetic Biology, University of California, San Francisco, San Francisco, CA, USA; Department of Bioengineering, Stanford University, Stanford, CA, USA; Department of Chemical and Systems Biology, Stanford University, Stanford, CA, USA; Stanford ChEM-H, Stanford, CA, USA*

Udi Qimron • *Department of Clinical Microbiology and Immunology, Sackler Faculty of Medicine, Tel Aviv University, Tel Aviv, Israel*

Lennart Randau • *Prokaryotic Small RNA Biology, Max-Planck Institute for Terrestrial Microbiology, Marburg, Germany*

Nalin Rastogi • *WHO Supranational TB Reference Laboratory, Tuberculosis & Mycobacteria Unit, Institut Pasteur de la Guadeloupe, Abymes, Guadeloupe, France*

Deepak Reyon • *Molecular Pathology Unit, Center for Cancer Research, Massachusetts General Hospital, Charlestown, MA, USA; Center for Computational and Integrative Biology, Massachusetts General Hospital, Charlestown, MA, USA; Department of Pathology, Harvard Medical School, Boston, MA, USA*

Ekaterina Savitskaya • *Skolkovo Institute of Science and Technology, Moscow, Russia; Institute of Molecular Genetics, Russian Academy of Sciences, Moscow, Russia*

Ekaterina Semenova • *Waksman Institute of Microbiology, Piscataway, NJ, USA*

Konstantin Severinov • *Skolkovo Institute of Science and Technology, Moscow, Russia; Institute of Gene Biology, Russian Academy of Sciences, Moscow, Russia; St. Petersburg State Polytechnical University, St. Petersburg, Russia; Waksman Institute of Microbiology, Piscataway, NJ, USA*

Cynthia M. Sharma • *Research Centre for Infectious Diseases (ZINF), University of Würzburg, Würzburg, Germany*

Virginijus Siksnys • *Institute of Biotechnology, Vilnius University, Vilnius, Lithuania*

Tomas Sinkunas • *Institute of Biotechnology, Vilnius University, Vilnius, Lithuania*

Christophe Sola • *Microbiology Department, Institut de Biologie Intégrative de la Cellule (I2BC), CEA, CNRS, Université Paris-Sud, Orsay, France*

ALEXANDRA STROTKSAYA • *Skolkovo Institute of Science and Technology, Moscow, Russia; St. Petersburg State Polytechnical University, St. Petersburg, Russia; Institute of Molecular Genetics, Russian Academy of Sciences, Moscow, Russia*

JÖRG VOGEL • *Institute for Molecular Infection Biology (IMIB), University of Würzburg, Würzburg, Germany*

SAKHARAM P. WAGHMARE • *Department of Chemical and Biological Engineering, ChELSI Institute, University of Sheffield, Sheffield, UK; Laboratory of Plant Physiology and Biophysics, University of Glasgow, Glasgow, UK*

FRANÇOIS-XAVIER WEILL • *Unité des Bactéries Pathogènes Entériques, Institut Pasteur, Paris, France*

EDZE R. WESTRA • *Biosciences, University of Exeter, Penryn, UK*

MALCOLM F. WHITE • *Biomedical Sciences Research Complex, University of St Andrews, St Andrews, UK*

JILL WILDONGER • *Department of Biochemistry, University of Wisconsin-Madison, Madison, WI, USA*

SPENCER WONG • *Medical Microbiology and Immunology, University of California, Davis, Davis, CA, USA; Department of Biology, Massachusetts Institute of Technology, Cambridge, MA, USA*

ALEXANDER F. YAKUNIN • *Department of Chemical Engineering and Applied Chemistry, University of Toronto, Toronto, ON, Canada*

JING-RUEY JOANNA YEH • *Cardiovascular Research Center, Massachusetts General Hospital, Charlestown, MA, USA; Department of Medicine, Harvard Medical School, Boston, MA, USA*

IDO YOSEF • *Department of Clinical Microbiology and Immunology, Sackler Faculty of Medicine, Tel Aviv University, Tel Aviv, Israel*

JING ZHANG • *Biomedical Sciences Research Complex, University of St Andrews, St Andrews, UK*

# Chapter 1

## Investigating CRISPR RNA Biogenesis and Function Using RNA-seq

Nadja Heidrich, Gaurav Dugar, Jörg Vogel, and Cynthia M. Sharma

### Abstract

The development of deep sequencing technology has greatly facilitated transcriptome analyses of both prokaryotes and eukaryotes. RNA-sequencing (RNA-seq), which is based on massively parallel sequencing of cDNAs, has been used to annotate transcript boundaries and revealed widespread antisense transcription as well as a wealth of novel noncoding transcripts in many bacteria. Moreover, RNA-seq is nowadays widely used for gene expression profiling and about to replace hybridization-based approaches such as microarrays. RNA-seq has also informed about the biogenesis and function of CRISPR RNAs (crRNAs) of different types of bacterial RNA-based CRISPR-Cas immune systems. Here we describe several studies that employed RNA-seq for crRNA analyses, with a particular focus on a differential RNA-seq (dRNA-seq) approach, which can distinguish between primary and processed transcripts and allows for a genome-wide annotation of transcriptional start sites. This approach helped to identify a new crRNA biogenesis pathway of Type II CRISPR-Cas systems that involves a trans-encoded small RNA, tracrRNA, and the host factor RNase III.

**Key words** RNA-sequencing (RNA-seq), Deep sequencing, cDNA library, Terminator exonuclease, Differential RNA-seq (dRNA-seq), PNK, Small RNA identification, CRISPR-Cas, tracrRNA, crRNA

## 1 Introduction

A transcriptome by simple definition is the collective knowledge of all RNA transcripts in an organism, both in identity and abundance. Unlike the genome, the transcriptome is highly variable and underlies rapid changes in gene expression defined by the state and environment of the organism. Hybridization-based technologies including microarrays and tiling arrays have played a vital role in gene expression analyses and transcriptome annotation over the last two decades. However, they suffer from some inherent limitations such as high amount of required input material, restricted dynamic range, poor availability of probes for nonmodel organisms, and cross-hybridization problems. The recent advent of deep sequencing technology has revolutionized the field of transcriptomics [1], greatly facilitating the identification of novel transcripts and transcriptome annotation of both prokaryotes and eukaryotes [2–4]. Several different sequencing technologies are on the market, i.e., Solexa (Illumina), 454 (Roche), and the SOLiD system (ABI). Whereas these technologies normally require an amplification step of the cDNA, several technologies also permit single-molecule sequencing such as SMRT sequencing (Pacific Biosciences) or nanopore sequencing (Oxford Nanopore Technologies). Nonetheless, Illumina sequencing of cDNA libraries is currently the most commonly used method for transcriptome analyses by RNA-seq and also allows for multiplexing of samples.

For a typical RNA-seq experiment, first total RNA or a fraction thereof is converted into a cDNA library, which is then analyzed by one of the current next generation sequencing platforms. One major challenge in the first bacterial RNA-seq studies was the abundance of ribosomal RNAs (rRNA) and transfer RNAs (tRNA) that constitute more than 95 % of the cellular RNA pool. In contrast, eukaryotic RNA-seq studies were facilitated through a specific capture of the poly(A)-tailed mRNA fraction using oligo-d(T) priming for cDNA synthesis [2]. However, bacterial mRNAs do not usually carry poly(A) tails, and if so they constitute a degradation signal. Therefore, most of the initial bacterial RNA-seq studies included an rRNA depletion step such as oligonucleotide-based removal of rRNAs with magnetic beads or size fractionation using gel electrophoresis (reviewed in refs. [3, 5]). Nowadays, with ever-growing sequencing depth, the rRNA fraction is no longer a problem as sequencing of total RNA fractions yields enough reads for the mRNA and ncRNA fractions, too. Moreover, this has the advantage that all transcripts in a cell are captured and species-specific capture-probes for rRNA removal become unnecessary.

An important step during an RNA-seq experiment is the construction of the cDNA library. Some of the initial studies used protocols that were not strand-specific due to random hexamer priming of cDNA synthesis and ligation of sequencing adapters to double-stranded cDNA. There are now diverse protocols available for a strand-specific transcriptome analysis. These use, for example, 5′-end linker ligation combined with poly(A)-tailing with *Escherichia coli* poly(A) polymerase [6–8], direct sequencing of first strand cDNA [9], template switching PCR [10], bisulfite-induced C to U conversions prior to cDNA synthesis [11], or incorporation of deoxyuridine into second strand cDNA, which renders it sensitive to uracil-*N*-glycosylase degradation [12]. Moreover, *E. coli* poly(A) polymerase preferentially polyadenylates mRNAs over rRNAs and can thus be used to deplete rRNAs [13].

It should be noted that the different protocols used for library preparation and sequencing can introduce certain biases due to different efficiencies in reverse transcription owing to RNA secondary structures or G/C-content, or different ligation efficiencies [14]. This can lead to aberrant amplification of certain transcripts, which may explain the variations observed among different studies of the same organism [15].

After completion and quality check of library preparation, the cDNAs are sequenced on one of the next generation sequencing platforms resulting in millions of short sequence reads (35–300 bp). After quality filtering and linker and/or poly(A)-tail trimming, these transcript fragments are computationally aligned and mapped to a reference genome. The number of mapped reads per nucleotide is then used to generate whole-genome cDNA coverage plots for visualization in a genome browser for gene expression profiling and transcript annotation.

Based on RNA-seq it is now possible to define in one sequencing experiment the exact boundaries and relative abundances of potentially all RNA molecules transcribed from a genome. This has improved our understanding of functional elements and facilitates better genome annotations. For example, RNA-seq has greatly facilitated the determination of transcriptional start sites (TSS) or processing sites on a global scale. Traditionally, 5′ ends of single genes were determined by primer extension [16] or 5′ RACE (rapid amplification of cDNA ends) [17–19], which is laborious and unsuitable for a global analysis. Several RNA-seq-based methods including a modified 5′ RACE protocol for combination with deep sequencing have been developed for global mapping of 5′ ends in prokaryotes [20–22]. However, most of these approaches cannot distinguish whether the 5′ ends correspond to TSS or processing sites of transcripts.

In this chapter, we describe a protocol for a differential RNA-sequencing (dRNA-seq) approach based on treatment with terminator exonuclease (TEX) [23], which we originally developed to study the primary transcriptome of the major human pathogen *Helicobacter pylori* [7]. This approach permits the distinction between primary transcripts marked by a 5′ tri-phosphate (5′PPP) and processed RNAs with a 5′ monophosphate (5′P) group. To generate dRNA-seq libraries, total bacterial RNA is split into two halves: one remains untreated; whereas, the second one is treated with TEX, which specifically degrades transcripts with a 5′P, whereas primary transcripts including sRNAs and mRNA 5′ ends are protected (Fig. 1). This leads to a relative enrichment of primary transcripts and the depletion of processed RNAs, such as the abundant rRNAs and tRNAs, which carry a 5′P. These enrichment patterns can then be used to define global maps of TSS, operon structure, and small RNA output.

Over the past few years, RNA-seq has been adapted to study the biogenesis of crRNAs and to highlight the roles of many Cas proteins involved in the process. These approaches range from global transcriptomics to more selective sequencing of RNAs associated with proteins after coimmunoprecipitation experiments. A variety of RNA-seq methods have identified CRISPR-derived RNAs (crRNAs) without phage infection in several prokaryotic

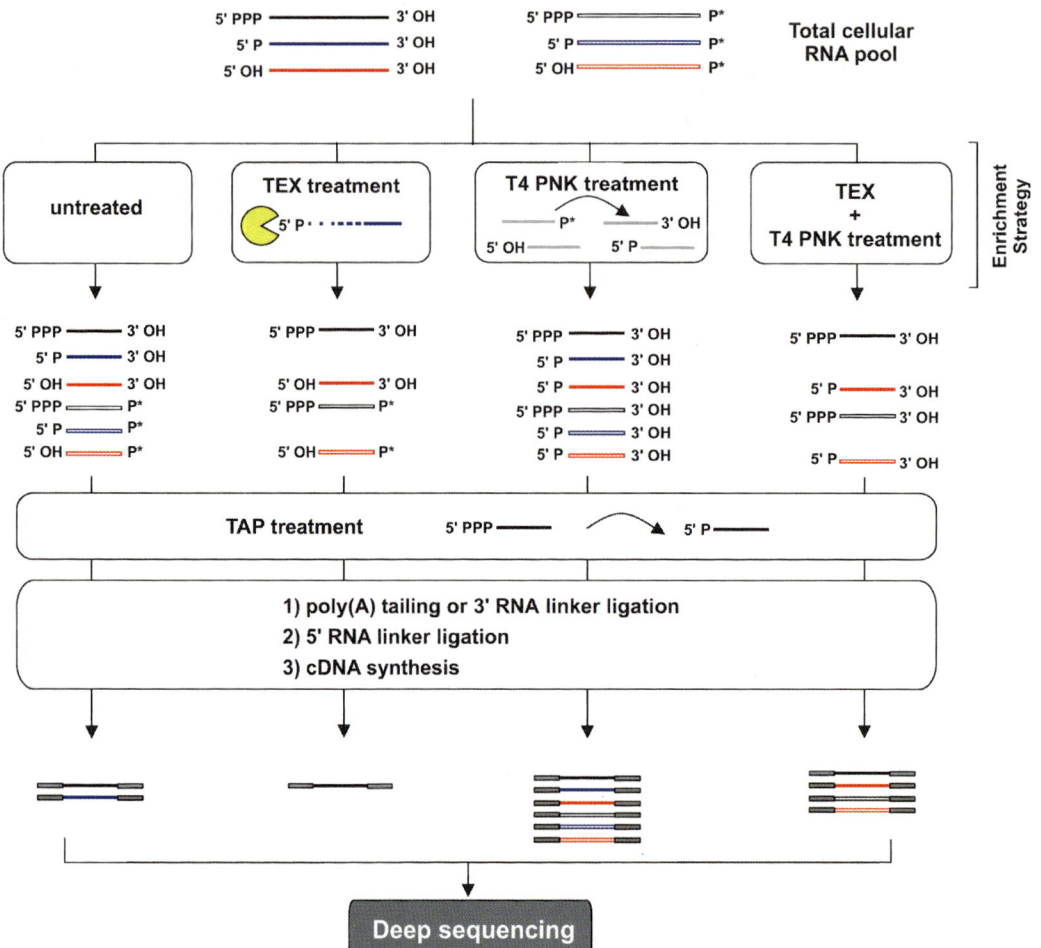

**Fig. 1** Schematic representation of different RNA sequencing approaches including TEX and PNK treatment and combinations of both treatments. The bacterial cellular RNA pool consists of primary transcripts (5′PPP) and processed transcripts (5′P or 5′OH). RNA species generated from multiple processing events may possess non-canonical groups at both 3′ and 5′ ends. RNAs represented by *filled bars* have canonical 3′OH ends, while the *unfilled bars* represent 3′ non-OH ends (2′P, 3′P or <2′, 3′P; shown as P*). Different RNA treatments or combinations thereof prior to library generation allow for enrichment and sequencing of cDNA reads from transcripts with original non-5′P ends and/or original non-3′OH ends. Primary transcripts will be enriched in TEX-treated libraries

organisms, such as *Thermus thermophilus* [24], *Neisseria meningitidis* [25], *Campylobacter jejuni* [26], and *Pyrococcus furiosus* [27, 28]. These prokaryotes all harbor different CRISPR-Cas systems of Type I, II, or III [29]. Identification of crRNAs in RNA-seq profiles provides a first support for functionality of a CRISPR-Cas locus, including information regarding where the crRNA transcripts start and end, and where and how precursor-CRISPR RNAs (pre-crRNAs) are processed into small crRNAs.

Besides annotation of transcript boundaries, RNA-seq can be used for expression analysis of individual crRNAs and reveal differences of abundance patterns for crRNAs within one locus or between different loci [30]. Deep sequencing analyses reported highly variable crRNA abundance patterns for CRISPR arrays from *Sulfolobus solfataricus*, *Pyrobaculum* species, and *Methanococcus maripaludis* [31–33]. The analysis of the abundance of crRNAs can reveal biological insights into (1) transcriptional polarity within the array, (2) efficiency of pre-crRNA processing for each individual crRNA, (3) stochastic termination of the pre-crRNA transcript, and (4) differential stability of mature crRNAs [32].

Various RNA-seq protocols have been used to study crRNA biogenesis (*see* Table 1). These studies have applied different protocols for the preparation of cDNA libraries to capture transcripts with different 5′ or 3′ ends (Fig. 1). For example, the above described dRNA-seq approach based on treatment with terminator exonuclease has been successfully used for the determination of TSS in a wide range of prokaryotic and eukaryotic organisms [7, 34–37]

Table 1
Summary of RNA-seq techniques used to study CRISPR RNAs

| Organism | CRISPR type | Treatment | Enrichment | Study |
|---|---|---|---|---|
| *Campylobacter jejuni* | II-C | −/+TEX | Primary transcripts | [26] |
| *Clostridium difficile* | I-B | −/+TAP | Primary transcripts | [49] |
| *Clostridium thermocellum* and *Methanococcus maripaludis* | I-B | T4 PNK | Mature crRNAs | [30] |
| *Haloferax volcanii* | I-B | T4 PNK | Mature crRNAs | [59] |
| *Methanopyrus kandleri* | III-A III-B | T4 PNK | Mature crRNAs | [48] |
| *Methanosarcina mazei* strain Gö1 | I-B III-B | PNK/TAP | Mature crRNAs | [45] |
| *Nanoarchaeum equitans* | I | T4 PNK | Mature crRNAs | [60] |
| *Neisseria meningitidis* | II-C | −/+TEX | Primary transcripts | [25] |
| crenarchaeal genus *Pyrobaculum* | I-A III-A III-B | size selection 15-70 nt | Small RNAs | [32] |
| *Pyrococcus furiosus* | III-B | size selection 18-65 nt | Small RNAs | [28] |
| *Pyrococcus furiosus* | III-B | coIP of Cmr complex + T4 PNK | Mature Cmr-associated RNAs | [27] |
| *Streptococcus pyogenes* | II-A | −/+TEX | Primary transcripts | [38] |
| *Synechocystis* sp. PCC6803 | I-D III | T4 PNK | Mature crRNAs | [46] |
| *Thermococcus kodakarensis* | I-A I-B | TSAP[a] | | [61] |
| *Thermus thermophilus* HB8 | I-E III-A III-B | size fractionation/ dephosphorylation/ T4 PNK | Mature crRNAs | [24] |

[a]Thermosensitive alkaline phosphatase

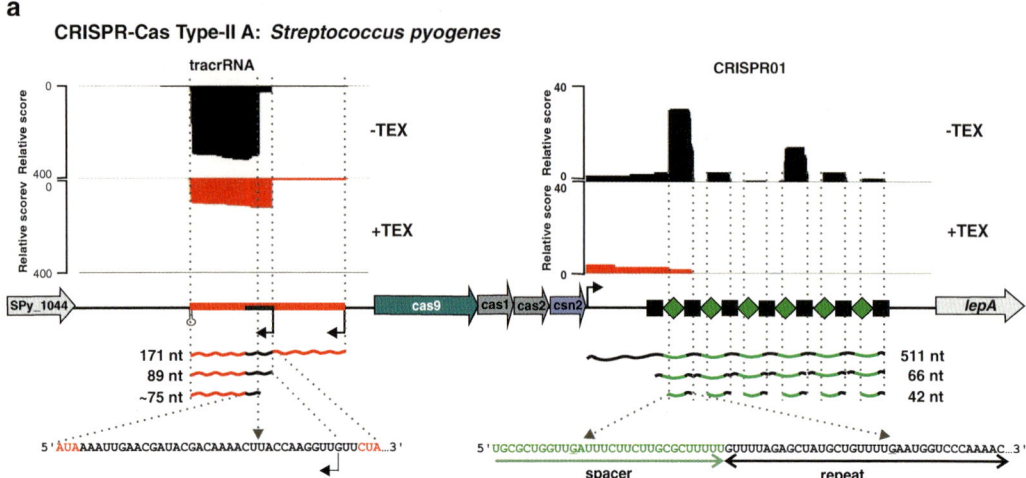

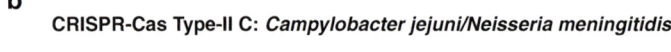

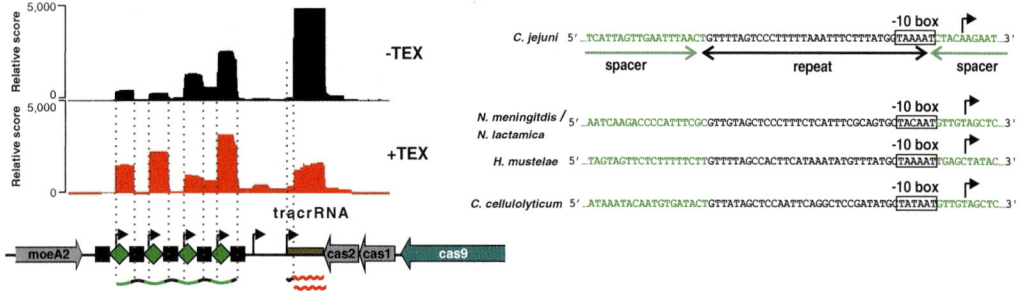

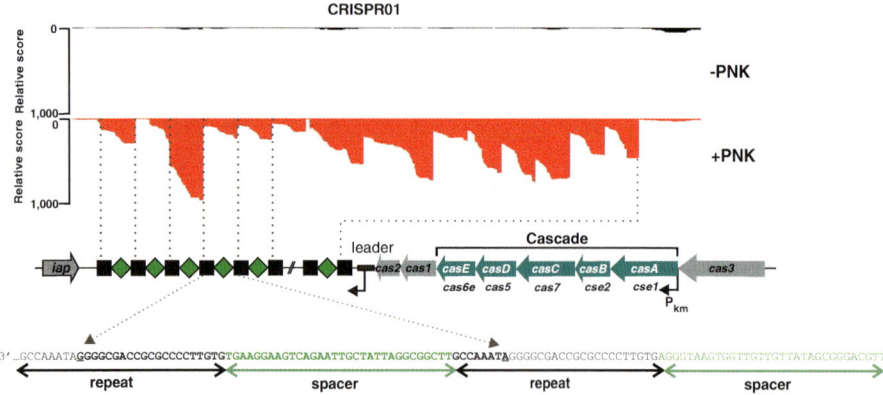

**Fig. 2** RNA-seq of CRISPR loci from different species. (**a**) Mapping of cDNA reads of libraries generated from −/+TEX treated RNA to the Type II CRISPR-Cas locus of *S. pyogenes* revealed coprocessing of a trans-encoded sRNA, tracrRNA, and pre-crRNA [38]. *Black arrows* indicate transcriptional start sites determined by dRNA-seq. The figure was adapted from ref. [38]. (**b**) cDNA reads of −/+TEX libraries mapped to the CRISPR loci in *C. jejuni* NCTC11168 revealed transcriptional start sites for crRNA-repeat units along with the processed tracrRNA [26].

and has also facilitated the study of the biogenesis and mechanisms of crRNAs from Type II CRISPR-Cas systems (Fig. 2a, b) [25, 26, 38]. Differential treatment of RNA libraries has been of particular interest for CRISPR-Cas research owing to the complex and unusual processing of CRISPR transcripts. The right choice of protocol for small RNA library preparation is crucial, since many protocols require small RNAs to have 5′P and 3′OH groups [39]. However, certain CRISPR-Cas subtypes produce crRNAs with different chemical groups at their 5′ and 3′ ends. For example, biochemical characterization of pre-crRNA cleavage products generated by Cas6 endoribonucleases, such as Cas6e and Cas6f (Type I) and Cas6 (Type III) revealed that they exhibit 5′ hydroxyl (5′OH) and 2–3′ cyclic phosphate ends [40–42]. Such pools of crRNAs with non-5′P ends and/or original non-3′OH ends are not accessible for 5′-linker ligation and cannot be extended by poly(A) polymerase. Therefore, they require a PNK-treatment prior to cDNA preparation to be captured in RNA-seq libraries. Here, we focus on protocols for dRNA-seq for primary transcriptome analyses including the determination of TSS and processing sites as well as for PNK-treatment to capture transcripts with 5′OH or 3′P groups. In this chapter, we primarily focus on preparation of RNA-seq libraries for maximum knowledge on crRNA biogenesis and functional elucidation.

## 2 Materials

All necessary solutions and reagents should be prepared prior to the experiment. Unless otherwise stated, autoclaved, deionized water (dH$_2$O) is used to make all solutions. Solutions and buffers should be stored at the proper conditions as indicated. Make sure that you work under RNase-free conditions and always keep the RNA samples on ice. Standard equipment used in molecular biology research is required for this protocol, e.g., a heat block, a gel electrophoresis apparatus, and a benchtop microcentrifuge.

### 2.1 Key Reagents, Components, and Equipment

1. DNase I (1 U/μL, Fermentas, Cat# EN0521, 1,000 U).
2. Superase-In RNase Inhibitor (20 U/μL, Ambion, Cat# AM2694, 2,500 U).
3. Terminator™ 5′-phosphate-dependent exonuclease (TEX) (1 U/μL, Epicentre, Cat# TER51020, 40 U).

**Fig. 2** (continued) In line with this, an extended −10 box promoter motif is found in each repeat unit in *C. jejuni*. The figure was adapted from ref. [26]. Sequences from other Type II-C subtype also show a conserved −10 box at the end of the repeats [25]. (**c**) cDNA enrichment in a PNK-treated (+PNK) library compared to a −PNK library reveals processed crRNAs with non-5′P/non-3′OH end architecture specific for Type I-E CRISPR-Cas systems in *Salmonella enterica* (Heidrich and Vogel, unpublished)

4. T4 Polynucleotide Kinase (T4 PNK) (10 U/μL, NEB, Cat# M0201S, 500 U) and ATP.
5. Tobacco acid pyrophosphatase (TAP) (10 U/μL, Epicentre, Cat# T19100, 100 U).
6. Roti-Aqua-P/C/I (phenol/chloroform/isoamylalcohol) (Carl Roth GmbH, Cat# X985.1).
7. GlycoBlue™ (15 mg/mL; Ambion, Cat# AM9515).
8. Stains-all (Sigma-Aldrich, Cat# E9379-1G).
9. RNA RiboRuler™ High Range RNA Ladder (Fermentas, Cat# SM1821).
10. Phase-Lock tubes (VWR, Cat# 713-2536, 2 mL).
11. Eppendorf tubes (Eppendorf Vertrieb Deutschland GmbH, #30120086).
12. Nanodrop spectrophotometer (NanoDrop Technologies).
13. Heated shaker (thermo-shaker).
14. Reagents for high-fidelity polymerase chain reaction (PCR).
15. Agarose gels and electrophoresis equipment.
16. Agilent Bioanalyzer (2100) or MultiNA microchip electrophoresis (Shimadzu).
17. Polyacrylamide/Urea gels and electrophoresis equipment.
18. Poly(A) polymerase and a suitable RNA adapter.
19. M-MLV reverse transcriptase and an oligo(dT)-adapter primer.
20. Agencourt AMPure XP kit (Beckman Coulter Genomics).

## 2.2 Selected Solutions and Buffers for RNA Preparation and RNA-seq Library Preparation

1. RNA precipitation mix (30:1 EtOH:3 M NaOAc, pH 6.5): Add 1 mL of 3 M NaOAc (pH 6.5) to 29 mL of 100 % Ethanol. The buffer can be stored at room temperature between usages.
2. Loading buffer II: 95 % Formamide; 18 mM EDTA; and 0.025 % each of sodium dodecyl sulfate (SDS), Xylene Cyanol, and Bromophenol Blue. Store at −20 °C.
3. Stains-all stock solution: Add 0.1 g Stains-all to 100 mL of formamide and mix. The solution should be stored at 4 °C in dark to avoid fading of the dye (*see* **Note 1**).
4. Stains-all working solution: Mix 30 mL of Stains-all stock solution with 90 mL formamide. Add water to bring the final volume to 200 mL. Store in dark at 4 °C.
5. Other buffers and solutions: 3 M NaOAc (pH 6.5), 75 % EtOH, 0.5 M EDTA (pH 8.0) along with ultrapure water should also be prepared for general use.

# 3 Methods

## 3.1 Overview of the dRNA-seq Method (See also Fig. 1)

1. Isolation of high-quality RNA samples (*see* Subheading 3.2).
2. DNase I treatment of RNA samples (*see* Subheading 3.3).
3. Enrichment of primary transcripts using TEX (*see* Subheading 3.4). One half of the DNA-free RNA sample is treated with TEX, which degrades the processed RNAs and, thereby, enriches primary transcripts. The other half of the RNA sample is mock-treated (includes both primary and processed RNAs).
4. Optional step: The RNAs with 3′P or 5′OH ends can be enriched using PNK treatment (*see* Subheading 3.5).
5. Both of the TEX−/+ samples are treated with TAP, which cleaves the 5′PPP group of primary transcripts leaving a 5′P that is required for 5′ linker ligation (*see* Subheading 3.6).
6. After TEX and TAP treatment, visual quality control of the RNA samples on a polyacrylamide gel followed by direct staining of the RNA is recommended (*see* Subheading 3.7).
7. Next, strand-specific cDNA libraries are constructed from both RNA samples by the following steps: (1) poly(A) tailing at the 3′ end with *E. coli* poly(A) polymerase, (2) 5′ linker ligation, (3) first-strand cDNA synthesis using an oligo(dT)-adapter primer and an RNaseH⁻ (minus) reverse transcriptase, and (4) PCR amplification with primers containing barcodes for the designated sequencing platform (*see* Subheading 3.8).
8. After sequencing of cDNAs on the designated high-throughput sequencing platform, cDNA reads are mapped to a reference genome and transcript levels are quantified (*see* Subheading 3.9).

## 3.2 Isolation of High-Quality RNA Samples

To construct dRNA-seq libraries, ideally ~15 μg of high quality RNA is required (*see* **Note 2**). However, we advise the preparation of larger RNA amounts from the same sample to have enough material for subsequent confirmation experiments, such as 5′ RACE, Northern blot analysis, or quantitative RT-PCR. It is important to use an RNA preparation method that yields high-quality RNA to avoid extensive sequencing of rRNA degradation fragments. For bacterial total RNA preparation, we recommend RNA isolation methods based on hot-phenol extraction [43, 44]. Note that for Gram-positive bacteria, a more extensive lysis method, e.g., using lysozyme or glass-beads, prior to RNA extraction may be required. With the current state of sequencing technology, removal of rRNA is an optional parameter, which could easily be avoided by sequencing a little deeper (*see* **Note 3**).

## 3.3 DNase Treatment of RNA Samples

Prior to any optional RNA enrichment procedure and subsequent cDNA library preparation, residual genomic DNA should be removed from RNA samples using DNase I digestion.

1. Dissolve RNA sample (~20 µg) in 39.5 µL dH$_2$O and denature for 5 min at 65 °C.

2. After cooling on ice for 5 min, add 5 µL 10× DNase I buffer including MgCl$_2$, 0.5 µL Superase-In RNase Inhibitor (20 U/µL) (*see* **Note 4**), and 5 µl DNase I (1 U/µL). This will give a total reaction volume of 50 µL.

3. Perform DNase I digestion by incubation for 30–45 min at 37 °C.

4. Remove DNase I by phenol–chloroform extraction. First adjust the reaction volume to 100 µL by addition of 50 µL dH$_2$O to each tube. Next, add 100 µL Roti-Aqua-P/C/I to a 2 mL Phase-Lock tube followed by DNase I digested RNA samples. Mix for 15 s by shaking the tube (do not vortex!), followed by centrifugation for 12 min at 15 °C and 16,000 × $g$. Transfer upper phase to a fresh 1.5 mL Eppendorf tube.

5. To precipitate RNA, add 2.5 volumes (~300 µL) 30:1 RNA precipitation mix (EtOH: 3 M NaOAc, pH 6.5) and incubate at least 1 h or preferably overnight at −20 °C. Afterwards, centrifuge samples for 30 min at 4 °C and 16,000 × $g$. Discard supernatant and wash the pellet with 350 µL of 75 % EtOH. Centrifuge again for 10 min at 4 °C and 16,000 × $g$. Discard supernatant and air-dry pellet.

6. Add 20 µL dH$_2$O (do not pipet up and down!). Dissolve pellet by 5 min incubation at 65 °C and ~800 rpm on a thermoshaker, vortex two to three times in between. Check RNA concentration on NanoDrop (final RNA concentration ~1 µg/µL).

7. Efficient removal of genomic DNA contamination is checked using a control PCR. Use any primers that yield a product of ~500–1,000 bp. Test 100 ng RNA before and after DNase I digestion as input and 100 ng genomic DNA as positive control, 40 cycles PCR amplification.

8. Lastly, the integrity of the DNase-free RNA should be analyzed by visual inspection of the 23S and 16S bands on an agarose gel or on a Bioanalyzer.

## 3.4 Enrichment of Primary Transcripts Using Terminator Exonuclease (TEX)

The dRNA-seq approach aims to analyze primary transcriptomes and distinguish between 5′PPP- and 5′P-ends of transcripts. The workflow for preparing dRNA-seq libraries is illustrated in Fig. 1 (first two columns). It is based on the preparation of two differential cDNA libraries, one from untreated RNA and one from RNA that was pretreated with TEX, which specifically degrades RNAs with a 5′P. After TEX treatment, both RNA samples are treated with Tobacco Acid Pyrophosphatase (TAP) to convert 5′-PPP ends into 5′P for 5′end linker ligation (*see* Subheading 3.6). Note that unless a PNK treatment is included, RNAs with a 5′-OH end are not captured in the cDNA library since they are not accessible for linker ligation (*see* Subheading 3.5).

1. Separate DNase I-treated RNA sample in two 1.5 mL tubes (7 μg each) with 37.5 μL dH$_2$O and denature for 2 min at 90 °C (*see* **Note 4**).

2. After cooling on ice for 5 min, add 5 μL 10× TEX buffer and 0.5 μL Superase-In RNase inhibitor (20 U/μL) to each tube. Add 7 μL terminator exonuclease (1 U/μL) to one tube (TEX+) and 7 μL dH$_2$O to the other (TEX−).

3. Incubate for 60 min at 30 °C and afterwards stop the reaction by addition of 0.5 μL 0.5 M EDTA, pH 8.0 on ice.

4. Afterwards add 50 μL dH$_2$O to each tube to fill the volume to 100 μL. Purify and concentrate the RNA using P/C/I treatment by addition of 100 μL Roti-Aqua-P/C/I in 2 mL Phase-Lock gel tube to the reaction samples, vigorous mixing and inverting of the tubes for 15 s, and centrifugation for 12 min at 15 °C and 16,000 ×*g*.

5. After transferring the upper phase to a fresh 1.5 mL Eppendorf tube, add 2 μL GlycoBlue followed by 300 μL of 30:1 RNA precipitation mix (EtOH: 3 M NaOAc, pH 6.5).

6. After precipitation overnight at −20 °C, centrifuge for 30 min at 4 °C and 16,000 ×*g*, discard supernatant and wash the pellet with 90 μL 75 % EtOH. Centrifuge again for 10 min at 4 °C and 16,000 ×*g*, discard supernatant and air-dry pellet.

7. Add 11 μL dH$_2$O (do not pipet up and down!). Dissolve pellet by 5 min incubation at 65 °C and ~800 rpm on a thermo-shaker, vortex two to three times in between. Check RNA concentration on a NanoDrop (the TEX treated sample should contain less RNA due to removal of processed RNAs).

### 3.5 Enrichment of RNAs with 3′P or 5′OH Ends Using PNK Treatment

The current dRNA-seq method is limited to the detection of transcripts with 5′PPP and 5′P ends, respectively. Processed RNAs carrying a 5′OH group are not digested by TEX (as are primary transcripts with a 5′PPP), but they are not accessible for 5′ RNA linker ligation either and, thus, will not be represented in the cDNA library. To capture these RNAs, an additional 5′ phosphorylation step using T4 polynucleotide kinase is required (Fig. 1).

RNA-seq approaches that included such PNK-treatment successfully resolved crRNAs in cDNA libraries (Fig. 2c) and led to the identification of 5′ and 3′ groups of processed crRNAs from different Types of CRISPR-Cas systems, e.g., in *Salmonella enterica* (Heidrich and Vogel, unpublished), *Pyrococcus furiosus* [27], *Methanosarcina mazei* strain Göl [45], *Synechocystis* sp. PCC6803 [46], *Clostridium thermocellum* and *Methanococcus maripaludis* [30], and *Thermus thermophilus* HB8 [24]. Furthermore, the application of such a terminal-phosphate-dependent small RNA cDNA library preparation method revealed mechanistic key features of the processing enzyme. For example, in *Sulfolobus solfataricus*, CMR-complex mediated cleavage of target RNAs seem to have parallels with a guide RNA loaded endoribonuclease (Piwi) of the piRNA pathway, that targets mobile mRNA and mediates dsRNA cleavage [47]. However, a functional homology between Cmr proteins and Piwi can be excluded since Piwi recognizes the 5′ phosphate of the guide RNA while crRNAs utilized by CMR-complexes lack a 5′ phosphate and have an essential 5′ tag [31]. In *Thermus thermophilus*, which carries multiple CRISPR loci, RNA-seq revealed two kinds of mature crRNAs with different 3′ termini suggesting that multiple crRNA processing systems operate in parallel in a single bacterial strain [24].

To detect all combinations of CRISPR-derived transcripts, RNA-seq analysis that combines TEX and PNK treatment for cDNA library preparations would be the ultimate strategy. The following PNK treatment protocol is based on the study by Su et al. [48].

1. Separate the DNase I treated RNA sample in two tubes (10 μg each in 42.5 μL dH$_2$O) and denature for 2 min at 90 °C.

2. After cooling on ice for 5 min, add 5 μL 10× T4 PNK buffer and 0.5 μL Superase-In RNase inhibitor (20 U/μL) to each tube. Add 2 μL T4 PNK (10 U/μL) to one tube (PNK+) and 2 μL dH$_2$O to the other (PNK−).

3. Incubate for 6 h at 37 °C. Subsequently, add an additional 1 μL of T4 PNK (10 U/μL) and ATP (2 mM final) to the PNK+ sample. Add an equal amount of dH$_2$O to the PNK− sample. Incubate the samples for an additional 1 h at 37 °C.

4. Afterwards remove T4 PNK and precipitate RNA by phenol–chloroform extraction as described in Subheading 3.3, **steps 4** and **5**.

5. Add 11 μL dH$_2$O to the precipitated RNA (do not pipet up and down!). Dissolve pellet by 5 min incubation at 65 °C and ~800 rpm on a thermo-shaker, vortex two to three times in between. Check RNA concentration on a NanoDrop (the PNK treated sample should contain an equal RNA amount compared to the untreated one).

## 3.6 Generation of 5′P for RNA Linker Ligation Using Tobacco Acid Pyrophosphatase (TAP) Treatment

Another strategy to enrich primary transcripts uses treatment of RNA with TAP. This enzyme removes pyrophosphates (PP) from RNAs that carry a 5′PPP, leaving a 5′P end. This step is required prior to the RNA linker ligation step during cDNA library preparation to ensure linker-ligation of the 5′PPP ends of primary transcripts. Therefore, in the dRNA-seq approach both samples (TEX− and TEX+) are treated with TAP. Sequencing of TAP−/

TAP+ samples without TEX treatment has been used for global identification of sRNAs and their TSS as well as Type-I B derived crRNAs in *Clostridium difficile* [49]. Hereby, a comparison of libraries constructed from one RNA sample treated with TAP (TAP+) and one without TAP treatment (TAP−) identifies reads associated with the TSS due to an enrichment of primary transcripts in the TAP+ library versus the TAP− library. In contrast, in the dRNA-seq approach, primary transcripts are represented in both the TEX− and TEX+ libraries but TSS are identified through a depletion of processed RNAs in the TEX+ library.

1. Denature the remaining 10 μL of the −/+TEX RNA samples from Subheading 3.4 or −/+PNK treated RNA samples from Subheading 3.5 for 1 min at 90 °C.

2. After cooling on ice for 5 min, add 7 μL dH$_2$O, 2 μL 10× TAP buffer, 0.5 μL Superase-In RNase inhibitor (20 U/μL) and 0.5 μL TAP (10 U/μL) to each tube and mix well by pipeting up and down.

3. Incubate for 60 min at 37 °C.

4. Afterwards, adjust the volume to 100 μL by adding 80 μL dH$_2$O and remove TAP by phenol–chloroform extraction (*see* Subheading 3.4, **steps 4** and **5**).

5. Add 20 μL dH$_2$O to the precipitated RNA (do not pipet up and down!). Dissolve pellet by 5 min incubation at 65 °C and ~800 rpm on a thermo-shaker, vortex two to three times in between. Check RNA concentration on a NanoDrop.

## 3.7 Quality Control of TEX and TAP Treatment

The quality of RNA samples should be analyzed after the different enzymatic treatments on a denaturing polyacrylamide (PAA) gel, prior to cDNA synthesis. This will enable confirmation of successful TEX treatment, since it should lead to a depletion of the 23S and 16S rRNA bands. In the TEX− sample, intact 23S and 16S rRNA bands should be visible on the gel.

1. Prepare a 4 % polyacrylamide/8.3 M urea gel (~10 cm × 10 cm).

2. Add 5 μL loading buffer II to 5 μL of the TEX− and TEX+ samples, denature samples 1–2 min at 95 °C and load on gel. Denature also an RNA ladder and load 7.5 μL RNA RiboRuler™ High Range RNA Ladder.

3. Run gel at 150 V for approx. 1–1.5 h and place the gel in 100 mL stains-all working solution for 20 min in the dark while shaking. De-stain the gel with dH$_2$O in light.

## 3.8 cDNA Library Preparation

After a successful RNA enrichment as well as TAP treatment and quality control, cDNA libraries are constructed according to the designated sequencing platform in a strand-specific manner. For example, in our dRNA-seq analyses of *C. jejuni* and *N. meningitidis* transcriptomes [25, 26], cDNA library construction was performed following a previously established protocol for the identification of eukaryotic microRNAs [50], but omitting size-fractionation of RNA prior to cDNA synthesis. Afterwards, cDNA libraries were sequenced on a HiSeq2000 (Illumina) system.

1. Poly(A)-tail equal amounts (about 200 ng) of −/+TEX-treated RNA using poly(A) polymerase using 2.5 U *E. coli* poly(A) polymerase (NEB) for 5 min at 37 °C.

2. Ligate an RNA adapter (5′ Illumina sequencing adapter, 5′-UUUCCCUACACGACGCUCUUCCGAUCU-3′) to the 5′-phosphate of the RNAs (30 min, 25 °C).

3. Perform first-strand cDNA synthesis using an oligo(dT)-adapter primer (3′ for Illumina sequencing adapter, 5′-GTGACTGGAGTTCAGACGTGTGCTCTTCCGATCTTTTTTTTTTTTTTTTTTTTTTTTTVN-3′) and M-MLV reverse transcriptase (AffinityScript, Agilent). Incubation at 42 °C for 20 min, ramp to 55 °C followed by 55 °C for 5 min.

4. PCR-amplify the final cDNAs with a high-fidelity DNA polymerase (e.g., Herculase II Fusion DNA Polymerase from Agilent) to a final concentration of 20–30 ng/μL (initial denaturation at 95 °C for 2 min, 10–15 cycles 95 °C for 20 s and 68 °C for 2 min). The primers for PCR amplification should be designed for the chosen sequencing platform (e.g., TruSeq sequencing according to the instructions of Illumina). *See* **Note 5** for examples of TruSeq primers.

5. Purify PCR products using the Agencourt AMPure XP kit (Beckman Coulter) (1,8 × sample volume).

6. Examine cDNA sizes by capillary electrophoresis on a MultiNA microchip electrophoresis system (Shimadzu) (*see* **Note 6**).

*3.9 cDNA Library Sequencing and Analysis*

The cDNA libraries are sequenced following the Illumina standard protocols. Note that there are now many suitable commercially available kits for cDNA library preparation. Methodological details of the sequencing and data analysis are beyond the scope of this chapter, but the broader concepts of data interpretation are discussed below.

Sequencing of differential dRNA-seq libraries leads to a characteristic enrichment pattern reflected by a redistribution of a gene's cDNAs towards a sawtooth-like profile with an elevated sharp 5′ flank, which can be used to annotate transcriptional start sites. This can be done manually by visual inspection [7] or through algorithms that allow for automated TSS annotation [26]. Application of the dRNA-seq approach to total RNA of the human pathogen *Streptococcus pyogenes* revealed transcriptional start sites and processing patterns of CRISPR-derived RNAs (Fig. 2a) [38]. Closer inspection of the dRNA-seq data indicated that processing of pre-crRNA into mature crRNA forms (39–42 nt) relies on a trans-encoded small RNA (tracrRNA) and the host factor RNase III in the first step of CRISPR RNA processing that was verified subsequently by a genetic and biochemical analysis. This pioneering study resulted in the classification of so-called Type II CRISPR-Cas systems that are marked by a particular maturation pathway and that are generally associated with vertebrate pathogens and commensals [29, 51].

Two other studies using dRNA-seq uncovered a remarkably distinct crRNA biogenesis mechanism in *Campylobacter jejuni* and *Neisseria meningitidis*, leading to a new subtype C of Type II CRISPR-Cas systems [25, 26]. Remember that treatment of RNA with terminator exonuclease (TEX) causes a relative enrichment of primary transcripts with 5′PPP ends in the TEX+ library (Figs. 1 and 2a). In *C. jejuni* and *Neisseria*, the crRNAs were enriched rather than depleted in the TEX+ libraries, revealing that the crRNAs are transcribed from promoters that are embedded within each repeat. Thus, here the crRNA 5′ ends are formed by transcription and not by processing from a longer precursor transcript (Fig. 2b).

In addition to the identification of transcript boundaries, dRNA-seq can be applied to study expression and processing patterns of crRNAs under different conditions, as done in *Methanosarcina mazei* strain Göl [45] or among different strains (*see* **Note** 7). Regarding the latter, a comparative transcriptome view based on RNA-seq across multiple strains of the genus *Campylobacter* or *Pyrobaculum* revealed in vivo evidence for the presence or absence of functional CRISPR-Cas loci [26, 32]. Moreover, such a comparative transcriptional analysis can provide a genus-level evolutionary perspective on bacterial and archaeal CRISPR-Cas systems [32]. In the same line, RNA-seq can be applied to compare various CRISPR-Cas systems, as performed for five bacterial strains that harbor Type II systems [51]. This study revealed the tracrRNA family as an unusual sRNA family with no obvious conservation of structure, sequence or location of tracrRNA homologs within Type II CRISPR-Cas loci. Moreover, RNA-seq-based validation of Type II CRISPR-Cas expression and processing will underpin the design of RNA-programmable Cas9 as genome editing tool [52].

## 4 Outlook: Analyses of Ribonucleoprotein Complexes Using RNA-seq

Many noncoding RNAs reside within discrete ribonucleoprotein (RNP) complexes and require proteins for their stability, processing, or function. A variety of RNA-seq-based methods have been developed to study RNA–protein interactions and often involve a coimmunoprecipitation (coIP) step with or without prior cross-linking [53]. For example, a combination of coIP with RNA-seq revealed the direct sRNA and mRNA binding partners of the RNA

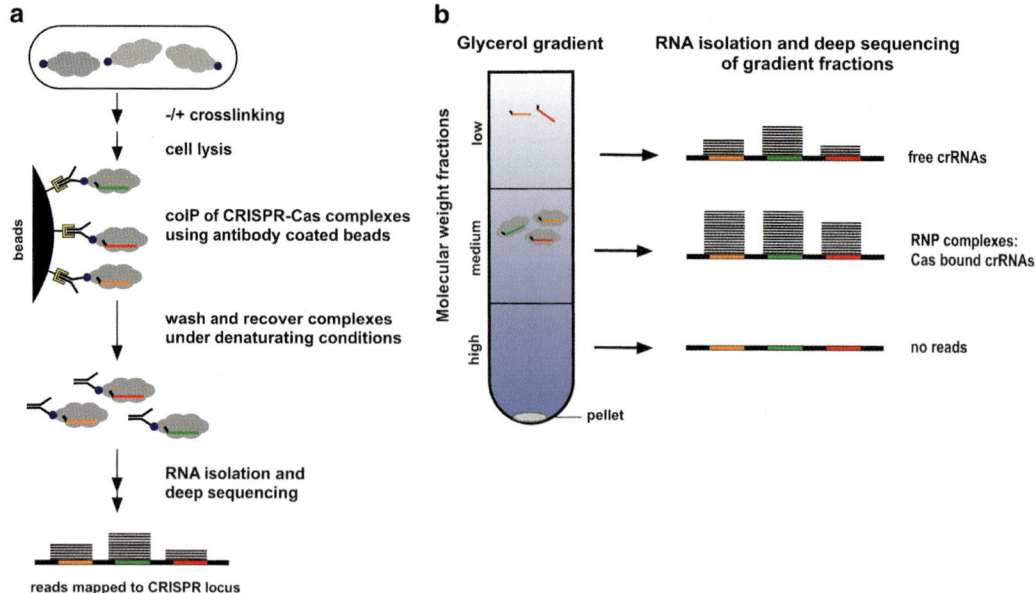

**Fig. 3** Novel applications of RNA-seq for crRNA analysis. (**a**) Analysis of RNAs associated with CRISPR complexes by coimmunoprecipitation (coIP) of a desired Cas protein with or without crosslinking. Upon coIP and RNA isolation by phenol-chloroform extraction from the complex, RNA binding partners are analyzed by RNA-seq and subsequent cDNA mapping. (**b**) Detection of CRISPR complex associated and nonassociated crRNAs based on fractionation of cell lysates by glycerol gradients followed by RNA-seq of different gradient fractions

chaperone Hfq in *Salmonella* on a global scale [6, 54]. Since crRNAs also act as part of RNP complexes [55, 56], the composition of CRISPR-derived RNP complexes can be determined by such coIP methods (Fig. 3a). Selected *cas* genes could be epitope-tagged in the chromosome or on a plasmid of a desired strain using either a FLAG- or a His-tag. Afterwards, active Cas complexes from bacterial protein lysates can be purified through coIP with a respective antibody (FLAG tag) and or through affinity chromatography (His tag) to look for Cas-associated RNA or DNA by deep sequencing. To determine conserved RNA-binding motifs in crRNAs a UV-crosslinking step can be included in this assay; UV covalently links one nucleotide of each RNA molecule to its respective protein binding partner. Such a coIP-approach successfully copurified active Cmr-crRNA complexes in *P. furiosus* [27]. Comparison of deep sequencing profiles of crRNAs found in total *P. furiosus* RNA and immunopurified Cmr complexes revealed essential features of the crRNAs associated with the Cmr effector. This included the unexpected cleavage of an endogenous target RNA by the Cmr complex in vivo. A similar coIP-approach with antibodies against Cmr7 was applied in *S. solfataricus* to characterize CMR complexes [31]. As for *P. furiosus*, RNA-seq revealed important features of the CMR complex associated crRNAs that informed reconstitution experiments of the Cmr–crRNA complexes for cleavage activity assays.

Recently, a function in targeting of endogenous genes was demonstrated for a Type II CRISPR-Cas system in *Francisella novicida* [57]. Thus, in addition to providing insights into mechanistic features, RNA-seq analysis of Cas protein-associated RNAs promises to identify other classes of non CRISPR-RNAs as novel targets, possibly extending the functional role of CRISPR-Cas to a mediator of RNA surveillance. The sequencing of Cas-associated mRNAs will help understand whether this alternative function of Cas proteins in endogenous gene regulation is widespread [58].

Another strategy to gain insight into the functional components of CRISPR-Cas complexes is based on gradient fractionation of whole cell lysates coupled with high-throughput analysis of RNAs by deep sequencing and proteins by mass-spectrometry (Fig. 3b). This method can identify intermediates of assembled Cas-RNPs that may contain larger CRISPR-derived RNAs as well as other cellular RNAs. In *P. furiosus* active CMR complexes were purified from S100 cellular extracts on the basis of crRNA fractionation profiles and Cas proteins were identified by tandem mass spectrometry of a particular gradient fraction [28]. Deep sequencing of RNA extracted from the same fraction revealed two size classes of mature crRNAs as active components of the CMR complex.

Overall, RNA-seq-based methods provide a powerful tool for the identification of crRNAs and their biogenesis as well as functional analyses of RNP complexes. Future technologies like single-cell or single-molecule sequencing will provide further insight into spatial and temporal transcriptome changes of bacteria and will allow for population-wide studies of CRISPR-Cas function and silencing of foreign or endogenous nucleic acids prior and after infecting the bacteria with phages.

## 5 Notes

1. Stains-all working solution can be reused 20–30 times when stored properly in dark at 4 °C

2. The amount of RNA used as input in dRNA-seq can also be scaled down.

3. Ribosomal RNA depletion or size selection could be performed prior to TEX or TAP treatments and cDNA library preparation for increasing coverage of non-ribosomal RNAs.

4. RNA samples should never be left out at room temperature or 37 °C in the absence of Superase-In RNase inhibitor.

5. The following adapter sequences flank the cDNA inserts: TruSeq_Sense_primer; 5′-AATGATACGGCGACCACCGAG ATCTACACTCTTTCCCTACACGACGCTCTTCCG ATCT-3′. TruSeq_Antisense_primer including 6-mer barcode sequence; 5′-CAAGCAGAAGACGGCATACGAGAT-NNN NNN-GTGACTGGAGTTCAGACGTGTGCTCTTC

CGATC(dT25)-3'. The combined length of the flanking sequences is 146 bases.

6. If the cDNA library contains many cDNAs in the size range of the 5' end and 3' end adaptors, a size-fractionation of the cDNA library on a gel is required to avoid extensive sequencing of self-ligation products of the two adaptors.

7. Different abundance patterns of crRNAs from different spacers within one species can also be caused due to selective recovery and amplification of certain crRNA sequences during library preparation.

## Acknowledgements

CRISPR work in the Vogel lab is funded by DFG Grant Vo875/7-1 and the Bavarian BioSysNet program. Work in the Sharma lab is supported by the ZINF Young Investigator program at the Research Center for Infectious Diseases (ZINF) in Würzburg, Germany, the Bavarian BioSysNet program, DFG Grant Sh580/1-1 and the Daimler and Benz foundation. GD is supported by the Graduate School for Life Sciences (GSLS) Würzburg.

## References

1. Mutz KO, Heilkenbrinker A, Lonne M, Walter JG, Stahl F (2013) Transcriptome analysis using next-generation sequencing. Curr Opin Biotechnol 24:22–30
2. Wang Z, Gerstein M, Snyder M (2009) RNA-Seq: a revolutionary tool for transcriptomics. Nat Rev Genet 10:57–63
3. Croucher NJ, Thomson NR (2010) Studying bacterial transcriptomes using RNA-seq. Curr Opin Microbiol 13:619–624
4. van Vliet AH (2010) Next generation sequencing of microbial transcriptomes: challenges and opportunities. FEMS Microbiol Lett 302:1–7
5. Sorek R, Cossart P (2010) Prokaryotic transcriptomics: a new view on regulation, physiology and pathogenicity. Nat Rev Genet 11:9–16
6. Sittka A, Lucchini S, Papenfort K, Sharma CM, Rolle K, Binnewies TT, Hinton JC, Vogel J (2008) Deep sequencing analysis of small noncoding RNA and mRNA targets of the global post-transcriptional regulator, Hfq. PLoS Genet 4:e1000163
7. Sharma CM, Hoffmann S, Darfeuille F, Reignier J, Findeiss S, Sittka A, Chabas S, Reiche K, Hackermuller J, Reinhardt R, Stadler PF, Vogel J (2010) The primary transcriptome of the major human pathogen *Helicobacter pylori*. Nature 464:250–255
8. McGrath PT, Lee H, Zhang L, Iniesta AA, Hottes AK, Tan MH, Hillson NJ, Hu P, Shapiro L, McAdams HH (2007) High-throughput identification of transcription start sites, conserved promoter motifs and predicted regulons. Nat Biotechnol 25:584–592
9. Croucher NJ, Fookes MC, Perkins TT, Turner DJ, Marguerat SB, Keane T, Quail MA, He M, Assefa S, Bahler J, Kingsley RA, Parkhill J, Bentley SD, Dougan G, Thomson NR (2009) A simple method for directional transcriptome sequencing using Illumina technology. Nucleic Acids Res 37:e148
10. Cloonan N, Forrest AR, Kolle G, Gardiner BB, Faulkner GJ, Brown MK, Taylor DF, Steptoe AL, Wani S, Bethel G, Robertson AJ, Perkins AC, Bruce SJ, Lee CC, Ranade SS, Peckham HE, Manning JM, McKernan KJ, Grimmond SM (2008) Stem cell transcriptome profiling via massive-scale mRNA sequencing. Nat Methods 5:613–619
11. He Y, Vogelstein B, Velculescu VE, Papadopoulos N, Kinzler KW (2008) The antisense transcriptomes of human cells. Science 322:1855–1857

12. Parkhomchuk D, Borodina T, Amstislavskiy V, Banaru M, Hallen L, Krobitsch S, Lehrach H, Soldatov A (2009) Transcriptome analysis by strand-specific sequencing of complementary DNA. Nucleic Acids Res 37:e123
13. Frias-Lopez J, Shi Y, Tyson GW, Coleman ML, Schuster SC, Chisholm SW, Delong EF (2008) Microbial community gene expression in ocean surface waters. Proc Natl Acad Sci U S A 105:3805–3810
14. Raabe CA, Tang TH, Brosius J, Rozhdestvensky TS (2014) Biases in small RNA deep sequencing data. Nucleic Acids Res 42:1414–1426
15. t Hoen PA, Friedlander MR, Almlof J, Sammeth M, Pulyakhina I, Anvar SY, Laros JF, Buermans HP, Karlberg O, Brannvall M, den Dunnen JT, van Ommen GJ, Gut IG, Guigo R, Estivill X, Syvanen AC, Dermitzakis ET, Lappalainen T (2013) Reproducibility of high-throughput mRNA and small RNA sequencing across laboratories. Nat Biotechnol 31:1015–1022
16. Thompson JA, Radonovich MF, Salzman NP (1979) Characterization of the 5′-terminal structure of simian virus 40 early mRNA's. J Virol 31:437–446
17. Argaman L, Hershberg R, Vogel J, Bejerano G, Wagner EG, Margalit H, Altuvia S (2001) Novel small RNA-encoding genes in the intergenic regions of *Escherichia coli*. Curr Biol 11:941–950
18. Vogel J, Bartels V, Tang TH, Churakov G, Slagter-Jager JG, Huttenhofer A, Wagner EG (2003) RNomics in *Escherichia coli* detects new sRNA species and indicates parallel transcriptional output in bacteria. Nucleic Acids Res 31:6435–6443
19. Bensing BA, Meyer BJ, Dunny GM (1996) Sensitive detection of bacterial transcription initiation sites and differentiation from RNA processing sites in the pheromone-induced plasmid transfer system of *Enterococcus faecalis*. Proc Natl Acad Sci U S A 93:7794–7799
20. Wurtzel O, Sapra R, Chen F, Zhu Y, Simmons BA, Sorek R (2010) A single-base resolution map of an archaeal transcriptome. Genome Res 20:133–141
21. Mendoza-Vargas A, Olvera L, Olvera M, Grande R, Vega-Alvarado L, Taboada B, Jimenez-Jacinto V, Salgado H, Juarez K, Contreras-Moreira B, Huerta AM, Collado-Vides J, Morett E (2009) Genome-wide identification of transcription start sites, promoters and transcription factor binding sites in *E. coli*. PLoS One 4:e7526
22. Cho BK, Zengler K, Qiu Y, Park YS, Knight EM, Barrett CL, Gao Y, Palsson BO (2009) The transcription unit architecture of the *Escherichia coli* genome. Nat Biotechnol 27:1043–1049
23. Sharma CM, Vogel J (2014) Differential RNA-seq: the approach behind and the biological insight gained. Curr Opin Microbiol 19:97–105
24. Juranek S, Eban T, Altuvia Y, Brown M, Morozov P, Tuschl T, Margalit H (2012) A genome-wide view of the expression and processing patterns of *Thermus thermophilus* HB8 CRISPR RNAs. RNA 18:783–794
25. Zhang Y, Heidrich N, Ampattu BJ, Gunderson CW, Seifert HS, Schoen C, Vogel J, Sontheimer EJ (2013) Processing-independent CRISPR RNAs limit natural transformation in *Neisseria meningitidis*. Mol Cell 50:488–503
26. Dugar G, Herbig A, Forstner KU, Heidrich N, Reinhardt R, Nieselt K, Sharma CM (2013) High-resolution transcriptome maps reveal strain-specific regulatory features of multiple *Campylobacter jejuni* Isolates. PLoS Genet 9:e1003495
27. Hale CR, Majumdar S, Elmore J, Pfister N, Compton M, Olson S, Resch AM, Glover CV 3rd, Graveley BR, Terns RM, Terns MP (2012) Essential features and rational design of CRISPR RNAs that function with the Cas RAMP module complex to cleave RNAs. Mol Cell 45:292–302
28. Hale CR, Zhao P, Olson S, Duff MO, Graveley BR, Wells L, Terns RM, Terns MP (2009) RNA-guided RNA cleavage by a CRISPR RNA-Cas protein complex. Cell 139:945–956
29. Makarova KS, Haft DH, Barrangou R, Brouns SJ, Charpentier E, Horvath P, Moineau S, Mojica FJ, Wolf YI, Yakunin AF, van der Oost J, Koonin EV (2011) Evolution and classification of the CRISPR-Cas systems. Nat Rev Microbiol 9:467–477
30. Richter H, Zoephel J, Schermuly J, Maticzka D, Backofen R, Randau L (2012) Characterization of CRISPR RNA processing in *Clostridium thermocellum* and *Methanococcus maripaludis*. Nucleic Acids Res 40:9887–9896
31. Zhang J, Rouillon C, Kerou M, Reeks J, Brugger K, Graham S, Reimann J, Cannone G, Liu H, Albers SV, Naismith JH, Spagnolo L, White MF (2012) Structure and mechanism of the CMR complex for CRISPR-mediated antiviral immunity. Mol Cell 45:303–313
32. Bernick DL, Cox CL, Dennis PP, Lowe TM (2012) Comparative genomic and transcriptional analyses of CRISPR systems across the genus *Pyrobaculum*. Front Microbiol 3:251
33. Richter H, Lange SJ, Backofen R, Randau L (2013) Comparative analysis of Cas6b processing

and CRISPR RNA stability. RNA Biol 10:700–707

34. Kroger C, Colgan A, Srikumar S, Handler K, Sivasankaran SK, Hammarlof DL, Canals R, Grissom JE, Conway T, Hokamp K, Hinton JC (2013) An infection-relevant transcriptomic compendium for *Salmonella enterica* Serovar Typhimurium. Cell Host Microbe 14:683–695

35. Jager D, Sharma CM, Thomsen J, Ehlers C, Vogel J, Schmitz RA (2009) Deep sequencing analysis of the Methanosarcina mazei Go1 transcriptome in response to nitrogen availability. Proc Natl Acad Sci U S A 106:21878–21882

36. Zhelyazkova P, Sharma CM, Forstner KU, Liere K, Vogel J, Borner T (2012) The primary transcriptome of barley chloroplasts: numerous noncoding RNAs and the dominating role of the plastid-encoded RNA polymerase. Plant Cell 24:123–136

37. Albrecht M, Sharma CM, Reinhardt R, Vogel J, Rudel T (2010) Deep sequencing-based discovery of the Chlamydia trachomatis transcriptome. Nucleic Acids Res 38:868–877

38. Deltcheva E, Chylinski K, Sharma CM, Gonzales K, Chao Y, Pirzada ZA, Eckert MR, Vogel J, Charpentier E (2011) CRISPR RNA maturation by trans-encoded small RNA and host factor RNase III. Nature 471:602–607

39. Ghildiyal M, Zamore PD (2009) Small silencing RNAs: an expanding universe. Nat Rev Genet 10:94–108

40. Carte J, Wang R, Li H, Terns RM, Terns MP (2008) Cas6 is an endoribonuclease that generates guide RNAs for invader defense in prokaryotes. Genes Dev 22:3489–3496

41. Haurwitz RE, Jinek M, Wiedenheft B, Zhou K, Doudna JA (2010) Sequence- and structure-specific RNA processing by a CRISPR endonuclease. Science 329:1355–1358

42. Jore MM, Lundgren M, van Duijn E, Bultema JB, Westra ER, Waghmare SP, Wiedenheft B, Pul U, Wurm R, Wagner R, Beijer MR, Barendregt A, Zhou K, Snijders AP, Dickman MJ, Doudna JA, Boekema EJ, Heck AJ, van der Oost J, Brouns SJ (2011) Structural basis for CRISPR RNA-guided DNA recognition by Cascade. Nat Struct Mol Biol 18:529–536

43. Blomberg P, Wagner EG, Nordstrom K (1990) Control of replication of plasmid R1: the duplex between the antisense RNA, CopA, and its target, CopT, is processed specifically in vivo and in vitro by RNase III. EMBO J 9:2331–2340

44. Mattatall NR, Sanderson KE (1996) *Salmonella typhimurium* LT2 possesses three distinct 23S rRNA intervening sequences. J Bacteriol 178:2272–2278

45. Nickel L, Weidenbach K, Jager D, Backofen R, Lange SJ, Heidrich N, Schmitz RA (2013) Two CRISPR-Cas systems in *Methanosarcina mazei* strain Go1 display common processing features despite belonging to different types I and III. RNA Biol 10:779–791

46. Scholz I, Lange SJ, Hein S, Hess WR, Backofen R (2013) CRISPR-Cas systems in the cyanobacterium *Synechocystis* sp. PCC6803 exhibit distinct processing pathways involving at least two Cas6 and a Cmr2 protein. PLoS One 8:e56470

47. Aravin AA, Hannon GJ, Brennecke J (2007) The Piwi-piRNA pathway provides an adaptive defense in the transposon arms race. Science 318:761–764

48. Su AA, Tripp V, Randau L (2013) RNA-Seq analyses reveal the order of tRNA processing events and the maturation of C/D box and CRISPR RNAs in the hyperthermophile *Methanopyrus kandleri*. Nucleic Acids Res 41:6250–6258

49. Soutourina OA, Monot M, Boudry P, Saujet L, Pichon C, Sismeiro O, Semenova E, Severinov K, Le Bouguenec C, Coppee JY, Dupuy B, Martin-Verstraete I (2013) Genome-wide identification of regulatory RNAs in the human pathogen *Clostridium difficile*. PLoS Genet 9:e1003493

50. Berezikov E, Thuemmler F, van Laake LW, Kondova I, Bontrop R, Cuppen E, Plasterk RH (2006) Diversity of microRNAs in human and chimpanzee brain. Nat Genet 38:1375–1377

51. Chylinski K, Le Rhun A, Charpentier E (2013) The tracrRNA and Cas9 families of type II CRISPR-Cas immunity systems. RNA Biol 10:726–737

52. Fonfara I, Le Rhun A, Chylinski K, Makarova KS, Lecrivain AL, Bzdrenga J, Koonin EV, Charpentier E (2014) Phylogeny of Cas9 determines functional exchangeability of dual-RNA and Cas9 among orthologous type II CRISPR-Cas systems. Nucleic Acids Res 42:2577–2590

53. Konig J, Zarnack K, Luscombe NM, Ule J (2012) Protein-RNA interactions: new genomic technologies and perspectives. Nat Rev Genet 13:77–83

54. Chao Y, Papenfort K, Reinhardt R, Sharma CM, Vogel J (2012) An atlas of Hfq-bound transcripts reveals 3′ UTRs as a genomic reservoir of regulatory small RNAs. EMBO J 31:4005–4019

55. Brouns SJJ, Jore MM, Lundgren M, Westra ER, Slijkhuis RJH, Snijders APL, Dickman MJ, Makarova KS, Koonin EV, van der Oost J

(2008) Small CRISPR RNAs guide antiviral defense in prokaryotes. Science 321:960–964
56. Jinek M, Chylinski K, Fonfara I, Hauer M, Doudna JA, Charpentier E (2012) A programmable dual-RNA-guided DNA endonuclease in adaptive bacterial immunity. Science 337: 816–821
57. Sampson TR, Saroj SD, Llewellyn AC, Tzeng YL, Weiss DS (2013) A CRISPR/Cas system mediates bacterial innate immune evasion and virulence. Nature 497:254–257
58. Heidrich N, Vogel J (2013) CRISPRs extending their reach: prokaryotic RNAi protein Cas9 recruited for gene regulation. EMBO J 32:1802–1804
59. Maier LK, Lange SJ, Stoll B, Haas KA, Fischer S, Fischer E, Duchardt-Ferner E, Wohnert J, Backofen R, Marchfelder A (2013) Essential requirements for the detection and degradation of invaders by the *Haloferax volcanii* CRISPR/Cas system I-B. RNA Biol 10:865–874
60. Randau L (2012) RNA processing in the minimal organism *Nanoarchaeum equitans*. Genome Biol 13:R63
61. Elmore JR, Yokooji Y, Sato T, Olson S, Glover CV 3rd, Graveley BR, Atomi H, Terns RM, Terns MP (2013) Programmable plasmid interference by the CRISPR-Cas system in *Thermococcus kodakarensis*. RNA Biol 10: 828–840

# Chapter 2

## In Vitro Co-reconstitution of Cas Protein Complexes

André Plagens and Lennart Randau

### Abstract

CRISPR-Cas systems employ diverse and often multimeric CRISPR-associated (Cas) protein effector complexes to mediate antiviral defense. The elucidation of the mechanistic details and the protein interaction partners requires production of recombinant Cas proteins. However, these proteins are often produced as inactive inclusion bodies. Here, we present a detailed protocol for the isolation and purification of insoluble Cas proteins. Guidelines for their solubilization via co-reconstitution strategies and procedures to upscale the production of soluble multimeric Cas protein complexes are provided.

**Keywords** CRISPR-Cas system, Cas protein, Cascade, Inclusion body, Solubilization, Co-reconstitution

## 1 Introduction

The production of recombinant proteins in a chosen microbial host system is frequently used to investigate protein function and interaction partners, and to engineer proteins for biotechnological applications. These types of studies and applications require the production of soluble proteins. However, in many cases the recombinant protein accumulates intracellularly as inclusion bodies (IB), i.e., insoluble aggregates of inactive and denatured protein [1]. Numerous studies have addressed the isolation of IBs from different gene expression systems [2, 3], theoretical and mechanical aspects of protein folding [4, 5], and application strategies for protein renaturation [6–9]. The formation of IB is commonly observed for the production of recombinant Cas proteins in *Escherichia coli* strains; this has impeded the biochemical analysis of Cas protein function and complex formation. Possible reasons for IB formation include the notion that the activities of individual Cas proteins (e.g., helicase, nucleic-acid binding, nuclease) cause cellular toxicity and that the folding of the large multimeric ribonucleoprotein complexes (e.g., the Cascade interference complex) requires a coordinated and stoichiometrically balanced assembly strategy. Here, we present a technique for the in vitro co-reconstitution of interacting Cas proteins based on a model archaeal multimeric Cascade complex. The expression of recombinant proteins in IBs offers advantages in comparison to the time-consuming purification of proteins that exhibit limited solubility and stability. The IB proteins can be obtained in high amounts and purity even in small-scale expression systems. The proteins are usually protected from proteolytic degradation and protein or nucleic acid contamination can be minimized [10]. Our protocol details the isolation (*see* Subheading 3.1) and solubilization (*see* Subheading 3.2) of IBs of individual Cas proteins in vitro [11]. The major task is to recover biologically active and soluble protein with sufficient yield. Typically, each protein exhibits specific characteristics during the reconstitution procedure and requires individual cofactors for stability and activity. To establish a strategy for the refolding of a Cas protein effector complex, the interacting proteins need to be defined. This can be addressed experimentally or deduced from the computational classification of Cas proteins into different types and subtypes and the identification of gene co-occurrence rules [12, 13]. General conditions for the refolding procedure require testing and adjustment for individual Cas protein complexes. The possible adjustments include pH value and buffer system, temperature, ionic conditions, low-molecular-weight additives, and addition of functional nucleic acids (e.g., CRISPR RNAs (crRNAs)). Here we provide a general strategy for the co-reconstitution of Cas protein complexes with a focus on the successful assembly of a co-refolded archaeal Cascade complex from *Thermoproteus tenax* (subtype I-A, [14]). The provided co-refolding assembly strategy uses four individual solubilized IB proteins (Cas5a, Cas3′, Cas3″, and Cas8a2) together with two purified soluble proteins (Csa5 and Cas7) (*see* Subheading 3.3). In addition, a procedure to upscale the production and purification of soluble multimeric Cascade complex in a range of mg/mL is described (*see* Subheading 3.4).

## 2 Materials

Prepare all solutions using ultrapure water and analytical grade reagents. Prepare and store all reagents at room temperature (unless otherwise indicated).

### 2.1 Components for Purification and Solubilization of Inclusion Bodies

1. Homogenization buffer: 100 mM Tris–HCl pH 7, 1 mM EDTA pH 8. For the preparation of homogenization buffer, mix 5 mL of a 1 M Tris–HCl stock solution (pH 7) and 100 μL of a 500 mM EDTA stock solution (adjusted to pH 8 with NaOH) and add water to a final volume of 50 mL.

2. Cell lysis components: Lysozyme powder (Fluka), ultrasonic homogenizer (Branson, Sonifier 250 set to a duty cycle of 40 % and an output control setting of 5; use six cycles of 30-s on and 30-s off).

3. DNase I (Sigma, 400 U/mg): Dissolve 10 mg DNase I in homogenization buffer to a final volume of 1 mL. 1 M $MgCl_2$ ($MgCl_2 \cdot 6H_2O$, hexahydrate): Dissolve 2.03 g $MgCl_2$ in 10 mL $H_2O$.

4. Detergent buffer: 60 mM EDTA pH 8, 6 % Triton X-100, 1.5 M NaCl. Dissolve 8.77 g of NaCl in 12 mL of 500 mM EDTA pH 8 stock solution, 6 mL of Triton X-100 (Sigma), and water to a final volume of 100 mL.

5. Washing buffer: 100 mM Tris–HCl pH 7, 20 mM EDTA pH 8. Mix 20 mL of the 1 M Tris–HCl pH 7 stock solution and 8 mL of the 500 mM EDTA pH 8 stock solution and adjust the volume to 200 mL with water.

6. Solubilization buffer: 6 M guanidine-HCl (Roth), 100 mM Tris–HCl pH 8, 100 mM DTT (Roth), 1 mM EDTA pH 8. Dissolve 2.87 g guanidine-HCl and 77 mg DTT in 0.5 mL of a 1 M Tris–HCl stock solution (pH 8), 10 μL of a 500 mM EDTA stock, and water to a final volume of 5 mL.

7. 1 M HCl: Dilute 8.2 mL of concentrated HCl (37.5 %) into 91.8 mL of water.

8. Dialysis buffer 1: 4 M guanidine-HCl, 10 mM HCl (dialysis buffer 2: 4 M guanidine-HCl). Dissolve 382.12 g guanidine-HCl in 10 mL of the 1 M HCl stock solution and water to a final volume of 1 L. Use dialysis tubing with an appropriate molecular weight exclusion (ZelluTrans T2, MWCO: 6,000–8,000 Da, Roth), seal with clamps, and equilibrate on a magnetic stirrer (IKA® RCT classic).

9. SDS-PAGE components: Standard SDS-PAGE methodology can be employed. We used 2× SDS-loading buffer (150 mM Tris–HCl pH 6.8, 1.2 % SDS, 30 % glycerol, 15 % ß-mercaptoethanol (ß-Me), 0.2 mg bromophenol blue, 10 mL), the Mini-PROTEAN® Tetra Cell system (BioRad), the ColorPlus™ Prestained Protein Ladder (Broad Range, 10–230 kDa, NEB), and the following staining/destaining solutions (staining: 2.5 g Coomassie Brilliant Blue R-250, 45 % methanol, 10 % acetic acid, 1 L; destaining: same but without Coomassie stain).

10. Protein concentration estimation: Dissolve 2 mg BSA in 10 mL of 4 M guanidine-HCl (200 μg/mL BSA stock) and use for the calibration according to the BioRad Protein Assay [15].

## 2.2 Components for Co-refolding of Cas Complexes

1. Buffer combinations listed in Table 1 are prepared with stock solutions. Additional stock solutions (not covered in Subheading 2.1): 80 % glycerol stock: Mix 800 mL of glycerol (100 %) with 200 mL of water and autoclave. 5 M NaCl stock: Dissolve 29.22 g NaCl in water to a final volume of 100 mL and autoclave. 0.5 M CHES buffer: Dissolve 10.36 g CHES in 80 mL water, adjust to pH 9 with NaOH, and add water to

## Table 1
Scheme of initial buffer mixtures for the optimization of co-refolding conditions

|   | Stocks | μL | 1 | 2 | 3 | 4 | 5 | 6 | 7 | 8 | 9 | 10 | 11 | 12 | 13 | 14 | 15 | 16 |
|---|---|---|---|---|---|---|---|---|---|---|---|---|---|---|---|---|---|---|
| 5 % glycerol | 80 % | 62.5 | x | x | x | x | x | x | x | x | x | x | x | x | x | x | x | x |
| 0.3 M NaCl | 5 M | 60 | x | x | x | x | x | x | x | x | x | x | x | x | x | x | x | x |
| 10 mM ß-Me | 14.3 M | 0.7 | x | x |   |   | x | x |   |   | x | x |   |   | x | x |   |   |
| HEPES pH 7* | 1 M | 100 | x | x | x | x |   |   |   |   |   |   |   |   |   |   |   |   |
| Tris pH 7* | 1 M | 100 |   |   |   |   | x | x | x | x |   |   |   |   |   |   |   |   |
| Tris pH 8* | 1 M | 100 |   |   |   |   |   |   |   |   | x | x | x | x |   |   |   |   |
| CHES pH 9* | 0.5 M | 200 |   |   |   |   |   |   |   |   |   |   |   |   | x | x | x | x |
| RT | – | – | x |   | x |   | x |   | x |   | x |   | x |   | x |   | x |   |
| 4 °C | – | – |   | x |   | x |   | x |   | x |   | x |   | x |   | x |   | x |
| 10 mM Mg | 1 M | 10 | x | x |   |   | x | x |   |   | x | x |   |   | x | x |   |   |
| 0.3 mM GSSG | 10 mM | 30 |   |   | x | x |   |   | x | x |   |   | x | x |   |   | x | x |
| 3 mM GSH | 100 mM | 30 |   |   | x | x |   |   | x | x |   |   | x | x |   |   | x | x |
| 1 mM EDTA | 0.5 M | 2 |   |   | x | x |   |   | x | x |   |   | x | x |   |   | x | x |
| Add H₂O, vol. 1 mL |   |   |   |   |   |   |   |   |   |   |   |   |   |   |   |   |   |   |

Listed buffer ingredients marked by an asterisk are at a final concentration of 100 mM in the folding solution

a final volume of 100 mL. 10 mM GSSG/100 mM GSH: Dissolve 6.1 mg oxidized glutathione (GSSG) or 30.73 mg reduced glutathione (GSH) in a 1 mL volume of water.

2. Protein precipitation solutions: Trichloroacetic acid (TCA) 100 % (w/v; 500 g dissolved in 227 mL water) and cold acetone 100 %.

3. Co-refolding start buffer (15 mL): 3.5 M guanidine-HCl, 100 mM HEPES/KOH pH 7, 5 % glycerol, 300 mM NaCl, 10 mM $MgCl_2$, 10 mM ß-Me. Dissolve 5.02 g guanidine-HCl in 1.5 mL HEPES (1 M HEPES/KOH stock pH 7), 938 μL glycerol (80 % glycerol stock), 900 μL NaCl (5 M NaCl stock), 150 μL $MgCl_2$ (1 M $MgCl_2$ stock), 10.5 μL ß-Me, and water to a final volume of 15 mL.

4. Co-refolding dialysis buffer 1: 3 M guanidine-HCl, 100 mM HEPES/KOH pH 7, 5 % glycerol, 300 mM NaCl, 10 mM $MgCl_2$, 10 mM ß-Me. Dissolve 114.64 g guanidine-HCl, 7.01 g NaCl, and 0.81 g $MgCl_2$ and add 40 mL of the HEPES stock solution, 25 mL glycerol (80 % stock), and 280 μL ß-Me in water to a final volume of 400 mL.

5. Co-refolding dialysis buffer 2: 100 mM HEPES/KOH pH 7, 5 % glycerol, 300 mM NaCl, 10 mM MgCl$_2$, 10 mM ß-Me. Dissolve 43.83 g NaCl and 5.08 g MgCl$_2$ and add 250 mL of the HEPES stock solution, 156 mL glycerol (80 % stock), and 1.75 mL ß-Me in water to a final volume of 2.5 L.

6. Concentrating of Cascade proteins was carried out with Amicon® Ultra Centrifugal Filter Units (Millipore) used according to the manufacturer's instructions.

## 3 Methods

### 3.1 Isolation of Inclusion Bodies

1. Resuspend 5 g *E. coli* cell pellet in 25 mL in homogenization buffer and vortex or mix thoroughly until no cell clumps are visible (*see* **Note 1**). Add 5 mg lysozyme to the suspension, vortex, and incubate for 30 min at 4 °C.

2. Lyse the cells with an ultrasonic homogenizer (six cycles of 30-s on and 30-s break). During this procedure, cool the sample on ice to prevent foaming, which inhibits the efficiency of cell lysis (*see* **Note 2**).

3. Reduce the viscosity of the suspension, caused by the liberated DNA, by adding 125 µL of the DNase I stock (500 U) and 250 µL of the 1 M MgCl$_2$ (final concentration: 10 mM) stock. Mix the suspension and incubate at 25 °C for 30 min on a rotary shaker.

4. Add 12.5 mL of the detergent buffer to the cell suspension to maximize the efficiency of lysis and incubate for 30 min at 4 °C. This step is required to enhance the purity of the IB preparation.

5. Centrifuge the cell lysate at 3,700 × *g* for 15 min at 4 °C. The pellet is difficult to resuspend; using the ultrasonic homogenizer (two cycles of 30-s on and 30-s off) alleviates this problem. Repeat the centrifugation step and decant the supernatant.

6. Resuspend the pellet in 40 mL washing buffer as described in Subheading 3.1.5. Recentrifuge the suspension a third time (3,700 × *g*, 15 min, 4 °C), decant the supernatant, and repeat the centrifugation a fourth time with fresh 40 mL washing buffer (*see* **Note 3**). After decanting the wash buffer the obtained IB pellet can be stored for a few days at −20 °C (*see* **Note 4**).

### 3.2 Solubilization of Inclusion Bodies

1. To obtain a Cas protein concentration of ~5 mg/mL, add 5 mL fresh solubilization buffer to 500 mg IB pellet (*see* **Note 5**). Resuspend the pellet by pipetting the suspension up and down for a few minutes and then incubate for 3 h at 25 °C on rotary shaker. Typically, at the beginning there are small pellet clumps visible, which should be completely resuspended after 1–2 h of incubation.

2. Add 250 µL of a 1 M HCl stock solution and check to insure that pH of the protein solution is between pH 3 and 4. If this is not the case, titrate small amounts of HCl to reach the desired pH range. Subsequently, spin down the insoluble cell debris ($10,000 \times g$, 10 min, 4 °C).

3. To remove the DTT from the supernatant, add to dialysis tubes that have been thoroughly rinsed with water and then tightly closed with clamps. Dialyze for 2 h at room temperature against 1 L dialysis buffer 1 on a magnetic stirrer at 100 rpm. Take the tube out and dialyze overnight at 4 °C against 1 L dialysis buffer 2. After the final dialysis aliquot the Cas protein and store at −80 °C (see **Note 6**).

4. Check the purity of the protein by adding 10 µL protein solution directly into a tube with 1 mL of distilled water. The fast dilution in water leads to precipitation of the protein, which can be obtained as a white pellet after centrifugation ($16,000 \times g$, 10 min, 4 °C). Dissolve the pellet in 10 µL homogenization buffer, add 10 µL of 2× SDS loading buffer, and heat for 5 min at 95 °C. Load the sample in a lane next to 5 µL of the protein marker on a 10–15 % SDS-PAGE. After destaining of the SDS gel the isolated Cas protein should have a purity of at least 90 % (see **Note 7**). Figure 1 shows the purity of the four isolated Cascade subunits Cas5a, Cas3′, Cas3″, and Cas8a2 of *T. tenax*.

5. The protein concentration has to be estimated in comparison to the calibration of BSA dissolved in 4 M guanidine-HCl as the protein standard.

## 3.3 Optimization of Co-refolding Conditions

1. The rapid dilution of denatured proteins is helpful to test a large number of Cas protein combinations and buffer conditions (see **Note 8**). For example, for I-A Cascade we tested the refolding efficiency by simultaneous reconstitution of five or six proteins (Fig. 2a–g) [16]. Additionally, Table 1 shows a scheme for buffer mixtures that provide possible initial directions for the optimization of co-refolding conditions.

2. Prepare 1 mL of each buffer mixture to use for rapid dilution. Set up a mixture with 25 µg of each of the relevant Cas proteins to be tested (for I-A Cascade: 3.5–5 µL of the insoluble proteins Cas5a, Cas3′, Cas3″, and Cas8a2, 10–12.5 µL of the soluble and purified proteins Csa5 and Cas7).

3. Add the protein mixture drop by drop to 1 mL of the test buffer to allow rapid protein refolding. The protein concentration for each drop should be in a range of 1–3 µg (for I-A Cascade: 0.5 µL). After each drop close the tube, let it shake for 1 min at 500 rpm in a reaction tube mixer, and continue with the procedure until the complete protein solution is added to the 1 mL buffer mixture. In total, incubate for 1 h at room temperature with continuous mixing at 500 rpm (see **Note 9**).

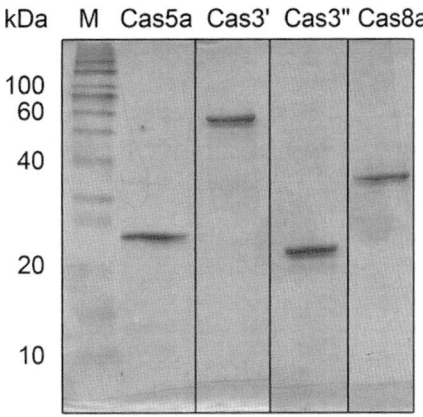

**Fig. 1** Purity of isolated and solubilized inclusion body Cas proteins. The inclusion bodies of the insoluble *T. tenax* I-A Cascade subunits Cas5a, Cas3', Cas3", and Cas8a2 were isolated and solubilized in guanidine-HCl and subsequently loaded on a 15 % SDS gel to check for the protein production and *E. coli* protein contaminations

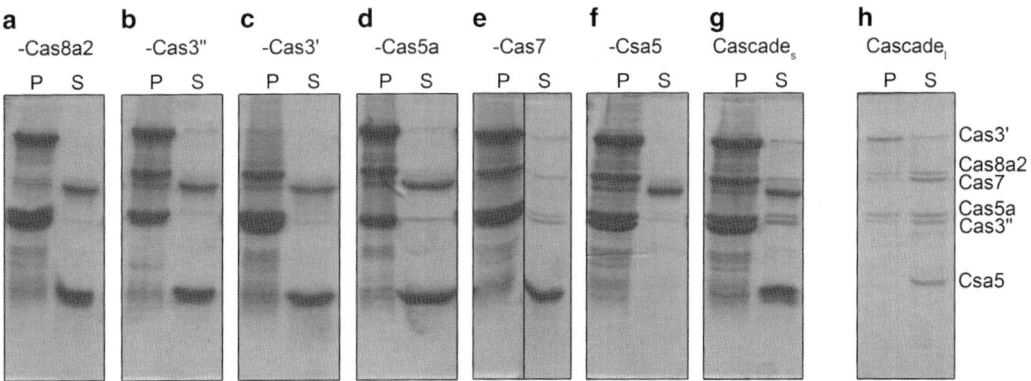

**Fig. 2** Establishment of a co-refolding strategy to obtain soluble I-A Cascade complex. Different combinations of five (**a–f**) or all six (**g**) *T. tenax* I-A Cascade subunits were mixed (25 µg of Cas5a, Cas3', Cas3", Cas8a2 in solubilized form, 25 µg of purified Csa5 and Cas7 via standard chromatography techniques) and refolded by rapid dilution in buffer combination 1 (Table 1). After separation of insoluble (P) and soluble protein (S) complexes, both fractions were loaded on a 15 % SDS gel. The co-refolding of all six I-A Cascade subunits (Cascade$_s$) results in the highest amount of soluble protein. These six protein subunits and buffer combination 1 were used for the upscale refolding procedure via stepwise dialysis (Cascade I, **h**), yielding sufficient amounts of pure I-A Cascade for biochemical activity assays [16]

4. Centrifuge each tube (13,000 ×$g$, 30 min, 4 °C) and decant the supernatant without disturbing the pellet to separate insoluble and refolded soluble protein. To compare the amount of soluble and insoluble protein, resuspend the protein pellet in 15 µL homogenization buffer and add 15 µL 2× SDS-loading buffer.

5. Set up a TCA precipitation of the 1 mL supernatant by adding 250 μL of the TCA stock, mix, incubate for 10 min at 4 °C, and centrifuge (16,000×g, 10 min, RT). Decant the supernatant, wash with 200 μL cold acetone, and centrifuge again (16,000×g, 2 min, RT). Repeat the acetone washing step, then resuspend the pellet in 15 μL homogenization buffer, and add 15 μL 2× SDS-loading buffer.

6. Load both the pellet and the supernatant fraction onto a 15 % SDS-PAGE alongside the protein marker and after staining of the gel compare the amount of denatured and refolded protein for the tested conditions (*see* **Note 10**).

### 3.4 Upscale of the Co-refolding Procedure

1. The tested buffer conditions and protein:protein co-refolding mixtures that exhibited the highest recovery of soluble protein are used for the following upscale procedure.

2. Use 12 mL of the fresh co-refolding start buffer and add 300 μg of each protein (for I-A Cascade: 20–40 μL of the insoluble proteins Cas5a, Cas3′, Cas3″, and Cas8a2, 125–150 μL of the soluble and purified proteins Csa5 and Cas7); bring the volume of the mix to 15 mL with the residual co-refolding start buffer (*see* **Note 11**). Transfer the protein mixture into a clean dialysis tube and tightly close with clamps.

3. Place the dialysis tube in 400 mL co-refolding dialysis buffer 1 in a 500 mL plastic beaker and check that the tubing is completely covered with liquid. Dialyze for 2 h at room temperature on the magnetic stirrer with 100 rpm.

4. Lower stepwise the concentration of the denaturant guanidine-HCl, by carefully pouring the 400 mL dialysis buffer 1 and the dialysis tube into a 1 L plastic beaker, add 200 mL of the prepared co-refolding dialysis buffer 2, and continue the dialysis for 2 h. Next add an additional 200 mL of the dialysis buffer 2 to the container and dialyze for 1.5 h at room temperature.

5. Transfer the 800 mL dialysis solution and the dialysis tube in a 2 L plastic beaker, add 400 mL of co-refolding dialysis buffer 2, and mix for 1.5 h on the magnetic stirrer.

6. Carefully pour 600 mL dialysis buffer out off the beaker, add 600 mL of dialysis buffer 2, and mix again for 1.5 h on the magnetic stirrer. To remove the last traces of guanidine-HCl, pour off the complete dialysis buffer, add 1 L of dialysis buffer 2, and dialyze for 1.5 h.

7. Typically, aggregates of misfolded protein or dead-end subcomplexes will appear during the co-refolding procedure. Therefore, clear the supernatant by spinning down the precipitate (13,000×g, 30 min, 4 °C) and concentrate the 15 mL protein solution with the Amicon unit to about 1 mL (Fig. 2h) (*see* **Note 12**).

## 4 Notes

1. A simple check for the presence of IBs in a small-scale (50 mL) *E. coli* expression culture can be achieved after standard cell lysis, ultracentrifugation (50,000 ×$g$, 1 h, 4 °C), and an SDS-PAGE of the supernatant and the pellet fraction. If the majority of the protein produced is visible in the pellet fraction, a 1 L expression culture usually delivers enough cell material to start the IB isolation.

2. Different methods of lysing *E. coli* cells also work efficiently for the isolation of IBs. Cell homogenization can be achieved with a French pressure cell disruptor or the freezing/grinding of cells in liquid nitrogen.

3. The intensive washing procedure of the isolated IB is very important to yield clean protein. In the beginning, the protein pellet is whitish smooth; it should have an ocher color and pasty texture in the end. Repeat the IB washing steps to get a cleaner IB pellet if necessary.

4. The storage of isolated IBs at −20 °C should be as short as possible to maximize the protein concentration. The best way is to continue directly with the solubilization of protein or the storage for 1 or 2 days at most. Longer storage times significantly reduce the yield of protein.

5. The typical IB yield of a 1 L *E. coli* expression culture is between 0.5 and 1 g depending on the Cas protein. Avoid using too much IB pellet for solubilization as it gets harder to dissolve the pellet in the solubilization buffer. An alternative weaker denaturant is 8 M urea which can also be used for the solubilization of proteins. From our experiences with all tested Cas proteins, urea did not solubilize reasonable amounts of protein.

6. Proteins solubilized in 4 M guanidine-HCl can be stored at −80 °C for several months or up to 1 year without any negative effect on stability or refolding efficiency.

7. If the purity of the obtained Cas protein is not high enough or a lot of contamination bands are visible on the SDS gel, an additional purification step can be integrated here. For His-tagged proteins, the solubilized protein can be loaded onto a 5 mL Ni-NTA column that is equilibrated and washed with dialysis buffer 2 and subsequently eluted with dialysis buffer 2 containing 500 mM imidazole. Untagged Cas proteins can also be purified via size-exclusion chromatography under denaturing conditions.

8. The folding of proteins with known disulfide bonds needs a special handling to form the covalent chemical bonds of the cysteine residues in the refolding process. Mixtures of oxidized

and reduced glutathione (GSSG/GSH) are used to facilitate the disulfide bond formation and increase the yield of refolded protein. In general, the buffer composition for the co-refolding of Cas proteins or complexes has to be examined experimentally. Several low-molecular-weight additives (L-arginine, lauryl maltoside, Triton X-100), ions, supporting proteins (BSA, chaperones), or functional nucleic acids, e.g., crRNAs, can help in folding proteins.

9. During the process of rapid dilution, it is very important to keep the protein concentration during each step as low as possible. High local concentrations of protein tend to cause the immediate precipitation in the native buffer.

10. The optimization of the co-refolding procedure only provides a qualitative clue to identify the buffer composition that works best for the examined Cas protein complex. Usually most of the protein still precipitates and is found in the pellet fraction.

11. In the process of stepwise dialysis for the recovery of higher concentrations of co-refolded Cas proteins, we obtained the best results for protein concentrations of 100–200 μg/mL in the beginning of the dialysis.

12. The refolded and concentrated Cas protein complex can now be used for measuring the protein concentration, SDS-PAGE analysis, or further purification via different chromatography techniques. Subsequently, biochemical assays should be employed to verify the correct folding of the Cas protein complex and its activity [16].

## Acknowledgement

The authors thank Patrick Dennis for critical reading of the manuscript. This work was supported by the Deutsche Forschungsgemeinschaft (DFG, FOR1680) and the Max Planck Society.

## References

1. Clark EDB (1998) Refolding of recombinant proteins. Curr Opin Biotechnol 9(2):157–163, doi:10.1016/S0958-1669(98)80109-2
2. Rudolph R, Lilie H (1996) In vitro folding of inclusion body proteins. FASEB J 10(1):49–56
3. Burgess RR (2009) Refolding solubilized inclusion body proteins. Methods Enzymol 463:259–282. doi:10.1016/S0076-6879(09)63017-2
4. Umetsu M, Tsumoto K, Ashish K, Nitta S, Tanaka Y, Adschiri T, Kumagai I (2004) Structural characteristics and refolding of in vivo aggregated hyperthermophilic archaeon proteins. FEBS Lett 557(1–3):49–56
5. Creighton TE (1988) Toward a better understanding of protein folding pathways. Proc Natl Acad Sci U S A 85(14):5082–5086
6. Burgess RR (1996) Purification of overproduced Escherichia coli RNA polymerase sigma factors by solubilizing inclusion bodies and refolding from Sarkosyl. Methods Enzymol 273:145–149, doi:10.1016/S0076-6879(96)73014-8

7. Misawa S, Kumagai I (1999) Refolding of therapeutic proteins produced in Escherichia coli as inclusion bodies. Biopolymers 51(4):297–307
8. Werner F, Weinzierl ROJ (2002) A recombinant RNA polymerase II-like enzyme capable of promoter-specific transcription. Mol Cell 10(3):635–646
9. Stempfer G, Holl-Neugebauer B, Rudolph R (1996) Improved refolding of an immobilized fusion protein. Nat Biotechnol 14(3):329–334. doi:10.1038/nbt0396-329
10. Lilie H, Schwarz E, Rudolph R (1998) Advances in refolding of proteins produced in E-coli. Curr Opin Biotechnol 9(5):497–501
11. Rudolph R, Böhm G, Lilie H, Jaenicke R (1997) Folding proteins. In: Creighton TE (ed) Protein function: a practical approach, 2nd edn. IRL Press, Oxford, pp 57–99
12. Makarova KS, Aravind L, Wolf YI, Koonin EV (2011) Unification of Cas protein families and a simple scenario for the origin and evolution of CRISPR-Cas systems. Biol Direct 6:38. doi:10.1186/1745-6150-6-38
13. Makarova KS, Haft DH, Barrangou R, Brouns SJ, Charpentier E, Horvath P, Moineau S, Mojica FJ, Wolf YI, Yakunin AF, van der Oost J, Koonin EV (2011) Evolution and classification of the CRISPR-Cas systems. Nat Rev Microbiol 9(6):467–477. doi:10.1038/nrmicro2577
14. Plagens A, Tjaden B, Hagemann A, Randau L, Hensel R (2012) Characterization of the CRISPR-Cas subtype I-A system of the hyperthermophilic crenarchaeon Thermoproteus tenax. J Bacteriol 194(10):2491–2500. doi:10.1128/JB.00206-12
15. Bradford MM (1976) Rapid and sensitive method for quantitation of microgram quantities of protein utilizing principle of protein-dye binding. Anal Biochem 72(1–2):248–254
16. Plagens A, Tripp V, Daume M, Sharma K, Klingl A, Hrle A, Conti E, Urlaub H, Randau L (2014) In vitro assembly and activity of an archaeal CRISPR-Cas type I-A Cascade interference complex. Nucleic Acids Res 42(8):5125-5138. doi: 10.1093/nar/gku120

# Chapter 3

## Analysis of CRISPR Pre-crRNA Cleavage

Erin L. Garside and Andrew M. MacMillan

### Abstract

We have examined the processing of precursor-clustered regularly interspaced short palindromic repeat (CRISPR) RNAs (pre-crRNAs) of the Type I CRISPR-Cas system by incubation of radiolabeled model RNAs with recombinant CRISPR-associated (Cas) endoribonucleases, followed by denaturing polyacrylamide gel electrophoresis (PAGE) of the products. Determination of cleavage position is based on comparison with RNase T1 digestion and base hydrolysis products. The mechanism of cleavage is investigated by chemical and enzymatic characterization of the reaction products as well as by the demonstration that a specific 2′-deoxy substitution 5′ to the scissile phosphate blocks endonucleolytic cleavage.

**Key words** Polyacrylamide gel electrophoresis, RNase T1 mapping, Base hydrolysis mapping, Periodate oxidation/base elimination

## 1 Introduction

Individual clustered regularly interspaced short palindromic repeat (CRISPR) loci in bacterial and archaeal genomes feature a cluster of repeats separated by short spacer elements; these also typically include a set of CRISPR-associated (Cas) genes coding for proteins necessary for CRISPR-based adaptation and nucleic acid interference [1–5]. Specific subtypes of CRISPR systems are defined by the structure and organization of the repeats as well as by the particular set of accompanying protein-coding *cas* genes [6–8]. Three major types of CRISPR-Cas systems that may be further divided into ten subtypes have been described [9]. In the type I and III CRISPR-Cas systems an initial pre-crRNA is cleaved to yield mature crRNA by a Cas6 endoribonuclease. The crRNA, along with additional Cas proteins, targets foreign nucleic acid complementary to the crRNA spacer for degradation (Fig. 1a).

In the CRISPR-Cas subtype I-E from *Escherichia coli* the endoribonuclease Cas6e (Cse3) cleaves the pre-crRNA at the base of a stem-loop structure formed by the repeat [2]. In order to elucidate the basis of pre-crRNA recognition and processing we

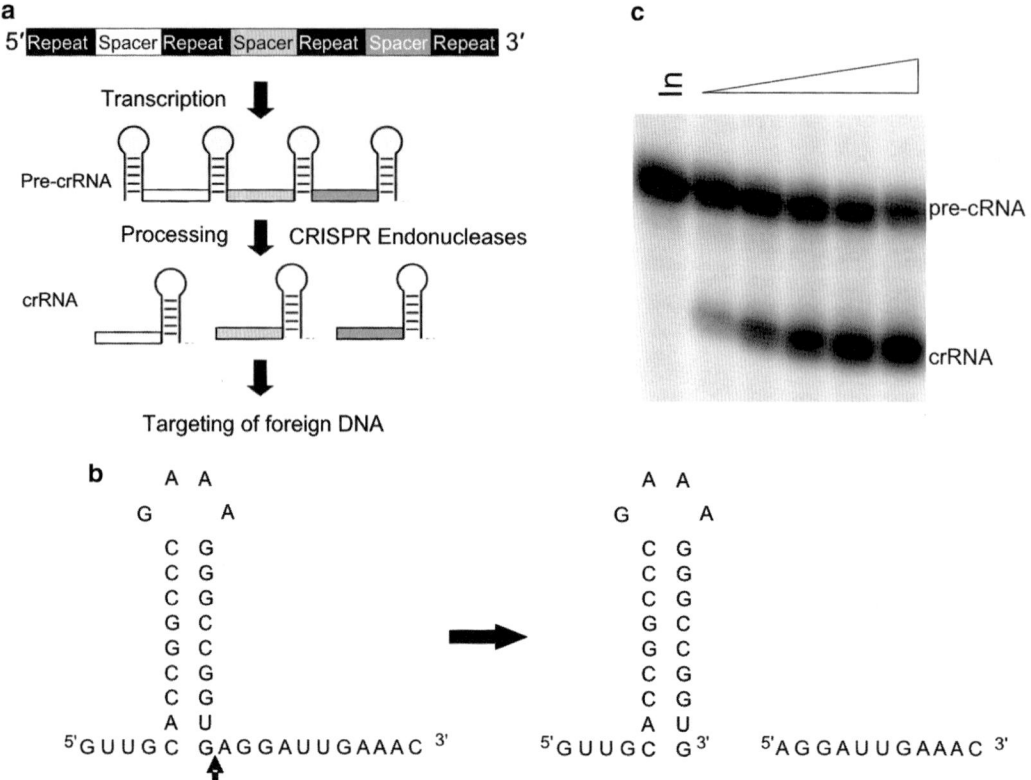

**Fig. 1** Processing of pre-crRNA by a CRISPR endoribonuclease. (**a**) Overview of the type I and III CRISPR-Cas defense mechanisms. The CRISPR DNA locus is transcribed to yield precursor CRISPR RNA (pre-crRNA), which is recognized and processed by a CRISPR endonuclease. The resulting CRISPR RNA (crRNA) activates cleavage of foreign DNA with a complementary sequence. (**b**) Sequence and cut site of a model pre-crRNA used in characterizing the CRISPR endonuclease from the I-C subtype. The location of cleavage is indicated with an arrow. (**c**) Cas5d cleaves the model pre-crRNA. Time course assay of 100 nM Cas5d incubation with 500 nM $^{32}$P 5′ end-labeled model pre-crRNA substrate shows the products of a specific cleaved crRNA increasing from 0 (*In* input) to 60 min

functionally and structurally characterized *T. thermophilus* Cas6e [10]. Using small model substrates, we were able to demonstrate RNA cleavage by Cas6e, identify the site of processing, and characterize the reaction mechanism by chemical and enzymatic characterization of the reaction products.

We have also recently characterized the activity of the CRISPR-Cas subtype I-C protein Cas5d from *T. thermophilus* [11]. A distinguishing feature of the I-C subtype is the lack of a gene coding for an ortholog of Cas6, the pre-crRNA processing RNA endonuclease in type I and III systems. It has been proposed, based on a bioinformatic analysis, that Cas5d might process the pre-crRNA in this system [9, 12]. Upon incubation of purified recombinant *T. thermophilus* Cas5d with RNA modeled on the pre-crRNA from CRISPR locus 5 on the TT27 plasmid we observed specific,

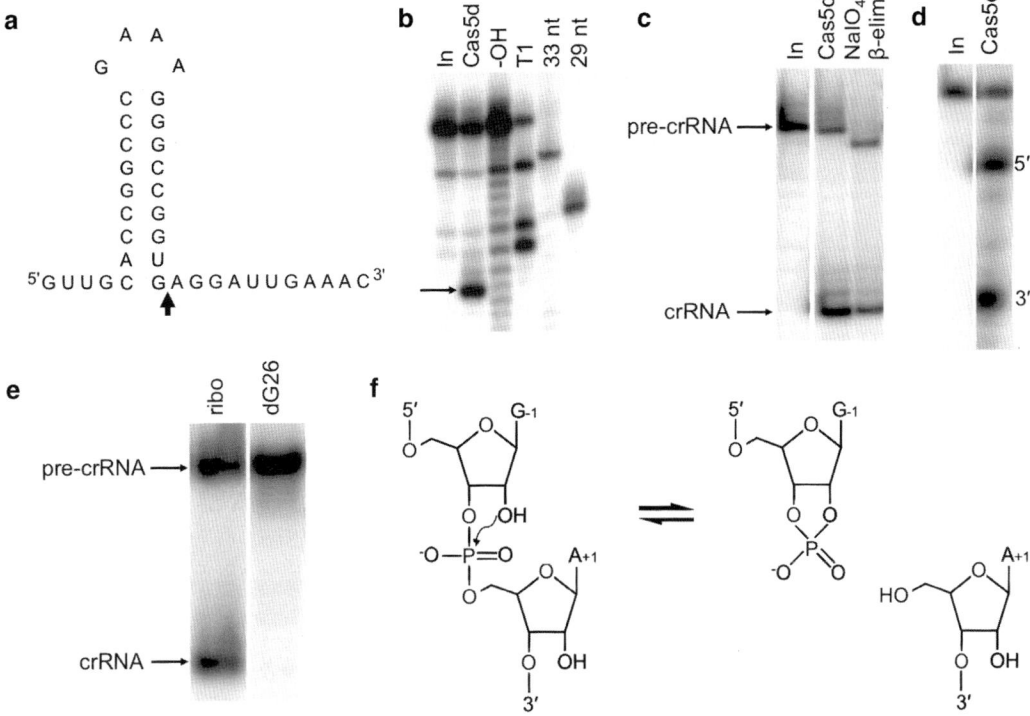

**Fig. 2** Characterization of CRISPR endonuclease Cas5d cleavage products. (**a**) Secondary structure of the model pre-crRNA hairpin used in these studies. The cleavage site, following G26, is indicated by an *arrow*. (**b**) RNase T1 treatment and base hydrolysis map the cleavage site to between G26 and A27. Cleaved crRNA is indicated by an *arrow*. RNAs 29 and 33 nucleotides in length were used for purposes of size comparison. (**c**) Periodate/base treatment of pre-crRNA with and without preincubation with Cas5d. Increased mobility of the full-length pre-crRNA but not the cleavage product indicates a modified 2′/3′ terminus on the latter. (**d**) $^{32}$P 5′ radiolabeling of the Cas5d cleavage products. The presence of a 5′ radiolabel on the 3′ cleavage product indicates a free 5′ hydroxyl on the initial cleavage product. (**e**) Processing of the model pre-crRNA substrate is inhibited by a deoxyG at position 26 consistent with the CRISPR endoribonuclease cleavage mechanism. (**f**) Proposed mechanism of pre-crRNA processing by the type I and III Cas endoribonucleases. The 2′ OH of the nucleotide 5′ to the cleavage site attacks the scissile phosphate generating cleavage products with a 2′–3′ cyclic phosphate on the 5′ cleavage product and a free 5′ hydroxyl on the 3′ cleavage product

Cas5d-dependent cleavage of the RNA (Fig. 1b, c). We then mapped the pre-crRNA cleavage site by comparison with RNase T1 and base hydrolysis ladders, and determined it to be 3′ to G26 at the base of a predicted RNA hairpin structure (Fig. 2a, b). The 5′ cleavage product was resistant to oxidation with periodate and base-mediated elimination, indicating a 3′ end lacking free 2′ and/or 3′ hydroxyls (Fig. 2c). The 3′ product could be 5′ end-labeled, indicating the presence of a 5′ hydroxyl (Fig. 2d). Additionally, substitution of a 2′-deoxy residue at the G26 position abolished cleavage in the presence of the enzyme (Fig. 2e), but the affinity of Cas5d for this RNA was unimpaired (data not shown). The generation of functionalized 2′/3′ and free 5′-hydroxyl termini on the longer and shorter products, respectively, and the disruption of processing by a deoxy residue 5′ to the cut site are consistent with a cleavage mechanism involving attack of the G26 2′-hydroxyl group on the scissile phosphate (Fig. 2f) as observed in both protein and RNA-catalyzed RNA cleavage reactions. As observed with the type I Cas6e [10, 13] and Cas6f (Csy4) [14] endoribonucleases, and the type III Cas6 endoribonuclease [3], pre-crRNA cleavage by Cas5d was found to be metal independent (data not shown). These results strongly suggest that Cas5d is a sequence-specific RNA endonuclease responsible for processing pre-crRNAs in the type I-C CRISPR-Cas pathway.

The previously characterized Cas6e-mediated pre-crRNA processing proceeds via a mechanism similar to that mediated by Cas5d as evidenced by chemical and enzymatic characterization of the products of Cas6e-dependent RNA cleavage. Additionally, high-resolution X-ray structural analysis of a Cas6e-RNA product complex unequivocally demonstrated the presence of a 2′–3′ cyclic phosphodiester consistent with the proposed mechanism. A number of other pre-crRNA processing endoribonucleases have been extensively characterized functionally [3, 13–27]. Despite differences in the structure and recognition of their pre-crRNA substrates evidence suggests that all type I and III Cas endoribonucleases proceed via a similar chemical mechanism.

The following protocol describes the characterization of the pre-crRNA cleavage activity of the Cas5d Cas endoribonuclease. The cleavage activity of other Cas endoribonucleases may be assayed by the same basic protocols.

## 2 Materials

Deionized water, purified to attain a sensitivity of 18.2 MΩ cm at 25 °C, is used to prepare all solutions. Unless otherwise indicated, prepare and store all reagents at room temperature.

### 2.1 CRISPR Pre-crRNA Cleavage Assay

1. Purified Cas endoribonuclease 1 µM (*see* **Note 1**).
2. $^{32}$P 5′ end-labeled model pre-CRISPR RNA (pre-crRNA) $50–100 \times 10^3$ cpm (*see* **Notes 2–5**).
3. Unlabeled CRISPR RNA (pre-crRNA) 5 µM.
4. NaCl 1 M.
5. Tris 100 mM pH 8: Dissolve 4.04 g Tris base and 10.5 g Tris–HCl in 1 L of water.
6. RNA loading dye (*see* Subheading 2.2, **item 4**).

### 2.2 Denaturing Acrylamide Gel

1. $15 \times 23$ cm $\times 1$ or 0.5 mm (*see* **Note 6**) 20 % 19:1 acrylamide:bisacrylamide gel in 8 M urea with wells 0.5 cm wide (*see* **Notes 7** and **8**) with associated vertical gel box.

2. Tris-borate-EDTA (TBE) buffer: To make 10× TBE: Dissolve 108 g Tris base, 55 g boric acid, and 7 g ethylenediaminetetraacetic acid (EDTA) in 1 L water. Dilute 10× TBE 100 mL in 1 L water to make 1× TBE.

3. 60 mL syringe with 18 G needle bent at 90°.

4. RNA loading dye (1× TBE, 9 M urea, 2 % SDS, 0.5 % bromophenol blue, and 0.5 % xylene cyanol in water) is prepared by mixing 1 mL 10× TBE, 5.4 g urea, 2 mL 10 % SDS, 0.05 g bromophenol blue, and 0.05 g xylene cyanol in 10 mL water (*see* **Note 9**).

5. Phosphor screen and phosphorimager (GE Healthcare).

## 2.3 Phenol/Chloroform Extraction and Ethanol Precipitation

1. Sodium acetate 0.3 M.

2. Buffer-saturated phenol: Phenol stored under 100 mM Tris–HCl pH 8 with antioxidant 8-hydroxyquinoline (0.1 %).

3. Chloroform:isopentyl alcohol (v/v 24:1) stored under 100 mM Tris–HCl pH 8.

4. Glycogen (Thermo Scientific molecular biology grade).

5. 100 % ethanol.

6. Speed vacuum concentrator (Savant).

## 2.4 Base Hydrolysis Mapping

1. Denaturing acrylamide gel from Subheading 2.2.

2. Sodium carbonate 1 M pH 9.0.

3. EDTA 10 mM (*see* **Note 10**).

4. Yeast tRNA (Sigma Aldrich) stored at 20 mg/mL stock in water.

5. $^{32}$P 5′ end-labeled model pre-crRNA $100-200 \times 10^3$ cpm.

## 2.5 RNase T1 Mapping

1. RNase T1 buffer (20 mM citrate, 7 M urea, 1 mM EDTA, 1 % concentrated HCl) is prepared by mixing 20 μL of 1 M citrate, 875 μL of 8 M urea, 2 μL 0.5 M EDTA, 93 μL water, and 10 μL concentrated HCl (*see* **Note 10**).

2. Yeast tRNA (Sigma Aldrich) stored as 20 mg/mL stock in water.

3. $^{32}$P 5′ end-labeled model pre-CRISPR RNA (pre-crRNA) $100-300 \times 10^3$ cpm.

4. RNase T1 (Invitrogen) 10 U/μL (*see* **Note 11**).

## 2.6 Oxidation and Base Elimination

1. $^{32}$P 5′ end-labeled pre-CRISPR RNA (pre-crRNA) $100-300 \times 10^3$ cpm cleaved as in Subheading 3.1 (*see* **Note 12**), phenol/chloroform extracted, and ethanol precipitated (*see* Subheading 3.3) but not resuspended.

2. Borate/boric acid 0.12 M pH 8.6.

3. Sodium periodate 200 mM.
4. Glycogen (Thermo Scientific molecular biology grade).
5. Lysine-HCl 1 M pH 9.3.

## 3 Methods

All procedures are carried out at room temperature unless otherwise indicated.

### 3.1 CRISPR RNA Cleavage Assay

1. Incubate a 10 µL cleavage reaction consisting of 100 nM purified Cas endoribonuclease, $50-100 \times 10^3$ cpm radiolabeled model pre-crRNA, 500 nM unlabeled model pre-crRNA, 100 mM NaCl, and 10 mM Tris–HCl pH 8.0 (*see* **Notes 13–15**) at room temperature for 30 min (*see* **Note 16**). Quench in an equal volume of RNA loading dye.
2. Run 10 µL of sample on a denaturing polyacrylamide gel for 2 h at 30 W (*see* Subheading 3.2 and **Note 17**) (Fig. 1c).

### 3.2 Denaturing Acrylamide Gel

1. Clamp gel into vertical gel box (*see* **Note 18**).
2. Fill top and bottom with 1× TBE:100 mL 10× TBE in 1 L of water.
3. Use the needle and syringe to remove air from the bottom of the gel and to rinse out the wells of the gel (*see* **Note 19**).
4. Add 5 µL RNA loading dye to the wells and pre-run the gel at 30 W for 20 min (*see* **Note 20**).
5. Rinse out the wells again (*see* **Note 19**).
6. Pipet 5–10 µL quenched cleavage reaction into each well. Also run an uncleaved sample (RNA without protein) as a size control.
7. Run the gel for 2 h at 30 W or until the fast dye (bromophenol blue, the darker dye) reaches the bottom of the gel (*see* **Note 21**).
8. Disassemble the gel rig. Dry off the glass plates, and carefully remove the top plate from the gel. Cover the gel with plastic wrap.
9. Expose the gel overnight in a phosphor cassette with a storage phosphor screen. Scan the phosphor screen in a phosphorimager (*see* **Note 22**).

### 3.3 Phenol/Chloroform Extraction and Ethanol Precipitation (See Note 23)

1. Make sample volume up to 400 µL in 0.3 M sodium acetate and 20 µg glycogen (*see* **Notes 24 and 25**).
2. Add 120 µL buffer-saturated phenol (*see* **Note 26**). Vortex the sample for 10 s to mix. Spin at least at $9.3 \times g$ for 2 min in microcentrifuge. Pipet off the aqueous upper layer and transfer to new centrifuge tube.

3. Add 120 μL chloroform (*see* **Note 27**). Vortex the sample for 10 s to mix. Spin at least at $9.3 \times g$ for 2 min in microcentrifuge. Pipette off the aqueous upper layer and transfer to a new centrifuge tube.

4. Add 1 mL ethanol and invert tube five times to mix. Freeze the sample at −20 °C or on dry ice for at least 20 min (*see* **Note 28**). Spin in cold microcentrifuge at $9.3 \times g$ for 20 min (*see* **Note 29**).

5. Carefully remove the ethanol so as not to disturb the RNA pellet. Wash the pellet in 120 μL ethanol. Spin in microcentrifuge for 2 min at $9.3 \times g$. Carefully remove the ethanol.

6. Dry the pellet in speed vacuum centrifuge, or leave tube open to air-dry.

### 3.4 Base Hydrolysis Mapping

1. Heat one 20 μL base hydrolysis reaction: 50 mM sodium carbonate pH 9.2, 1 mM EDTA, 2 μg yeast tRNA, and $100–200 \times 10^3$ cpm radiolabeled model pre-crRNA at 55 °C for 15 min (*see* **Note 30**).

2. Quench the reaction with a phenol/chloroform extraction followed by ethanol precipitation (*see* Subheading 3.3) and resuspend the resulting pellet in 50 μL of water.

3. Run 10 μL hydrolyzed RNA on denaturing gel (*see* Subheading 3.2) with cleaved CRISPR RNA (*see* Subheading 3.1) and RNase T1-treated RNA (*see* Subheading 3.5) to map the cleavage site on the RNA (*see* **Note 31**) (Fig. 2b).

### 3.5 RNase T1 Mapping

1. Incubate $100–300 \times 10^3$ cpm radiolabeled model pre-crRNA with 12 μg yeast tRNA in 70 μL RNase T1 buffer at 92 °C for 2 min, then directly on ice for 2 min, and then 55 °C for 2 min (*see* **Note 32**).

2. Add 3–9 μL RNase T1 (*see* **Notes 33** and **34**) and incubate at room temperature for 0.5–2 min (*see* **Note 35**).

3. Quench the reaction with phenol/chloroform extraction and ethanol precipitation (*see* Subheading 3.3). Resuspend the reaction in 50 μL water.

4. Run 10 μL digested RNA in 10 μL RNA loading dye on denaturing gel (*see* Subheading 3.2) with cleaved CRISPR RNA (*see* Subheading 3.1) and base-hydrolyzed RNA (*see* Subheading 3.4) to map the cleavage site on the RNA (*see* **Note 31**) (Fig. 2b).

### 3.6 Oxidation and Base Elimination (See Note 36)

1. Resuspend cleaved, precipitated RNA in a 90 μL reaction containing borate/boric acid (65 mM pH 8.6) and sodium periodate (27 mM). Incubate reaction at 0 °C in the dark for 1 h. Quench reaction with 200 μg glycogen.

2. Phenol/chloroform extract and ethanol precipitate the reaction (*see* Subheading 3.3 and **Note 37**).

3. Resuspend the pellet in 40 μL 1 M lysine-HCl. Incubate the reaction at 45 °C for 90 min.

4. Run 5 μL of reaction in 5 μL of RNA loading dye on denaturing acrylamide gel (*see* Subheading 3.2) with 0.5 mm thick spacers and comb alongside cleaved CRISPR RNA (*see* Subheading 3.1 and **Note 38**) (Fig. 2c).

# 4 Notes

1. This protocol is applicable to type I and III Cas endoribonucleases: any of the Cas6 enzymes (including Cas6f and Cas6e) as well as Cas5d.

2. We use a synthetic RNA modeled on the full-length CRISPR repeat.

3. To radiolabel RNA *see* ref. 28.

4. The pre-CRISPR RNA used in this protocol can be produced by chemical synthesis or in vitro transcription using T7 RNA polymerase. We use synthetic RNA (Integrated DNA Technologies) which comes bearing a free 5′ hydroxyl. In vitro-transcribed RNA must first be treated with alkaline phosphatase (Invitrogen) before 5′ radiolabeling.

5. We prepare our RNA by radiolabeling 1 μL of 10 μM RNA, followed by gel extraction, then ethanol precipitation, and resuspension of the resulting RNA pellet in 50 μL water. *See* ref. 29 for gel extraction protocol.

6. Use 1 mm wide spacers and combs for CRISPR cleavage assays, base hydrolysis, and RNase T1 mapping. Use 0.5 mm wide spacers and combs for oxidation/base elimination reactions as the thinner spacer/comb combination increases resolution.

7. To make 20 % 19:1 acrylamide:bisacrylamide gel in 8 M urea mix 100 mL 10× TBE, 480 g urea, 190 g acrylamide, and 10 g bisacrylamide with water to 1 L. Polymerize with 10 % fresh ammonium persulfate (APS) solution in water and $N,N,N,N'$-tetramethyl-ethylenediamine (TEMED). Unpolymerized acrylamide is a neurotoxin. Take care to avoid skin contact with acrylamide. Acrylamide solution can be stored in a container covered with aluminum foil for 2 months at room temperature.

8. To pour a polyacrylamide gel *see* ref. 30.

9. SDS in the dye can precipitate at room temperature. Store dye at 37 °C or heat to 37 °C prior to use to redissolve SDS.

10. To make 500 mM EDTA: Dissolve 186.1 g EDTA and 20 g NaOH in 700 mL water.

    EDTA dissolves only at basic pH, add NaOH as needed, and top up to 1 L water. A solution of 10 mM EDTA is made by diluting a stock of 500 mM EDTA in water.

11. Invitrogen RNase T1 is provided at 1,000 U/μL. Dilute 1 in 100 before use.

12. RNA cleavage may be carried out at a higher endoribonuclease concentration for a longer time to ensure maximum digestion.

13. *See* refs. 3, 4, 10, 13–27 for purification and cleavage conditions for characterized endoribonucleases.

14. The final reaction contains 100 mM NaCl and 10 mM Tris–HCl pH 8. Cas6e and Cas5d will cleave their respective pre-crRNA under these conditions [4, 10]. Additional conditions that have been reported in assays with other endoribonucleases include 100–400 mM KCl and 20 mM HEPES pH 7–7.5 [3, 13–27]. Cas endoribonucleases are metal independent, and will cleave in the absence of magnesium and in the presence of EDTA. Titrations of salt concentration and type, trials of buffer system and pH, and variations of enzyme concentration may be performed to determine the optimum conditions for previously unstudied endoribonucleases.

15. The reaction can be scaled up to perform a time-course analysis of cleavage. Quench 10 μL of reaction in 10 μL loading dye at the desired time points. Resolve the reactions on a denaturing polyacrylamide gel (Fig. 1c).

16. Different Cas endoribonucleases require incubation at different temperatures. Proteins from thermophilic organisms, such as *T. thermophilus* Cas6e or Cas5d, may require incubation at higher temperatures (55 °C) for optimal RNA cleavage.

17. The RNA does not need to be purified or precipitated before being run on the gel.

18. Use a metal plate clamped to the glass plates to cool the gel.

19. Use the syringe to remove the urea that settles into the wells. Urea in the wells will distort the bands in the gel.

20. Pre-running the gel will remove excess persulfate.

21. On these 20 % denaturing gels the bromophenol blue dye runs at a position equivalent to an 8-nucleotide RNA while xylene cyanol runs at a position equivalent to a 20-nucleotide RNA.

22. The resulting *.gel file can be quantified with ImageQuant (Molecular Dynamics).

23. Phenol/chloroform extraction followed by ethanol precipitation is used to purify a sample of nucleic acid from salts and proteins or to concentrate the sample.

24. Add 0.3 M sodium acetate directly to sample, or add 40 µL 3 M sodium acetate and top up final volume to 400 µL.
25. Glycogen acts as a carrier for the RNA and will increase the yield of purified RNA and makes the pellet easier to see.
26. Buffer-saturated phenol contains the antioxidant 8-hydroxyquinoline (yellow). The phenol should not be used if the antioxidant turns red (indicates oxidation).
27. Chloroform removes the phenol.
28. RNA is not soluble in ethanol. Freezing will help precipitate the RNA.
29. The RNA will pellet on the side of the tube facing the outside of the rotor. It is helpful to position the tubes with the hinge of the lid facing the outside of the rotor to make finding the RNA pellet easier.
30. Partial digestion of the RNA will produce a ladder of RNA with each subsequent band truncated by 1 base relative to the one above it. Do not over-digest the RNA.
31. Once the cleavage site has been identified, it can be confirmed by performing the CRISPR cleavage assay (*see* Subheading 3.1) on an RNA substrate containing a deoxy residue at the cleavage position. Disruption of cleavage by such a substitution is evidence that the cut site is properly identified (Fig. 2e).
32. The RNase T1 digest takes place under denaturing conditions. The heating and quick cooling of RNA assist in denaturing the RNA before digest.
33. RNase T1 cleaves RNA 3′ to a non-base-paired G. It is important to note that some very stable crRNA hairpins are not denatured even under the RNase T1 cleavage conditions resulting in the protection of their internal G residues from cleavage.
34. Partial digestion of the RNA will produce a ladder of RNAs ending in G. This provides a register for the base hydrolysis ladder. Do not over-digest the RNA.
35. Decrease the incubation time as amount of RNase T1 added is increased to prevent over-digestion of RNA. 3 µL of RNase T1 in a 2-min digestion or 9 µL of RNase T1 in a 1-min digestion is a useful initial test condition.
36. Oxidation/base elimination [31] tests for the existence of a 2′–3′ cyclic phosphate, consistent with the proposed Cas endoribonuclease cleavage mechanism involving attack of the 2′ hydroxyl on the scissile phosphate linkage. Disruption of cleavage of an RNA substrate with a deoxy residue at the cut site is also evidence of this cleavage mechanism (Fig. 2e).
37. There is no need to add the 1 µL of glycogen in **step 2** of Subheading 3.3.

38. The thinner spacers and comb allow for higher resolution in this gel. The difference between the cut sample and the untreated sample is 1 base. Let the gel run longer to increase separation between the two bands, taking care not to run the small products off the bottom of the gel.

**References**

1. Barrangou R, Fremaux C, Deveau H et al (2007) CRISPR provides acquired resistance against viruses in prokaryotes. Science 315:1709–1712
2. Brouns SJJ, Jore MM, Lundgren M et al (2008) Small CRISPR RNAs guide antiviral defense in prokaryotes. Science 321:960–964
3. Carte J, Wang R, Li H et al (2008) Cas6 is an endoribonuclease that generates guide RNAs for invader defense in prokaryotes. Genes Dev 22:3489–3496
4. Jansen R, Embden JD, Gaastra W et al (2002) Identification of genes that are associated with DNA repeats in prokaryotes. Mol Microbiol 43:1565–1575
5. Marraffini LA, Sontheimer EJ (2008) CRISPR interference limits horizontal gene transfer in staphylococci by targeting DNA. Science 322:1843–1845
6. Haft DH, Selengut J, Mongodin EF et al (2005) A guild of 45 CRISPR-associated (Cas) protein families and multiple CRISPR/Cas subtypes exist in prokaryotic genomes. PLoS Comput Biol 1:e60
7. Kunin V, Sorek R, Hugenholtz P (2007) Evolutionary conservation of sequence and secondary structures in CRISPR repeats. Genome Biol 8:R61
8. Makarova KS, Grishin N, Shabalina S et al (2006) A putative RNA-interference-based immune system in prokaryotes: computational analysis of the predicted enzyme machinery, functional analogies with eukaryotic RNAi, and hypothetical mechanisms of action. Biol Direct 1:7–33
9. Makarova KS, Haft DH, Barrangou R et al (2011) Evolution and classification of the CRISPR-Cas systems. Nat Rev Microbiol 9:467–477
10. Gesner E, Schellenberg MJ, Garside EL et al (2011) Recognition and maturation of effector RNAs in a CRISPR interference pathway. Nat Struct Mol Biol 18:688–692
11. Garside EL, Schellenberg MJ, Gesner EM et al (2012) Cas5d processes pre-crRNA and is a member of a larger family of CRISPR RNA endonucleases. RNA 18:2020–2028
12. Makarova KS, Aravind L, Wolf YI et al (2011) Unification of Cas protein families and a simple scenario for the origin and evolution of CRISPR-Cas systems. Biol Direct 6:38
13. Sashital DG, Jinek M, Dounda JA (2011) An RNA-induced conformational change required for CRISPR RNA cleavage by the endoribonuclease Cse3. Nat Struct Mol Biol 18:680–687
14. Haurwitz RE, Jinek M, Wiedenheft B et al (2010) Sequence- and structure-specific RNA processing by a CRISPR endonuclease. Science 329:1355–1358
15. Carte J, Pfister NT, Compton MM et al (2010) Binding and cleavage of CRISPR RNA by Cas6. RNA 16:2181–2188
16. Haurwitz RE, Sternberg SH, Ja D (2012) Csy4 relies on an unusual catalytic dyad to position and cleave CRISPR RNA. EMBO J 31:2824–2832
17. Koo Y, Ka D, Kim EJ et al (2013) Conservation and variability in the structure and function of the Cas5d endoribonuclease in the CRISPR-mediated microbial immune system. J Mol Biol 425:3799–3810
18. Nam KH, Haitjema C, Liu X et al (2012) Cas5d protein processes pre-crRNA and assembles into a cascade-like interference complex in subtype I-C/DvulgCRISPR-Cas system. Structure 20:1574–1584
19. Niewoehner O, Jinek M, Doudna JA (2013) Evolution of CRISPR RNA recognition and processing by Cas6 endonucleases. Nucleic Acids Res 42:1341–1353
20. Plagens A, Tjaden B, Hagemann A et al (2012) Characterization of the CRISPR/Cas subtype I-A system of the hyperthermophilic crenarchaeon *Thermoproteus tenax*. J Bacteriol 194:2491–2500
21. Przybilski R, Richter C, Gristwood T et al (2011) Csy4 is responsible for CRISPR RNA processing in *Pectobacterium atrosepticum*. RNA Biol 8:517–528
22. Reeks J, Sokolowski RD, Graham S et al (2013) Structure of a dimeric crenarchaeal Cas6 enzyme with an atypical active site for CRISPR RNA processing. Biochem J 452:223–230

23. Richter J, Zoephel J, Schermuly J et al (2012) Characterization of CRISPR RNA processing in *Clostridium thermocellum* and *Methanococcus maripaludis*. Nucleic Acids Res 40:9887–9896
24. Shao Y, Li H (2013) Recognition and cleavage of a nonstructured CRISPR RNA by its processing endoribonuclease Cas6. Structure 21: 385–393
25. Sternberg SH, Haurwitz RE, Dounda JA (2012) Mechanism of substrate selection by a highly specific CRISPR endoribonuclease. RNA 18:661–672
26. Wang R, Preamplume G, Terns MP et al (2011) Interaction of the Cas6 riboendonuclease with CRISPR RNAs: recognition and cleavage. Structure 19:257–264
27. Wang R, Zheng H, Preamplume G et al (2012) The impact of CRISPR repeat sequence of structures of a Cas6 protein-RNA complex. Protein Sci 21:405–417
28. Rio DC, Hannon GJ, Ares M Jr, Nilsen TW (2011) Labeling of oligonucleotide probes (DNA, LNA, RNA) by polynucleotide kinase and [γ-32P]ATP. In: RNA: a laboratory manual. Cold Spring Harbor, New York
29. Rio DC, Hannon GJ, Ares M Jr, Nilsen TW (2011) Gel purification of RNA. In: RNA: a laboratory manual. Cold Spring Harbor, New York
30. Rio DC, Hannon GJ, Ares M Jr, Nilsen TW (2011) Gel electrophoresis. In: RNA: a laboratory manual. Cold Spring Harbor, New York
31. Igloi GL, Kossel H (1985) Affinity electrophoresis for monitoring terminal phosphorylation and the presence of queuosine in RNA. Application of polyacrylamide containing a covalently bound boronic acid. Nucleic Acids Res 13:6881–6898

# Chapter 4

# Annotation and Classification of CRISPR-Cas Systems

## Kira S. Makarova and Eugene V. Koonin

### Abstract

The clustered regularly interspaced short palindromic repeats (CRISPR)-Cas (CRISPR-associated proteins) is a prokaryotic adaptive immune system that is represented in most archaea and many bacteria. Among the currently known prokaryotic defense systems, the CRISPR-Cas genomic loci show unprecedented complexity and diversity. Classification of CRISPR-Cas variants that would capture their evolutionary relationships to the maximum possible extent is essential for comparative genomic and functional characterization of this theoretically and practically important system of adaptive immunity. To this end, a multipronged approach has been developed that combines phylogenetic analysis of the conserved Cas proteins with comparison of gene repertoires and arrangements in CRISPR-Cas loci. This approach led to the current classification of CRISPR-Cas systems into three distinct types and ten subtypes for each of which signature genes have been identified. Comparative genomic analysis of the CRISPR-Cas systems in new archaeal and bacterial genomes performed over the 3 years elapsed since the development of this classification makes it clear that new types and subtypes of CRISPR-Cas need to be introduced. Moreover, this classification system captures only part of the complexity of CRISPR-Cas organization and evolution, due to the intrinsic modularity and evolutionary mobility of these immunity systems, resulting in numerous recombinant variants. Moreover, most of the *cas* genes evolve rapidly, complicating the family assignment for many Cas proteins and the use of family profiles for the recognition of CRISPR-Cas subtype signatures. Further progress in the comparative analysis of CRISPR-Cas systems requires integration of the most sensitive sequence comparison tools, protein structure comparison, and refined approaches for comparison of gene neighborhoods.

**Key words** CRISPR-Cas classification, CRISPR-Cas annotation, CRISPR-Cas evolution, CRISPR, Cas bioinformatics, Cas1, RAMPs

## 1 Introduction

The clustered regularly interspaced short palindromic repeats (CRISPR)-Cas (CRISPR-associated proteins) modules are adaptive antivirus immunity systems that are present in most archaea and many bacteria and function on the self-nonself discrimination principle [1]. These systems incorporate fragments of alien DNA (known as spacers) into CRISPR cassettes, then transcribe the CRISPR arrays including the spacers, and process them to make a guide crRNA (CRISPR RNA) which is employed to specifically target and cleave the genome of the cognate virus or plasmid [2–5]. Numerous, highly diverse Cas (CRISPR-associated) proteins are involved in different steps of the processing of CRISPR loci transcripts, cleavage of the target DNA or RNA, and new spacer integration [5–7].

The action of the CRISPR-Cas system is usually divided into three stages: (1) adaptation or spacer integration, (2) processing of the primary transcript of the CRISPR locus (pre-crRNA) and maturation of the crRNA which includes the spacer and variable regions corresponding to 5′ and 3′ fragments of CRISPR repeats, and (3) DNA (or RNA) interference [3, 8, 9]. Two proteins, Cas1 and Cas2, that are present in the great majority of the known CRISPR-Cas systems are sufficient for the insertion of spacers into the CRISPR cassettes [10]. These two proteins form a complex that is required for this adaptation process; the endonuclease activity of Cas1 is required for spacer integration whereas Cas2 appears to perform a nonenzymatic function [11, 12]. The Cas1-Cas2 complex represents the highly conserved "information processing" module of CRISPR-Cas that appears to be quasi-autonomous from the rest of the system (see below).

The second stage, the processing of pre-crRNA into the guide crRNAs, is performed either by a dedicated RNA endonuclease complex or via an alternative mechanism that involves bacterial RNase III and an additional RNA species [13]. The mature crRNA is bound by one (type II) or several (types I and III) Cas proteins that form the effector complex, which targets the cognate DNA or RNA [14–19]. The effector complex of type I systems is known as Cascade (CRISPR-associated complex for antiviral defense) [20].

Because of the enormous diversity of CRISPR-Cas, classification of these systems and consistent annotation of the Cas proteins are major challenges [5]. Considering the complexity of the composition and architecture of the CRISPR-Cas systems and the infeasibility of a single classification criterion, a "polythetic" approach based on a combination of evidence from phylogenetic, comparative genomic, and structural analysis has been proposed [5]. Three major types of CRISPR-Cas systems are at the top of the classification hierarchy. The three types are readily distinguishable by virtue of the presence of three unique signature genes: Cas3 in type I systems, Cas9 in type II, and Cas10 in type III [5]. With several exceptions, all three CRISPR-Cas types contain full complements of components that are required for the key steps of the defense mechanism. Recently, thanks to in-depth sequence analysis and structure of the effector complexes from different variants of CRISPR-Cas systems, common principles of organization and function of the complexes have been uncovered allowing for further generalization of the CRISPR-Cas classification, at least for the systems of type I and type III [21–24].

In this chapter we present an overview of the approaches, methods, and further challenges of CRISPR-Cas subtype classification and Cas protein nomenclature taking into account recent advances in the understanding of the mechanisms and organization of CRISPR-Cas systems.

## 2 Phylogenomic Analysis and Classification of CRISPR-Cas Systems

### 2.1 Annotation of cas Genes by Comparative Analysis of the Encoded Protein Sequences

The sequences of most Cas proteins, with only a few exceptions, such as Cas1 and Cas3, are highly diverged, presumably owing to the fast evolution that is typical of defense systems. Accordingly, classification of *cas* genes on the basis of protein sequence conservation is a nontrivial task that requires careful application of the most sensitive available sequence analysis methods that typically compare frequency profiles (position-specific scoring matrices, PSSM) generated from multiple alignments of the analyzed protein families, rather than individual sequences. The most complete available set of PSSMs corresponding to the latest accepted nomenclature and classification of the *cas* gene [5] is currently available through the CDD database [25]. The list of these PSSMs and the correspondence between the CDD PSSMs and the respective Pfam and TIGR families also can be found in [5] and at ftp://ftp.ncbi.nih.gov/pub/wolf/_suppl/CRISPRclass/crisprPro.html. Several servers, widely used for sequence similarity searches, such as HHpred [26], include mirrors of the CDD database and thus can also be used for Cas protein annotation. Usually, PSI-BLAST [27] or HHpred [26] is used to identify similarity of a sequence (e.g., the sequence of a particular Cas protein) to a library of PSSMs generated from the respective multiple alignments. Different Cas profiles and different programs vary with respect to the sensitivity and selectivity when used for searches with the aforementioned programs with default parameters.

Table 1 provides information on those signature *cas* genes for individual subtypes that can be reliably identified by any program. Unfortunately, search for similarity with many other Cas families, including several signature genes, may result in a considerable number of false positives and false negatives. Specifically, such proteins as Cas3′, Cas3″, Cas10, Cas4, and Cas9 can display similarity to related proteins or domains that are not linked to CRISPR-Cas system, e.g., diverse helicases in the case of Cas3 and various polymerases and cyclases in the case of Cas10. Furthermore, many RNA recognition motif-containing proteins that function as Cas effector complex subunits are similar to each other despite being present in the loci for distinct CRISPR-Cas subtypes (*see* Subheading 2.6.2). By contrast, profiles for the small and large Cas effector complex subunits are not sensitive enough due to extreme divergence of these proteins even within a single subtype.

**Table 1**
**Classification of CRISPR-Cas subtypes**

| System subtype | Mono-phyletic on Cas1 tree | Signature genes: strong/weak[a] (other name) | Comment |
|---|---|---|---|
| I-A | No | Cas8a2, Cas5 | Cas3 is often split into helicase domain Cas3' and HD nuclease domain Cas3" and a separate gene for small subunit Csa5 is often present. There are hybrid systems having a Cas1-Cas2-Cas4 module similar to I-A but a Cascade module more similar to type I-B. They are paraphyletic to I-A clade. The latter systems often have the large subunits described as Cas8a1; however, they should be rather classified as I-B. |
| I-B | No | Cas8b | I-B systems belong to two distinct clades on the Cas1 tree. One is paraphyletic group to I-A on the Cas1 tree, and another is paraphyletic to III-A. Usually the Cas3 gene is not split. |
| I-C | No | Cas8c | These systems usually do not have a *cas6* gene. Cas5 is catalytically active and replaces Cas6 function. |
| I-C variant | No | GSU0054 (Cas5 group RAMP) | These systems usually do not have *cas6*. Cas5 has several specific insertions or fusions, but is likely to be catalytically active. There are systems with different subfamilies of the large subunit, which are often severely deteriorated and sometimes even missing. |
| I-D | No | Cas10d | The HD domain is associated with the large subunit rather than with Cas3, although it does not have the circular permutation of the motifs like the HD domain fused with Cas10 in type III systems. |
| I-E | Yes | Cse1, Cse2 | The *cas4* gene is not associated with this system. |
| I-F | Yes | Csy1, Csy2, Csy3, Cas6f | The *cas4* gene is not associated with this system, and *cas2* is fused to *cas3*. There is no separate gene for a small subunit, which is either missing or fused to the large subunit. |
| I-F variant 1 | N/A | Csy1/Csy2 fusion | The *cas1-cas2-cas3* genes are not present. Usually three genes (*csy1/csy2* fusion, *csy3*, and *cas6f*) are present in an operon, which is often found next to *tniQ/tnsD* family genes. These are potentially mobile effector complexes. |
| I-F variant 2 | Yes | PBPRB1993 PBPRB1992 | A derived variant of I-F different from predicted group 5 (PBPRB1992) and group 7 genes (PBPRB1993). |

| Subtype | Strong/weak | Signature gene | Description |
|---|---|---|---|
| II-A | Yes | Csn2 | Monophyletic group on Cas9 tree. There are four genes in these operons with csn2 gene in addition to cas1_2_9. There are at least five distinct families of Csn2. |
| II-B | Yes | Cas9 (Csx12 subfamily) | Monophyletic group on Cas9 tree with four gene operons containing cas4 in addition to cas1_2_9. |
| II-C | No | N/A | Only three genes are present in the II-C operon—cas1_2_9. |
| III-A | No | Csm2 (small subunit) | Also known as the Csm module and Cas10 usually has active catalytic motifs. III-A is associated with several Cas7 group RAMPs and is often linked to csm6 which has CARF and C-terminal HEPN domain. Might be associated with Cas1-Cas2 pairs of different origin. |
| III-A variant | No | Csx10 all1473 | These have many modifications. Csx10 is a fusion of Cas5 and Cas7 proteins. The specific gene all1473 is likely to be a component of Cascade but is not similar to any known Cas proteins. The large subunit is often lacking the HD domain. Csx10 could be fused to the small subunit and Cas7 group RAMPs are often fused and have large insertions. |
| III-B | No | Cmr5 (small subunit) | Also known as the Cmr (or RAMP) module and Cas10 often has active catalytic motifs. These systems are usually associated with several Cas7 group RAMPs and are rarely present in a genome as a stand-alone system They are usually not linked to cas1-cas2 gene pair and Cmr1 has a duplication of RAMPs both from the Cas7 group. |
| III-B variant | No | MTH326 (Cas10 or Csx11) | The large subunit is often inactivated and some Cmr1 family proteins possess only one RAMP domain. |
| IV | N/A | DinG (Csf4) | These systems possess a gene for a very reduced large subunit csf1. This variant, in addition to predicted large subunit and Cas7 and Cas5 group RAMPs, often encodes a gene for a DinG-like helicase. |
| IV-variant | N/A | RHA1_ro10070 | These variants possess a gene for a very reduced large subunit csf1 and RHA1_ro10070-like proteins are predicted small subunits of effector complexes. These systems are found mostly in Actinobacteria and often on plasmids. |

*Strong/weak*—is the characteristic of the signature protein family with respect to subtype recognition/classification ability using the respective profile. Strong means that it has a relatively high specificity and high selectivity, i.e., a reliable signature, whereas weak means that search for this family yields a high level of either false positives or false negatives, but nevertheless the family remains the best available signature for a particular subtype

In order to increase sensitivity of the searches for these families, distinct profiles for each subfamily have to be generated. For this purpose, sequences that are not recognized as known *cas* genes but are present in the CRISPR-Cas loci have to be clustered using a clustering method such as BLASTCLUST [28] and then aligned using an appropriate multiple alignment. Alternatively or additionally, these sequences can be used as queries for similarity searches using PSI-BLAST and the closely related homologs found in this search can also be aligned to generate profiles that are then used as queries for a more sensitive sequence similarity search method, such as HHpred, to detect potential remote sequence similarity with known Cas families.

## 2.2 Classification of CRISPR-Cas Systems

Considering technical problems with the sensitivity and selectivity of Cas protein family profiles and uncertainties of Cas1 phylogeny and CRISPR-Cas subtype classification described below, fully automated identification of CRISPR-Cas subtypes in general is not currently feasible. The best approach to ensure the correct classification is to combine several sources of information such as Cas1 phylogeny, identification, and annotation of as many Cas proteins as possible in the locus in question and, for type II systems, identification of the trans-activating crRNA (tracrRNA) genes [29]. Table 1 provides a description of the key features of each subtype that help in the classification of the CRISPR-Cas systems.

Extra caution should be exercised when introducing new gene names and new subtypes, because, due to the often extreme sequence divergence of the Cas proteins, the similarity with already defined genes and subtypes can easily be overlooked. Furthermore, the abundance of associated genes that are likely to represent (quasi)independent immunity mechanism and are only loosely linked to CRISPR-Cas loci requires extra evidence to assign new *cas* names [30–32].

## 2.3 A Brief History of CRISPR-Cas System Classification and cas Gene Nomenclature

The original bioinformatic analysis that linked the CRISPR repeats and *cas* genes proposed four names for the most conserved and abundant *cas* genes and their products: Cas1, Cas2, Cas3, and Cas4 [33]. Subsequent analyses of proteins associated with these systems have shown that the genomes of various CRISPR-containing organisms encode approximately 65 distinct sets of orthologous Cas proteins which can be classified into either 23–45 families depending on the classification criteria (granularity of clustering) [6, 7]. Two additional core *cas* gene names were introduced at this stage, namely *cas5* and *cas6* [6].

Cas1 is the most conserved protein that is present in most of the CRISPR-Cas systems and evolves slower than other Cas proteins [34]. Accordingly, Cas1 phylogeny has been used as the guide for CRISPR-Cas system classification. Distinct operon organization of the CRISPR-Cas models in the genomes also was recognized as an important additional classification criterion [6, 7]. Eight distinct subtypes were originally proposed and named after the species whose genomes encoded a typical system of each subtype: Ecoli, Ypest, Nmeni, Dvulg, Tneap, Hmari, Apern, and Mtube [6]. In addition, the RAMP module (named after several proteins from the RAMP—repeat-associated mysterious proteins—superfamily of proteins containing the RNA recognition motif (RRM) domain) was described as a gene complex that is often present in the genomes along with CRISPR-Cas systems of one of the aforementioned subtypes but is not linked with a distinct *cas1-cas2* gene pair [6]. Consequently, additional protein families specific for each subtype or the RAMP module received names indicating subtype of their origin. For example, CRISPR subtype Apern named after the system present in *Aeropyrum pernix* genome encompasses several specific genes: csa1 (*C*RISPR *s*ystem *A*pern gene *1*), csa2, csa3, and csa4 [6]. Additionally, several core genes from the subtypes that are sufficiently distinct received a suffix letter, e.g., Cas5a (Cas5 superfamily genes of *A*pern subtype) [6]. Numerous gene families, for which no clear link to a particular subtype has been established, received gene symbols with the prefix "csx" [6].

The accumulating limitations and inconsistencies in the classification and nomenclature of CRISPR-Cas systems and *cas* genes, along with the pressing need to accommodate the rapidly growing data on sequence analysis, and structural and biochemical characterization of Cas proteins, prompted a team of CRISPR researchers to propose a revision of the aforementioned classification and nomenclature that is the current standard in the field [24] and is considered in detail below. However, it should be noted right away that the remarkable progress in the understanding of the molecular mechanisms of CRISPR-Cas and the structure of effector complexes as well as the growing number of genomes with numerous highly derived variants of CRISPR-Cas systems reveal further problems and challenges that call for the next update and improvement of this classification in the near future.

## 2.4 Three Major Types of CRISPR-Cas Systems, Their Subtypes, and cas Gene Nomenclature

The top level of the current CRISPR-Cas classification hierarchy includes the three major types (I, II, and III) [5] and the less common but clearly distinct type IV [30, 31]. The distinction between the CRISPR-Cas types is based on the respective signature genes and the typical organization of the respective loci. The current CRISPR-Cas classification is summarized in Table 1 and the description of structural and functional features of core Cas protein is given in Table 2.

### 2.4.1 Type I CRISPR-Cas Systems

All type I loci contain the signature gene *cas3* which encodes a large protein with a helicase possessing a single-stranded DNA (ssDNA)-stimulated ATPase activity coupled to unwinding of DNA-DNA and RNA-DNA duplexes [40]. Often, but not always, the helicase

## Table 2
**Structures, domain architectures, and functions of the core components of CRISPR-Cas systems**

| Family | Biochemical/*in silico* evidence | Examples of available structures and structural features |
|---|---|---|
| Cas1 | Metal-dependent deoxyribonuclease [35, 36]; deletion of Cas1 in *E. coli* results in increased sensitivity to DNA damage and impaired chromosomal segregation [37]. | PDB: 3GOD, 3LFX, 2YZS<br>Unique fold with two domains: N-terminal β-stranded domain and catalytic C-terminal α-helical domain |
| Cas2 | RNase specific to U-rich regions [38]. Double-stranded DNase [39]. | PDB: 2IVY, 2I8E, and 3EXC<br>RRM (ferredoxin) fold |
| Cas3 (helicase and HD domain) | Single-stranded DNA nuclease (HD domain) and ATP-dependent helicase [40]; required for interference [20]. | |
| Cas3″ stand-alone HD nuclease | Metal-dependent deoxyribonuclease specific for double-stranded oligonucleotides [41]. | PDB: 3S4L and 3SKD |
| Cas4 | PD-(DE)xK superfamily nuclease with three-cysteine C-terminal cluster [7]; possesses 5′-ssDNA exonuclease activity [42]. | PDB: 4IC1 |
| Cas5 | Subunit of Cascade complex interacting with large subunit and Cas7 subunit and binding the 5′-handle of crRNA [20, 22, 23, 43–45]. In the subtype I-C system Cas5 is the ribonuclease that replaces Cas6 function [46]. | PDB: 3KG4; 3VZI; 3VZH<br>Two domains of RRM (ferredoxin) fold, the C-terminal domain is deteriorated in many Cas5 protein of type I; RAMP superfamily |
| Cas6 | Metal-independent endoribonuclease that generates crRNAs [20, 44, 47–52]. | PDB: 2XLJ, 1WJ9,3I4H, 4C8Z, 4DZD<br>Two domains of RRM (ferredoxin) fold, RAMP superfamily |

| | | |
|---|---|---|
| Cas7 | Subunit of Cascade complexes binding crRNA [20, 22, 23, 43, 45]; often present in Cascade complexes in several copies. | PDB: 3PS0, 4N0L<br>RRM (ferredoxin) fold with subdomains, RAMP superfamily |
| Cas8abcef (large subunit) | Subunit of Cascade complex, involved in PAM recognition [16–18, 20, 43]. | PDB: 4AN8 |
| Cas10 (large subunit) | Subunit of Cascade (Cmr) complex [22, 23, 45, 47]. | PDB: 3UNG, 4DOZ<br>Two domains homologous to palm domain polymerases and cyclases, both belonging to RRM (ferredoxin) fold; Zn finger containing domain and C-terminal alpha helical domain [53]; Fusion: HD nuclease domain |
| Small subunit | Small, mostly alpha helical protein, subunit of Cascade complex [20, 22, 23, 44, 45, 47, 54, 55]. | PDB: 2ZCA (Cse2); 2ZOP, 2OEB (Cmr5); 3ZC4 (Csa5)<br>Cse2 has two alpha helical bundle-like domains; Cmr5 has a domain matching N-terminal domain of Cse1 and Csa5 has a domain matching C-terminal domain of Cse2 |
| Cas9 | In type II CRISPR-Cas systems, Cas9 is sufficient both to generate crRNA and to cleave the target DNA [56, 57], although it requires the help of a housekeeping gene coding for RNase III and a special gene tracrRNA encoded in the respective CRISPR-Cas locus [13]. Both the RuvC and HNH nuclease domains of Cas9 are involved in the cleavage of the target DNA [15, 58]. | PDB: 4OGC, 4OO8, 4CMP<br>Cas9 has several subdomains and adopts a two-lobed general structure. Beyond two catalytic nuclease domain its subdomains do not appear to be similar to other known protein structures [59, 60] |

domain is fused to an HD family domain which has an endonuclease activity and is involved in the cleavage of the targeted DNA [40, 61]. Exonuclease (3′–5′) activity on single-stranded DNAs and RNAs has also been reported for the HD domain from *Methanocaldococcus jannaschii* [62]. The HD domain is located at the N-terminus of Cas3 proteins or is encoded by a separate gene within the same locus as *cas3* helicase. In the latter case, the helicase is denoted *cas3′* and the HD nuclease is denoted *cas3″* (Fig. 1 and [5]). In type I-F systems, *cas3* is additionally fused to the *cas2* gene.

Usually type I systems are encoded by a single operon containing the *cas1* and *cas2* genes, genes for the subunits of the Cascade or effector complex, including large subunit, small subunit (often fused to the large subunit), *cas5* and *cas7* genes, and *cas6* gene that is directly responsible for pre-crRNA transcript processing. Each gene in the type I system operons is usually present in a single copy. Several exceptions for effector complex organization are described in Table 1 and below in the text. Type I systems are currently divided into six subtypes, I-A to I-F, each of which has its own signature gene and distinct features of operon organization (Table 1). Unlike other subtypes, I-E and I-F lack the *cas4* gene. These subtypes are related according to the Cas1 phylogeny (Figs. 1 and 2). Subtypes I-A, I-B, I-C, I-E, and I-F mostly correspond to the originally proposed ones [6], with the exception of Hmari and Tneap subtypes that were combined into subtype I-B [5]. Recently, other diverged variants of several subtypes have been identified; these, however, share several features with the existing subtypes and thus still could be described within existing classification, e.g., several type I-C variants and a derived type I-F variant [24] (Fig. 1 and Table 1). In addition, the number and diversity of stand-alone (not associated with *cas1-cas2* gene pair) effector complexes are growing. These "solo" effector complexes are often present on plasmids and/or associated with transposon-related genes, such as TniQ/TnsD, a DNA-binding protein required for transposition [63, 64]. Many such cases are derivatives of subtype I-F (Fig. 1) and some others (e.g., Ava_3490-Ava_3493 *Anabaena variabilis* ATCC 29413, with genes encoding Cas6, Cas8, Cas5, Cas7) are derivatives of subtype I-C. If a system includes a derived variant of a known Cas protein family, this family might have an optional suffix indicating the subtype to which this protein belongs (e.g., Cas6f is a highly derived member of Cas6 superfamily specific for subtype I-F). Notably, the phylogenetic tree of the type I signature protein Cas3 seems to accurately reflect the subtype classification [65].

### 2.4.2 Type II CRISPR-Cas Systems

The signature gene for type II CRISPR-Cas systems is *cas9*, which encodes a multidomain protein that combines all the functions of effector complexes and the target DNA cleavage and is essential for the maturation of the crRNA [15]. The type II systems are also known as the "HNH" systems, *Streptococcus*-like or Nmeni

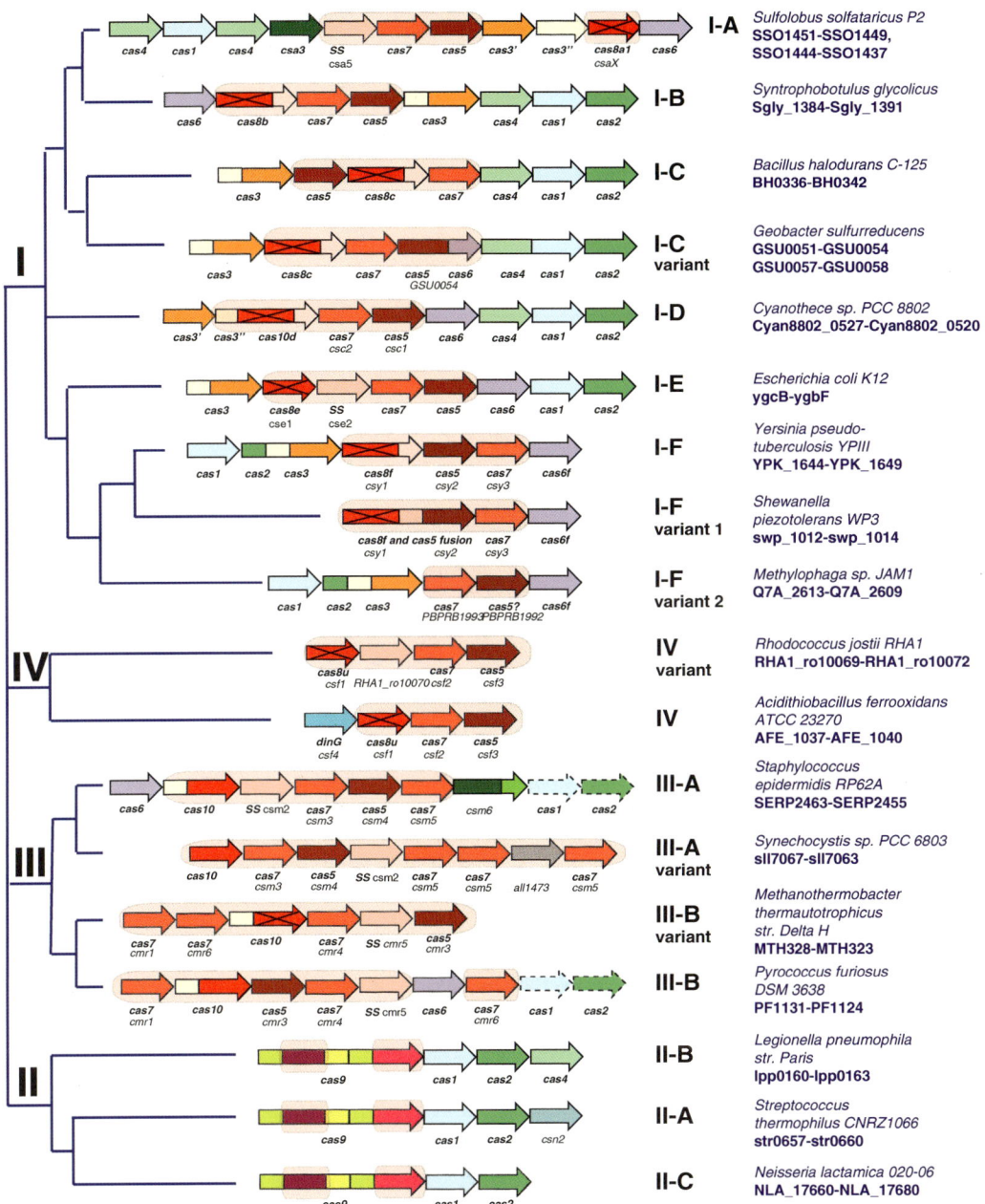

**Fig. 1** Classification and organization of CRISPR-Cas systems. Typical operon organization is shown for each CRISPR-Cas subtype. For each CRISPR-Cas subtype, a representative genome and the respective gene locus tag names are indicated. Homologous genes are *color coded* and identified by a family name. Names follow the classification from [5]. See also details in [30]. Names in *bold* are proposed systematic names; "legacy names" are in regular font. Abbreviations: *LS* large subunit (including subfamilies of Cas10, Cas8, Cse1, Csy1), *SS* small subunit (including Cmr2, Cmr5, Cse2). Genes coding for inactivated large subunits are indicated by *crosses*. Genes and domain components for effector complexes are highlighted by *pink background*

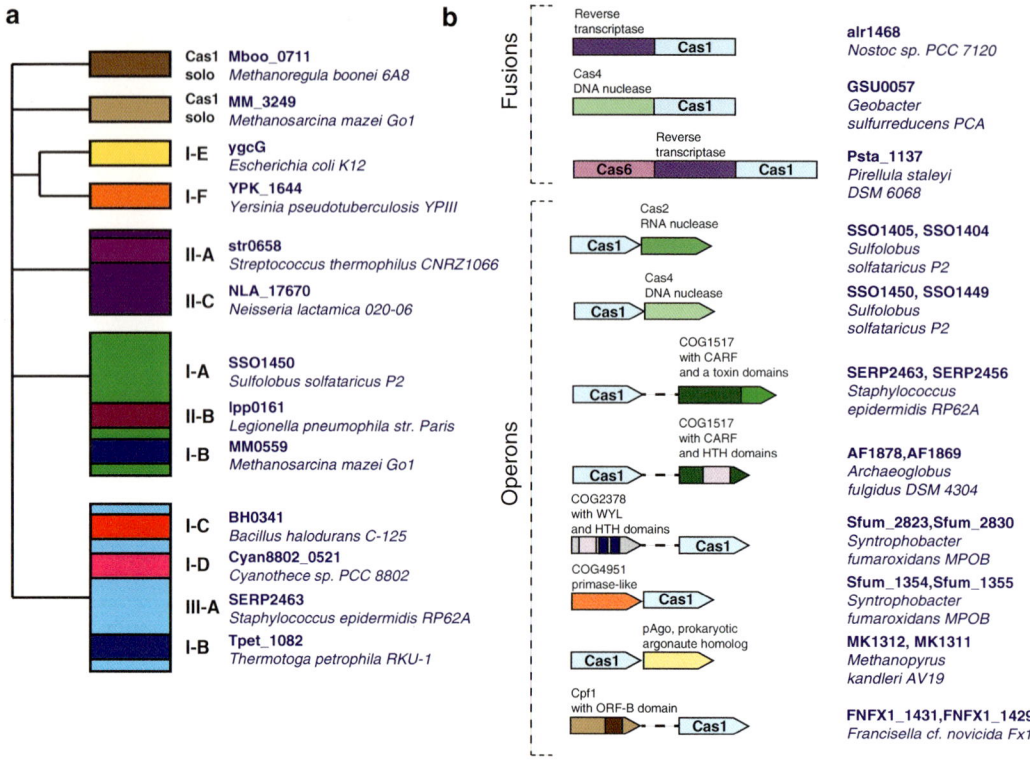

**Fig. 2** Phylogeny of Cas1 and its associations with other genes. (**a**) Schematic representation of Cas1 phylogeny (the complete tree and details of the tree reconstruction are available in [31]). The branches are *colored* according to the automatic assignment of *cas1* genes to CRISPR-Cas subtypes based on the analysis of ten up- and ten downstream genes. (**b**) Cas1 fusions and operonic associations

subtype. Every CRISPR-Cas locus of this subtype, in addition to the *cas9* gene, also contains the ubiquitous *cas1* and *cas2* genes. In addition to these three protein-coding genes, the vast majority of type II loci also encompass one or two genes for tracrRNA, an RNA that is partially homologous to the cognate CRISPR [29, 66]. These systems use cellular (not encoded within the CRISPR-Cas loci) RNase III and tracrRNA for the processing of pre-crRNA [13]. The large Cas9 protein (~800–1,400 amino acids) contains two nuclease domains, namely the RuvC-like nuclease (RNase H fold) and the HNH (McrA-like) nuclease domain that is located in the middle of the protein [24]. Both nucleases are required for target DNA cleavage [15, 58].

Recently, several crystal structures of Cas9 have been solved including one with an artificial single-guide RNA (sgRNA) and a target DNA [59, 60]. It has been shown that Cas9 forms a two-lobed structure, with the target DNA and sgRNA positioned in the interface between the two lobes. Two loops in both lobes contribute to the recognition of the PAM. A conserved arginine cluster at the N-terminus of Cas9 belongs to a bridge helix which is critical for sgRNA: DNA recognition [59]. Outside the RuvC and HNH domains, the Cas9 structure shows no apparent structural similarity to other proteins. The part of the Cas9 protein including both nuclease domains and the arginine-rich cluster probably originated from mobile genes that are not associated with CRISPR repeats [29]. These mobile genes themselves appear to descend from a transposon gene known as ORF-B whose role in the transposon life cycle remains unknown [29]. Due to the significant sequence similarity between Cas9 and its homologs that are unrelated to CRISPR-Cas, Cas9 cannot be used as the only marker for identification of type II systems.

Type II CRISPR-Cas systems are currently classified into three subtypes (II-A, II-B, and II-C), two of which were introduced in the updated classification [5] and one was proposed recently on the basis of the distinct operon organization [29, 66, 67] (Table 1). Type II-A systems encompass an additional gene, known as *csn2* (Fig. 1), which is considered a signature gene for this subtype. The Csn2 protein is not required for interference but apparently has an unclear role in spacer integration [56]. The Csn2 proteins form homotetrameric rings that bind linear double-stranded DNA through the central hole [68–71]. This protein has been shown to adopt a highly derived P-loop ATPase fold in which the ATP-binding site appears to be inactivated [68, 69, 71]. Several highly diverged Csn2 subfamilies have been identified [29], in particular short [68, 69] and long forms [71] for which structures and biochemical characterization are available [68–71]. Type II-B systems do not encode the *csn2* gene but possess a distinct fourth gene that belongs to the Cas4 family which is also associated with subtypes I-A to I-D (but not I-E and I-F) [5]. The Cas4 proteins possess 5′-single-stranded DNA exonuclease activity [42] and belong to the PD-EDxK family of nucleases [7]. The actual role of the Cas4 proteins in the CRISPR-Cas systems remains unknown. The recently proposed type II-C CRISPR-Cas systems possess only three protein-coding genes (*cas1*, *cas2*, and *cas9*) and are common in sequenced bacterial genomes [29, 30, 66]. Recently, type II systems have been developed into a powerful genome editing and engineering tool with a major biotechnological potential [72, 73].

### 2.4.3 Type III CRISPR-Cas Systems

All type III systems possess the signature gene *cas10* which encodes a multidomain protein containing a palm domain similar to that in cyclases and polymerases of the PolB family [74, 75]. Thus, this protein originally was predicted to be a polymerase [76]. Recently, the structure of Cas10 has been solved and four distinct domains have been identified [53, 77]: the N-terminal cyclase-like domain that adopts the same RRM fold as the palm domain but is not predicted to possess enzymatic activity, a helical domain containing the Zn-binding treble clef motif, the palm domain that retains the catalytic residues and is predicted to be active, and the C-terminal alpha helical domain resembling the thumb domain of A-family DNA polymerase and Cmr5, a small alpha helical protein present in some of the type III CRISPR-Cas systems. Cas10 is the large subunit of effector complexes of type III systems. Each type III locus also encodes other subunits of effector complexes such as one gene for the small subunit, one gene for a Cas5 group RAMP protein, and usually several genes for RAMP proteins of the Cas7 group (Fig. 1 and *see* Subheading 2.6.2). Often Cas10 is fused to an HD family nuclease domain that is distinct from the HD domains of type I CRISPR-Cas systems and, unlike the latter, contains a circular permutation of the conserved motifs [7, 76]. Type III CRISPR-Cas systems often do not encode their own *cas1* and *cas2* genes but use crRNAs produced from CRISPR arrays associated with type I or type II systems [14, 78]. Nevertheless, in many genomes that lack type I and type II systems, *cas1*, *cas2*, and *cas6* genes are linked to a type III system that accordingly is predicted to be fully functional [31]. Currently, there are two subtypes within type III, III-A (former Mtube subtype or Csm module), and III-B (former RAMP module or Cmr module), which are clearly related but could be distinguished by the presence of distinct genes for small subunits of effector complexes, *csm2* and *cmr5*, respectively (Fig. 1, Table 1, and Subheading 2.6.3). The subtype III-A loci often possess *cas1*, *cas2*, and *cas6* [31] and have been shown to target DNA [79], whereas most of the III-B systems lack these genes and therefore depend on other CRISPR-Cas systems present in the same genome. The type III-B CRISPR-Cas systems have been shown to target RNA [14, 23, 47].

The composition and organization of type III CRISPR-Cas systems are much more diverse compared with type I systems. The diversity is achieved by gene duplications and deletions, domain insertions and fusions, and the presence of additional, poorly characterized domains that presumably are involved in either effector complexes or associated immunity. At least two of the type III variants (one of type III-A and the other of type III-B) are relatively common (Fig. 1 and Table 1). The distinguishing feature of the type III-B variant is the apparent inactivation of the palm/cyclase domain of Cas10 whereas the type III-A variants typically encompass a Cas10 gene lacking the HD domain and additionally contain an uncharacterized gene homologous to all1473 from *Nostoc* sp. PCC 7120 [24]. Both type III variants are typically present in a genome along with other CRISPR-Cas systems.

### 2.4.4 Type IV CRISPR-Cas Systems

Type IV CRISPR-Cas systems, found in several bacterial genomes, often on plasmids, can be typified by the CRISPR-Cas locus in *Acidithiobacillus ferrooxidans* ATCC 23270 (operon AFE_1037-AFE_1040). Similar to subtype III-A, this system lacks *cas1* and *cas2* genes and often is not associated with CRISPR arrays.

Moreover, in many bacteria, this is the only CRISPR-Cas system, with no CRISPR cassette detectable in the genome. The type IV systems possess an effector complex that consists of a highly reduced large subunit (*csf1*), two genes for RAMP proteins of the Cas5 (*csf3*) and Cas7 (*csf2*) groups, and, in some cases, a gene for a predicted small subunit [24]. The *csf1* gene could be considered a signature gene for this system (Fig. 1 and Table 1). There are two distinct subtypes of type IV systems, one of which contains a DinG family helicase *csf4* [80], whereas the second subtype lacks DinG but typically contains a gene for a small alpha helical protein, presumably a small subunit [24]. Type IV CRISPR-Cas systems could be mobile modules that, similar to type III systems, could utilize crRNA from different CRISPR arrays once these become available. However, other mechanisms such as generation of crRNA directly from alien RNA, without incorporation of spacers in CRISPR cassettes, cannot be ruled out.

The classification of CRISPR-Cas systems outlined above more or less adequately covers the representation of these systems in sequenced bacterial and archaeal genomes. However, considering the rapid evolution of CRISPR-Cas, these variants might represent only the proverbial a tip of the iceberg with respect to the true diversity of prokaryotic adaptive immunity. As a case in point, two novel CRISPR-Cas systems have been recently identified in the genomes of *Thermococcus onnurineus* and *Ignisphaera aggregans* [81]. Based on some marginal similarities, these loci could be tentatively assigned to type I and type III, respectively; however, they do not contain any signature genes described above that would allow one to classify them into any known subtype. Similarly, classification of certain type I systems, such as the one from *Microcystis aeruginosa* (MAE_30760-MAE_30790) and several other species [82], is hampered by the apparent absence of signature genes of the known type I subtypes. Accumulation of such "unclassifiable" variants raises the possibility that the current principles of CRISPR-Cas system classification might have to be reconsidered to take into account the challenge of the ever-increasing diversity.

## 2.5 Phylogeny and Genomic Associations of Cas1

The endonuclease Cas1 is an essential Cas protein that ensures the unique ability of CRISPR systems to keep memory of previous encounters with infectious agents. Cas1 and Cas2 form a heterohexameric complex that is necessary and sufficient for spacer integration [10–12, 83]. However, only the enzymatic activity of Cas1 is required for spacer integration by the Cas1-Cas2 complex whereas the activity of Cas2 is dispensable indicating that this protein has a structural role in spacer acquisition [11].

To date, three Cas1 proteins, from *Escherichia coli*, *Pseudomonas aeruginosa*, and *Archaeoglobus fulgidus*, have been experimentally characterized and their structures have been solved [35, 37, 84]. It has been shown that Cas1 protein forms a homodimer and is a metal-dependent nuclease that cleaves ssDNA and dsDNA. The Cas1 monomer consists of two domains, with the C-terminal α-helical catalytic domain and the mostly beta-stranded N-terminal domain that is probably involved in dimerization and interaction with other proteins, in particular Cas2 [35, 37, 84].

Cas1 is the most conserved Cas protein and its phylogeny generally correlates with the organization of CRISPR-Cas system loci; accordingly, until recently, Cas1 has been considered the signature for the presence of CRISPR-Cas systems in a genome [5–7]. However, as pointed out above, recently it has been found that many genomes that lack a *cas1* gene possess Cas loci that encode apparently active effector complexes and thus might function in a Cas1-independent fashion. The examples of systems lacking *cas1* include the type IV systems, described above, subtype III-B, and a variant of subtype I-F (Fig. 1).

Conversely, it has been recently shown that *cas1* is a component of predicted self-synthesizing transposable elements, dubbed casposons, where it is always associated with a DNA polymerase of the B family and variable sets of diverse genes [85]. Furthermore, in some other archaeal genomes from the Methanomicrobiales lineage, *cas1* is linked neither to casposons nor to CRISPR-Cas system and its function in these organisms remains obscure [31, 85]. Figure 2a presents a scheme of the Cas1 phylogeny published before [31]. The two groups of Cas1 that are not associated with CRISPR-Cas systems form two separate branches deep in the Cas1 tree. Their relationships with branches that correspond to Cas1 groups associated with known CRISPR-Cas systems are not resolved. Consistent with previous analyses, most of the known type I and type II subtypes form distinct branches. However, only subtypes I-E, I-F, II-A, and II-B are strictly monophyletic whereas the other subtypes show multiple deviations from the classification scheme [31]. In contrast, Cas1 proteins associated with both type III subtypes do not form monophyletic groups suggesting that these systems are compatible with a wide range of Cas1 proteins acquired from other CRISPR-Cas types. The number of genomes that possess only subtype III-A CRISPR-Cas is growing fast whereas subtype III-B systems associated with Cas1 and Cas2 remain rare [31]. Accordingly, much of the diversity of Cas1 is concentrated within subtype III-A (Fig. 2a) [31]. Another important observation is the polyphyly of the type II systems whereby Cas1 sequences of type II-B form a clade within the type I-A branch. The origin of the other type II-B genes from within type I is also supported by phylogenetic analysis of Cas2 and Cas4 proteins [29]. Generally, these findings confirm that effector complexes of CRISPR-Cas can function in association with "information processing" modules of different origin.

The *cas1* gene is often found either fused or located in the same predicted operons with a number of enzymatically active domains or predicted transcriptional regulators (Fig. 2b).

Many enzymatic domains linked to Cas1 do not belong to any Cas families and are known components of various defense systems that possess either RNase or DNase activity (*see* Subheading 2.5). Thus, it appears that the expression and activity of Cas1 proteins are tightly controlled and coupled to programmed cell death/dormancy mechanisms [1, 30, 86].

One of the most puzzling connections of Cas1 is to a gene called *cpf1* (see description at http://www.jcvi.org/cgi-bin/tigrfams/HmmReportPage.cgi?acc=TIGR04330), which encodes a large protein (about 1,300 amino acids), an uncharacterized protein. This gene is found in several diverse bacterial genomes, typically in the same locus with *cas1*, *cas2*, and *cas4* genes and a CRISPR cassette (for example, FNFX1_1431-FNFX1_1428 of *Francisella* cf. *novicida* Fx1). Thus, the layout of this putative novel CRISPR-Cas system appears to be similar to that of type II-B. Furthermore, similar to Cas9, the Cpf1 protein contains a readily identifiable C-terminal region that is homologous to the transposon ORF-B and includes an active RuvC-like nuclease, an arginine-rich region, and a Zn finger (absent in Cas9). However, unlike Cas9, Cpf1 is also present in several genomes without a CRISPR-Cas context and its relatively high similarity with ORF-B suggests that it might be a transposon component. If however this is a genuine CRISPR-Cas system and Cpf1 is a functional analog of Cas9 it would be a novel CRISPR-Cas type, namely type V. Hopefully this interesting system will be experimentally studied in the near future.

## 2.6 Principles of Organization of CRISPR-Cas Surveillance and Effector Complexes

### 2.6.1 General Features of Effector Complex Organization

The effector (surveillance) complex of CRISPR-Cas systems is involved in pre-crRNA processing (except in type III) and crRNA-guided targeting of foreign DNA or RNA. This complex contains from one (type II CRISPR-Cas systems) to several proteins and binds crRNA and DNA. Specific recognition of the DNA sequence matching the spacer (termed the protospacer) within the respective crRNA is necessary for the Cascade to form an R-loop and recruit a nuclease to cleave the target DNA. In addition, in type I and II systems recognition of a flanking PAM (protospacer adjacent sequence) is required. To date, the structure and organization of several effector complexes from different CRISPR-Cas systems of both type I and type III have been studied in detail. These include the Cascade complex from *E. coli* (subtype I-E) [20, 43, 44], Csy complex (subtype I-F) from *P. aeruginosa* [19], a(rchaeal) Cascade from *S. solfataricus* (subtype I-A)[87], subtype I-C complex *from Bacillus halodurans* [39], Csm-complex from *S. solfataricus* (subtype III-A) [45], and Cmr complexes (subtype III-B) from *Thermus thermophilus* [22] and *Pyrococcus furiosus* [23]. The analysis of these complexes revealed striking similarities in their organization despite the absence of sequence similarity between the majority of the constituents. These findings are consistent with previous predictions that have been made using comparative genomic methods [24, 30].

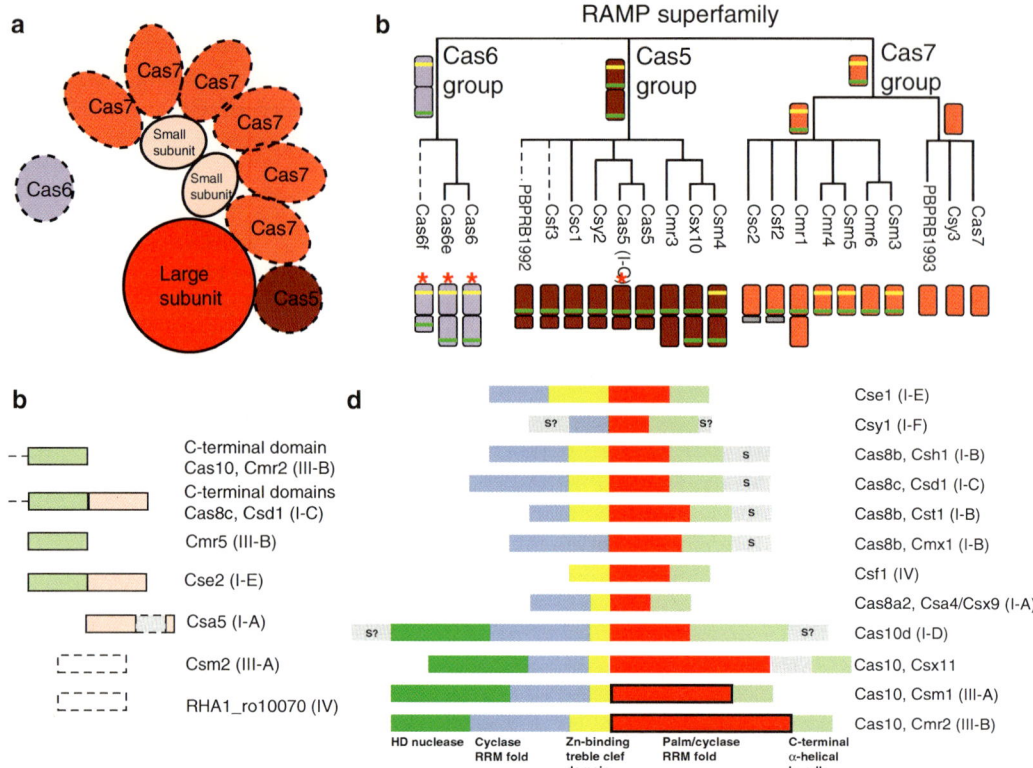

**Fig. 3** General organization of effector complexes in different types of CRISPR-Cas systems. (**a**) The generalized model of subunit composition of effector complexes of type I, III, and IV systems. Color coding is the same as in Fig. 2. The subunits that belong to the RAMP superfamily are shown by *dashed circles*. (**b**) Classification of the RAMP superfamily into three families. The tree-like scheme of RAMP relationships is based on the sequence similarity, structural features, and neighborhood analysis. Unresolved relationships are shown as multifurcations and tentative assignments are shown by *broken lines*. Glycine-rich loops are shown by *green lines*. The conserved histidines, suggesting catalytic activity of some of the RAMP proteins, are shown by *yellow lines*. Protein families shown to be active ribonucleases are marked by an *asterisk*. The deteriorated RRM domain is shown by the *gray rectangle* (i.e., Csf2 and Csc2). The predicted ancestral domain configuration is shown for each major node. (**c**) Domain organization of the small subunits of different subtypes of type I, III, and IV CRISPR-Cas systems. Homologous domains are color coded. *Dashed outline empty boxes* show that the structure similarity of these small subunits is unknown. The *gray box* shows a unique beta-stranded insertion. (**d**) Domain organization of the large subunits of different type I and III CRISPR-Cas systems. The palm-like domains of Cas10 proteins with intact cyclase/polymerase catalytic motifs are shown with a *black outline*. The letter "S" marks the regions that could be homologous to small subunits of Cascade complexes encoded as separated genes in type III systems, I-E subtype, and some systems of the I-A subtype

A general scheme of effector complex organization related to type I and type III systems is shown in Fig. 3a and reflects the following observations. Effector complexes consist of one large subunit, several small subunits, one Cas5 family protein, and several Cas7 family proteins. Cas5 and large and small subunits are usually encoded by a single gene each in both type I and type III systems, although large and small subunits are likely fused in several type I subtypes (Fig. 1) [24]. The Cas7 group proteins are encoded by a single gene in the respective type I system loci and by several separate genes in type III systems (Fig. 1). Functionally, Cas7 is involved in crRNA binding, and Cas5 in binding the 5′-handle of crRNA and interaction with the large subunit and the proximal Cas7 protein. The large subunit participates in DNA binding and recognition of the PAM sequence [17, 18, 22, 23, 43, 45, 88]. The Cas6 proteins, which are directly involved in pre-crRNA processing, usually do not belong to the effector complex but could be loosely associated with some of them [20]. A strong interaction has been detected between the Cas7 and Cas5 proteins and loose association has been identified between the large and small subunits when encoded by separate genes [17, 19, 45, 88]. It has been proposed that, in addition to crRNA-guided DNA targeting, type I-E Cascade can migrate along the DNA molecule, facilitating the selection of fragments to be incorporated into the CRISPR locus [83].

Four distinct subunits of the effector complexes of type I and type III CRISPR-Cas contain RRM domains, which consists of a four-stranded antiparallel β-sheet (arranged as β4β1β3β2), with two α-helices located after β1 and β3 in a βαββαβ ferredoxin-like fold [7, 54]. The type III large subunit, Cas10, contains two RRM domains, one of which is a polymerase/cyclase palm domain predicted to be active, whereas the other one is an inactivated version of the palm domain [53, 77]. The Cas5 proteins typically contain two RRM domains, with the C-terminal domain degraded in several subfamilies; the Cas7 proteins possess a single RRM domain; and the Cas6 proteins encompass two RRM domains [24, 30]. It has been hypothesized that the RAMP proteins evolved from the large subunit by duplication and specialization [24, 31]. The type II effector complex consists of a single multidomain protein, Cas9, that binds crRNA and tracrRNA. Similarly to the type I and type III complexes, the type II effector complex (Cas9) scans DNA, recognizes PAM, and forms an R-loop (*see* Subheading 2.4.2).

### 2.6.2 Three Major Families of RAMPs

Exhaustive sequence analysis, supported by the analysis of the growing collection of structural data, indicates that RAMPs can be classified into three families, the largest of which includes the Cas7 proteins (Fig. 3b). In the majority of CRISPR-Cas systems, processing of pre-crRNA is catalyzed by dedicated endoribonucleases that belong to the Cas6 family of RAMPs. Among all RAMP families, the Cas6 family has been characterized in most detail, both structurally and biochemically. This protein shows remarkable plasticity of the catalytic mechanism and RNA recognition modes [48–50]. The type member of the Cas6 family is the protein from the archaeon *Pyrococcus furiosus* [47, 51, 54]. The *P. furiosus* Cas6 contains two RRM domains with a G-rich loop located at the

C-terminus of the second RRM domain and the catalytic triad consisting of histidine, tyrosine, and lysine located within the first, N-terminal RRM domain [51]. The conserved catalytic histidine is located within the alpha helix that follows the first core beta strand of the N-terminal RRM domain. Many Cas6 subfamilies contain the catalytic histidine in the same position but other arrangements of catalytic residues have been detected as well [48, 49, 54]. The cleavage of the pre-crRNA occurs within a CRISPR repeat at the 5′ side of the phosphodiester bond, generating a 5′ end hydroxyl group and either a 3′ phosphate (Cas6 from *Pseudomonas aeruginosa*) or a 2′, 3′ end cyclic phosphate group (Cas6e), and yields a crRNA of approximately 60 nt in size [19, 52, 89]. The majority of the Cas6 proteins show substantial sequence conservation and belong to the core of the Cas6 family (COG1853/COG5551) but several are highly divergent, e.g., those associated with I-E (Cas6e) and I-F (Cas6f) CRISPR-Cas subtypes (Fig. 3b). The latter is the most derived Cas6 protein with a severely degraded C-terminal RRM domain [24, 52, 54].

The Cas7 proteins form the backbone of effector complexes that play the key role in binding and protecting the crRNA guide sequence. The Cas7 family proteins typically contain one RRM domain that is structurally similar to the N-terminal RRM domain of Cas6 and two distinct, albeit highly variable, subdomains [54, 87, 90]. The majority of the Cas7 family proteins associated with type III systems contain the characteristic G-rich loop, the structural marker of the RAMP superfamily (Fig. 3b). These proteins are diverse and could be present in several copies in the type III loci (Fig. 1). The Cas7 family proteins associated with type III systems are apparently prone to aggregation, forming multidomain proteins (e.g., Cmr1 family or Psta_1142 from *Pirellula staleyi* or HMPREF9137_2396 *Prevotella denticola*). Several subfamilies of the Cas7 family (Cmr4, Csm5, and Csm3) possess a conserved histidine that is structurally equivalent to the catalytic histidine of Cas6 required for pre-crRNA cleavage, suggesting that these Cas7 proteins are active RNases (Fig. 3b) [24]. This hypothesis is compatible with the demonstration of the RNA cleavage activity of the Cmr complex of *T. thermophilus* [22].

The Cas5 family proteins bind the 5′-handle of crRNA and provide the interaction interface for the large subunit and the Cas7 proteins. Similarly to the Cas6 family, most of the Cas5 proteins contain two RRM domains [24, 46, 54, 91] although the C-terminal domain is severely deteriorated in many Cas5 proteins associated with type I systems (Fig. 3b) [24]. The G-rich loop is easily detectable in the first RRM domain and often is present also in the second RRM domain in those Cas5 group RAMPs that are associated with type III systems (Fig. 3b) [24, 54]. Usually, only one Cas5 protein is encoded in a CRISPR-Cas locus (Fig. 1). These proteins, especially those associated with type III systems, are prone to structural rearrangements and fusions (e.g., Rcas_3293 from *Roseiflexus castenholzii* represents fusion of Cas5 and Cas7 group RAMPs). Some proteins are assigned to the Cas5 family provisionally, based on the general principles of organization of effector complexes because they do not share any similarity with known Cas proteins (Fig. 3b) [24]. Among the Cas5 family members, there are also proteins with a conserved N-terminal histidine (e.g., Csm4 subfamily). However, RNase activity has been experimentally demonstrated only for the Cas5 proteins that are associated with type I-C systems; these proteins have a different set of catalytic residues compared with Cas6 and are directly involved in pre-crRNA processing [39, 46].

### 2.6.3 Small Subunits

Typically, the small subunit is an alpha-helical protein containing up to eight predicted alpha helices. The small subunits are encoded by a separate gene in all type III CRISPR-Cas systems, in some type I systems, such as I-A and I-E, and one variant of type IV (Fig. 1). Analogous to the large subunit, the small subunits are highly diverse, such that different families often show no detectable sequence similarity to each other. In the majority of type I systems, large subunits contain a 4–6 alpha helical C-terminal extension that appears to complement the absence of a small subunit gene. Accordingly, it has been hypothesized that these proteins represent a fusion of the large and small subunits (Figs. 1 and 3b, d) [24]. In the effector complexes that include the small subunit as a separate component, there are usually several small subunit genes [17, 18, 20, 22, 23, 43, 45, 87, 92]. Recently, the structure of the small subunit Csa5 of type I-A from *S. solfataricus* has been reported [54, 55]. Comparison of these structures sheds new light on the evolution of the small subunits by demonstrating the evolutionary connection between small subunits of type I and type III systems [54, 55]. Structural comparison revealed that Cmr5, the small subunit of subtype III-B, corresponds to the N-terminal domain of Cse2, the small subunit of type I-E, whereas Csa5 partially corresponds to the C-terminal domain of Cse2 [54, 55]. In addition, Csa5 has a unique beta-stranded extension which is absent even in the proteins that belong to the same protein family (Fig. 3b). The relationships of these proteins to other distinct families of predicted small subunits, such as Csm2 (type III-A) and RHA1_ro10070 (type IV), remain to be elucidated (Fig. 3b).

### 2.6.4 Large Subunits

Multiple lines of evidence coming from *in silico* analysis of Cas proteins suggest that, the absence of significant sequence similarity notwithstanding, the large subunits present in most of the type I CRISPR-Cas systems could be homologous to Cas10 proteins, which contain two palm/cyclase domains, one of which is predicted to be enzymatically active (Fig. 3d) [24, 31]. Recent structures of effector and surveillance complexes from both major subtypes are compatible with this inference [19, 22, 23, 39, 45, 87, 88]. The crystal structures of two distinct large subunits have been solved, namely those of Cas10, the large subunit of type III systems [53, 77], and CasA (Cse1 or Cas8e) of subtype I-E [16]. The core domains of Cas10 are the N-terminal cyclase domain, Zn finger containing the treble clef domain, cyclase (palm) domain with the characteristic catalytic motif "GGDD," and the C-terminal alpha helical bundle (Fig. 3d). This arrangement is reminiscent to "Fingers," "Palm," and "Thumb" domains present in DNA polymerases of different families [93]. In Cas10, these four regions are arranged into four distinct domains, whereas Cse1 displays a much compact architecture where traces of the putative ancestral domain architecture are barely identifiable [31]. Major structural rearrangements of this type could have been anticipated even before these structures became available because the large subunits of different type I systems substantially differ in size and even seem to be missing in some systems (e.g., I-C variant, Table 1 and Fig. 3d) [24, 31].

Variations of the domain architecture and inactivation of the catalytic palm domain could be identified even within the Cas10 family. Some representatives lack the HD domain, a nuclease that is typically fused to Cas10 at the N-terminus (e.g., Caur_2291 from *Chloroflexus aurantiacus*), and the palm domain appears to be inactivated in many subtype III-B variants, e.g., MTH326 from *Methanothermobacter thermautotrophicus*) (see details in [24]).

An exhaustive comparison of multiple alignments and predicted secondary structures of the large subunits of type I and type III systems revealed several shared features, such as a Zn finger in the middle of the protein sequence in the Cas8a, Cas8b, Cse1, and Csf1 families, conserved beta-hairpin in the region roughly corresponding to the palm domain, and an alpha helical region at the C-terminus of the proteins compatible with the alpha helical bundle of Cas10 (Fig. 3d) [24]. However, direct comparisons detect no sequence similarity between Cas10 and any large subunits of type I systems, and moreover, many large type I subunits share no significant sequence similarity with each other, suggestive of extremely fast divergence of these proteins. An unusual variant of the large subunit denoted that Cas10d is associated with type I-D. This protein appears to be structurally similar to Cas10 and even contains an N-terminal HD domain but the latter has a circular permutation of the catalytic motifs which seems suggestive of its origin from the HD domain of Cas3 protein (Fig. 2) [24].

## 2.7 Potential Associated Immunity Genes and Regulatory Components

Many DNA-targeting defense systems contain previously overlooked components that are implicated in programmed cell death (PCD)/dormancy [1, 30, 31, 86, 94]. Experimental data on coupling between immunity and PCD is scarce, having been demonstrated only for the *Escherichia coli* anticodon nuclease (ACNase)

PrrC which contributes to the T4 phage exclusion mechanism as a component of the RM type Ic system PrrI [95]. A general hypothesis for the apparent integration of the two defense strategies has been proposed [1, 30, 31, 86, 94]. Specifically, a toxin associated with an immunity system, such as CRISPR-Cas, could act either as a dormancy inducer, which prevents fast virus propagation and could "buy the time" for the activation of the primary immune system, or, alternatively, as a toxin that causes altruistic suicide when immunity fails [86]. CRISPR-Cas systems are especially rich in genes encoding proteins associated with PCD [30, 31]. In particular, two core Cas proteins, Cas2 and Cas4, belong to families of nucleases that commonly function as toxins in toxin-antitoxin systems which are responsible for PCD in prokaryotes [1, 30, 86]. It has been shown that Cas2 forms a specific complex with Cas1 that is required for spacer acquisition by CRISPR-Cas but mutation of a residue predicted to be required for nuclease activity of Cas2 did not affect this step [11]. Thus, it appears likely that Cas1 and Cas2 modulate each other's activities, where the putative toxic nuclease activity of Cas2 is unleashed only under genotoxic stress caused by infection that is not controlled by immunity. Both Cas2 and Cas4 are fused with several other Cas proteins but not with effector complex components (Fig. 4). In several bacteria, an apparently inactivated Cas2 is fused to a 3'–5' exonuclease of the DEDDh family [7], suggesting that the lost nuclease activity of Cas2 could be replaced by an unrelated enzyme. Cas4 proteins of two distinct families are often present in the same operon (e.g., PAE0079-PAE0082 in *Pyrobaculum aerophilum* str. IM2) suggestive of functional differentiation (Fig. 4). A variety of other Cas proteins and proteins that are sporadically present within CRISPR-Cas loci are fused to or encoded next to Cas1, indicating that there are multiple ways to control the activity of this key Cas protein (Fig. 2b) [31]. One of such Cas1 fusions involves a reverse transcriptase for which cell toxicity has been recently demonstrated [96] (Fig. 2b).

Many more genes seem to be specifically associated with type III CRISPR-Cas systems. However, all these families have been identified in other genomic contexts as well. The largest superfamily of such proteins includes members of COG1517 that typically consist of a CARF domain (CRISPR-Cas-associated Rossmann-fold domain), an HTH domain [97, 98], and various effector domains, most of which are predicted to be active RNases and DNases [5, 7, 32] (Fig. 4). The HEPN domain containing a characteristic RxxxxH motif, the most abundant effector domain present in such families as Csm6 and Csx1, is a predicted ribonuclease [94]. Notably, the great majority of the COG1517 members associated with type III CRISPR-Cas systems contain effector domains [32]. The COG1517-related genes are often found in the same operon with other genes encoding uncharacterized proteins that

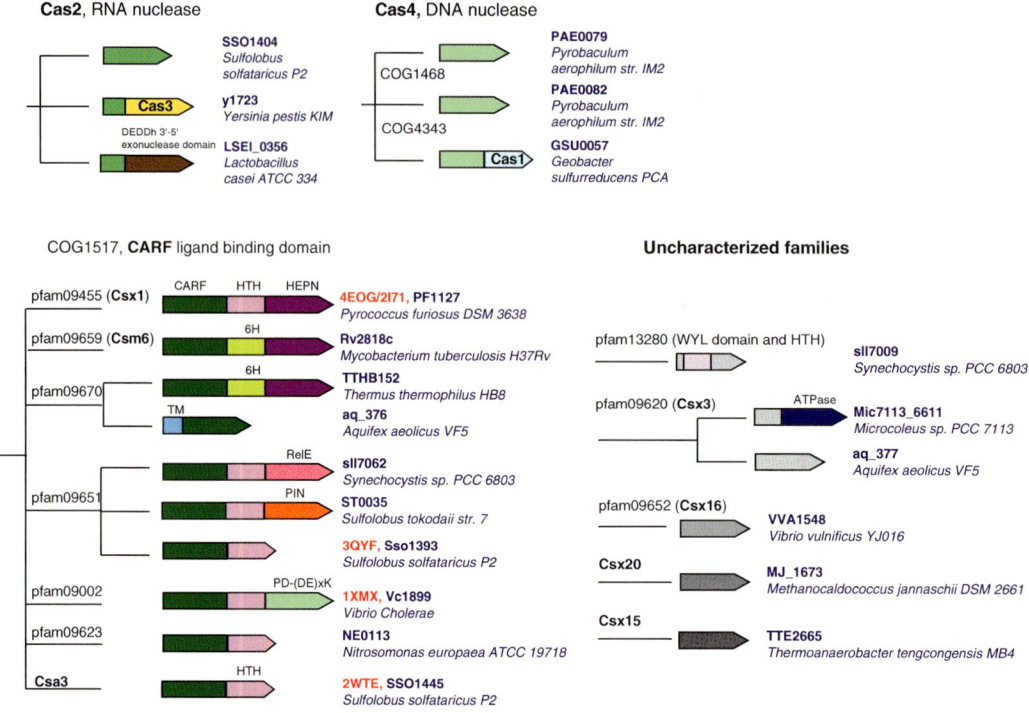

**Fig. 4** Associated immunity components of CRISPR-Cas systems. CRISPR-Cas gene names follow the nomenclature and classification from [5] and are shown in *bold*. An identifier of the sequence profile from COG and PFAM databases is provided when available. Examples of proteins for each distinct family are provided in the form of locus tag and organism name. The PDB code is shown in *red* when available. The genes are depicted as *block arrows* and domains as *block squares*. Homologous genes and domains are shown by *arrows* of the same color. The following domain names are indicated above the corresponding shape when shown for the first time: PD-(D/E)xK, restriction endonuclease superfamily protein, predicted DNA nuclease; HTH, DNA-binding helix-turn-helix domain; HEPN, HEPN domain, see details in [94]; RelE, RelE superfamily protein, predicted ribonuclease; PIN, PIN superfamily ribonuclease; 6H, helical middle domain in some of the COG1517 superfamily proteins

contain conserved potential catalytic residues and could represent novel families of nucleases (Fig. 4) [32]. Another abundant uncharacterized family typically contains the WYL domain (named after the respective conserved amino acid residues) and an HTH domain and thus reminds the core domain organization of COG1517. Both WYL and CARF domains are predicted to bind yet unidentified ligands, most likely nucleotides, and thus regulate the expression and/or activity of CRISPR-Cas systems via an allosteric mechanism [32]. Indeed, one of the WYL domain-containing proteins has been shown to regulate the expression of the CRISPR-Cas locus in *Synechocystis* sp. PCC6803 [99]. A plausible possibility seems to be that these regulators mediate the functional coupling of the CRISPR-Cas immunity with dormancy induction and PCD.

## 3 Conclusions and Outlook

The advances of comparative genomic analysis reveal unprecedented complexity of the CRISPR-Cas systems. The classification of CRISPR-Cas systems into three types and ten subtypes introduced some order into this striking diversity and provides the essential template for genome annotation and evolutionary studies. However, it is already perfectly clear that new types and subtypes of CRISPR-Cas have to be introduced. Moreover, this classification system, however refined and improved, can capture only part of the complexity of CRISPR-Cas organization and evolution, due to the intrinsic modularity and evolutionary mobility of these immunity systems, resulting in numerous recombinant variants. In particular, although Cas1 is the most conserved Cas protein, in terms of both presence in the great majority of CRISPR-Cas loci and sequence conservation, Cas1 phylogeny is of limited utility of CRISPR-Cas classification because of the extensive shuffling of the "informational" and "executive" modules. One possible way to achieve greater flexibility in CRISPR-Cas classification is to analyze these modules separately and explicitly recognize recombinants. However, gene and domain shuffling is extensive also within the modules so that we expect CRISPR classification to remain challenging for the foreseeable future. Above and beyond this organizational complexity of CRISPR-Cas systems, most of the *cas* genes evolve rapidly, which complicates the family assignment for many Cas proteins and the use of family profiles for the recognition of CRISPR-cas subtype signatures. Clearly, to achieve progress in the comparative analysis of CRISPR-Cas systems integration of the most sensitive sequence comparison tools with protein structure comparison is essential.

## References

1. Makarova KS, Wolf YI, Koonin EV (2013) Comparative genomics of defense systems in archaea and bacteria. Nucleic Acids Res 41: 4360–4377
2. Barrangou R, Horvath P (2012) CRISPR: new horizons in phage resistance and strain identification. Annu Rev Food Sci Technol 3:143–162
3. Wiedenheft B, Sternberg SH, Doudna JA (2012) RNA-guided genetic silencing systems in bacteria and archaea. Nature 482:331–338
4. van der Oost J, Jore MM, Westra ER, Lundgren M, Brouns SJ (2009) CRISPR-based adaptive and heritable immunity in prokaryotes. Trends Biochem Sci 34:401–407
5. Makarova KS, Haft DH, Barrangou R, Brouns SJ, Charpentier E, Horvath P, Moineau S, Mojica FJ, Wolf YI, Yakunin AF, van der Oost J, Koonin EV (2011) Evolution and classification of the CRISPR-Cas systems. Nat Rev Microbiol 9:467–477
6. Haft DH, Selengut J, Mongodin EF, Nelson KE (2005) A guild of 45 CRISPR-associated (Cas) protein families and multiple CRISPR/Cas subtypes exist in prokaryotic genomes. PLoS Comput Biol 1:e60
7. Makarova KS, Grishin NV, Shabalina SA, Wolf YI, Koonin EV (2006) A putative RNA-interference-based immune system in prokaryotes: computational analysis of the predicted enzymatic machinery, functional analogies with eukaryotic RNAi, and hypothetical mechanisms of action. Biol Direct 1:7
8. Barrangou R (2013) CRISPR-Cas systems and RNA-guided interference. Wiley Interdiscip Rev RNA 4:267–278

9. Westra ER, Swarts DC, Staals RH, Jore MM, Brouns SJ, van der Oost J (2012) The CRISPRs, they are a-changin': how prokaryotes generate adaptive immunity. Annu Rev Genet 46:311–339

10. Yosef I, Goren MG, Qimron U (2012) Proteins and DNA elements essential for the CRISPR adaptation process in *Escherichia coli*. Nucleic Acids Res 40:5569–5576

11. Nunez JK, Kranzusch PJ, Noeske J, Wright AV, Davies CW, Doudna JA (2014) Cas1-Cas2 complex formation mediates spacer acquisition during CRISPR-Cas adaptive immunity. Nat Struct Mol Biol 21:528–534

12. Richter C, Gristwood T, Clulow JS, Fineran PC (2012) In vivo protein interactions and complex formation in the *Pectobacterium atrosepticum* subtype I-F CRISPR/Cas System. PLoS One 7:e49549

13. Deltcheva E, Chylinski K, Sharma CM, Gonzales K, Chao Y, Pirzada ZA, Eckert MR, Vogel J, Charpentier E (2011) CRISPR RNA maturation by trans-encoded small RNA and host factor RNase III. Nature 471:602–607

14. Hale CR, Majumdar S, Elmore J, Pfister N, Compton M, Olson S, Resch AM, Glover CV III, Graveley BR, Terns RM, Terns MP (2012) Essential features and rational design of CRISPR RNAs that function with the Cas RAMP module complex to cleave RNAs. Mol Cell 45:292–302

15. Jinek M, Chylinski K, Fonfara I, Hauer M, Doudna JA, Charpentier E (2012) A programmable dual-RNA-guided DNA endonuclease in adaptive bacterial immunity. Science 337:816–821

16. Sashital DG, Wiedenheft B, Doudna JA (2012) Mechanism of foreign DNA selection in a bacterial adaptive immune system. Mol Cell 46:606–615

17. van Duijn E, Barbu IM, Barendregt A, Jore MM, Wiedenheft B, Lundgren M, Westra ER, Brouns SJ, Doudna JA, van der Oost J, Heck AJ (2012) Native tandem and ion mobility mass spectrometry highlight structural and modular similarities in clustered-regularly-interspaced shot-palindromic-repeats (CRISPR)-associated protein complexes from escherichia coli and pseudomonas aeruginosa. Mol Cell Proteomics 11:1430–1441

18. Zhang J, Rouillon C, Kerou M, Reeks J, Brugger K, Graham S, Reimann J, Cannone G, Liu H, Albers SV, Naismith JH, Spagnolo L, White MF (2012) Structure and mechanism of the CMR complex for CRISPR-mediated antiviral immunity. Mol Cell 45:303–313

19. Wiedenheft B, van Duijn E, Bultema JB, Waghmare SP, Zhou K, Barendregt A, Westphal W, Heck AJ, Boekema EJ, Dickman MJ, Doudna JA (2011) RNA-guided complex from a bacterial immune system enhances target recognition through seed sequence interactions. Proc Natl Acad Sci U S A 108:10092–10097

20. Brouns SJ, Jore MM, Lundgren M, Westra ER, Slijkhuis RJ, Snijders AP, Dickman MJ, Makarova KS, Koonin EV, van der Oost J (2008) Small CRISPR RNAs guide antiviral defense in prokaryotes. Science 321:960–964

21. Zhang Y, Heidrich N, Ampattu BJ, Gunderson CW, Seifert HS, Schoen C, Vogel J, Sontheimer EJ (2013) Processing-independent CRISPR RNAs limit natural transformation in *Neisseria meningitidis*. Mol Cell 50:488–503

22. Staals RH, Agari Y, Maki-Yonekura S, Zhu Y, Taylor DW, van Duijn E, Barendregt A, Vlot M, Koehorst JJ, Sakamoto K, Masuda A, Dohmae N, Schaap PJ, Doudna JA, Heck AJ, Yonekura K, van der Oost J, Shinkai A (2013) Structure and activity of the RNA-targeting Type III-B CRISPR-Cas complex of *Thermus thermophilus*. Mol Cell 52:135–145

23. Spilman M, Cocozaki A, Hale C, Shao Y, Ramia N, Terns R, Terns M, Li H, Stagg S (2013) Structure of an RNA silencing complex of the CRISPR-Cas immune system. Mol Cell 52:146–152

24. Makarova KS, Aravind L, Wolf YI, Koonin EV (2011) Unification of Cas protein families and a simple scenario for the origin and evolution of CRISPR-Cas systems. Biol Direct 6:38

25. Marchler-Bauer A, Anderson JB, Chitsaz F, Derbyshire MK, DeWeese-Scott C, Fong JH, Geer LY, Geer RC, Gonzales NR, Gwadz M, He S, Hurwitz DI, Jackson JD, Ke Z, Lanczycki CJ, Liebert CA, Liu C, Lu F, Lu S, Marchler GH, Mullokandov M, Song JS, Tasneem A, Thanki N, Yamashita RA, Zhang D, Zhang N, Bryant SH (2009) CDD: specific functional annotation with the conserved domain database. Nucleic Acids Res 37:D205–D210

26. Soding J, Biegert A, Lupas AN (2005) The HHpred interactive server for protein homology detection and structure prediction. Nucleic Acids Res 33:W244–W248

27. Altschul SF, Madden TL, Schaffer AA, Zhang J, Zhang Z, Miller W, Lipman DJ (1997) Gapped BLAST and PSI-BLAST: a new generation of protein database search programs. Nucleic Acids Res 25:3389–3402

28. Wheeler D, Bhagwat M (2007) BLAST QuickStart: example-driven web-based BLAST tutorial. Methods Mol Biol 395:149–176

29. Chylinski K, Makarova KS, Charpentier E, Koonin EV (2014) Classification and evolution of type II CRISPR-Cas systems. Nucleic Acids Res 42:6091–6105

30. Koonin EV, Makarova KS (2013) CRISPR-Cas: evolution of an RNA-based adaptive immunity system in prokaryotes. RNA Biol 10:679–686

31. Makarova KS, Wolf YI, Koonin EV (2013) The basic building blocks and evolution of CRISPR-CAS systems. Biochem Soc Trans 41:1392–1400

32. Makarova KS, Anantharaman V, Grishin NV, Koonin EV, Aravind L (2014) CARF and WYL domains: ligand-binding regulators of prokaryotic defense systems. Front Genet 5:102

33. Jansen R, Embden JD, Gaastra W, Schouls LM (2002) Identification of genes that are associated with DNA repeats in prokaryotes. Mol Microbiol 43:1565–1575

34. Takeuchi N, Wolf YI, Makarova KS, Koonin EV (2012) Nature and intensity of selection pressure on CRISPR-associated genes. J Bacteriol 194:1216–1225

35. Wiedenheft B, Zhou K, Jinek M, Coyle SM, Ma W, Doudna JA (2009) Structural basis for DNase activity of a conserved protein implicated in CRISPR-mediated genome defense. Structure 17:904–912

36. Han D, Lehmann K, Krauss G (2009) SSO1450–a CAS1 protein from Sulfolobus solfataricus P2 with high affinity for RNA and DNA. FEBS Lett 583:1928–1932

37. Babu M, Beloglazova N, Flick R, Graham C, Skarina T, Nocek B, Gagarinova A, Pogoutse O, Brown G, Binkowski A, Phanse S, Joachimiak A, Koonin EV, Savchenko A, Emili A, Greenblatt J, Edwards AM, Yakunin AF (2011) A dual function of the CRISPR-Cas system in bacterial antivirus immunity and DNA repair. Mol Microbiol 79:484–502

38. Beloglazova N, Brown G, Zimmerman MD, Proudfoot M, Makarova KS, Kudritska M, Kochinyan S, Wang S, Chruszcz M, Minor W, Koonin EV, Edwards AM, Savchenko A, Yakunin AF (2008) A novel family of sequence-specific endoribonucleases associated with the clustered regularly interspaced short palindromic repeats. J Biol Chem 283:20361–20371

39. Nam KH, Haitjema C, Liu X, Ding F, Wang H, DeLisa MP, Ke A (2012) Cas5d protein processes pre-crRNA and assembles into a cascade-like interference complex in subtype I-C/Dvulg CRISPR-Cas system. Structure 20:1574–1584

40. Sinkunas T, Gasiunas G, Fremaux C, Barrangou R, Horvath P, Siksnys V (2011) Cas3 is a single-stranded DNA nuclease and ATP-dependent helicase in the CRISPR/Cas immune system. EMBO J 30:1335–1342

41. Han D, Krauss G (2009) Characterization of the endonuclease SSO2001 from *Sulfolobus solfataricus* P2. FEBS Lett 583:771–776

42. Zhang J, Kasciukovic T, White MF (2012) The CRISPR associated protein Cas4 Is a 5′ to 3′ DNA exonuclease with an iron-sulfur cluster. PLoS One 7:e47232

43. Wiedenheft B, Lander GC, Zhou K, Jore MM, Brouns SJ, van der Oost J, Doudna JA, Nogales E (2011) Structures of the RNA-guided surveillance complex from a bacterial immune system. Nature 477:486–489

44. Jore MM, Lundgren M, van Duijn E, Bultema JB, Westra ER, Waghmare SP, Wiedenheft B, Pul U, Wurm R, Wagner R, Beijer MR, Barendregt A, Zhou K, Snijders AP, Dickman MJ, Doudna JA, Boekema EJ, Heck AJ, van der Oost J, Brouns SJ (2011) Structural basis for CRISPR RNA-guided DNA recognition by Cascade. Nat Struct Mol Biol 18:529–536

45. Rouillon C, Zhou M, Zhang J, Politis A, Beilsten-Edmands V, Cannone G, Graham S, Robinson CV, Spagnolo L, White MF (2013) Structure of the CRISPR interference complex CSM reveals key similarities with cascade. Mol Cell 52:124–134

46. Koo Y, Ka D, Kim EJ, Suh N, Bae E (2013) Conservation and variability in the structure and function of the Cas5d endoribonuclease in the CRISPR-mediated microbial immune system. J Mol Biol 425:3799–3810

47. Hale CR, Zhao P, Olson S, Duff MO, Graveley BR, Wells L, Terns RM, Terns MP (2009) RNA-guided RNA cleavage by a CRISPR RNA-Cas protein complex. Cell 139:945–956

48. Niewoehner O, Jinek M, Doudna JA (2014) Evolution of CRISPR RNA recognition and processing by Cas6 endonucleases. Nucleic Acids Res 42:1341–1353

49. Reeks J, Sokolowski RD, Graham S, Liu H, Naismith JH, White MF (2013) Structure of a dimeric crenarchaeal Cas6 enzyme with an atypical active site for CRISPR RNA processing. Biochem J 452:223–230

50. Richter H, Lange SJ, Backofen R, Randau L (2013) Comparative analysis of Cas6b processing and CRISPR RNA stability. RNA Biol 10:700–707

51. Carte J, Wang R, Li H, Terns RM, Terns MP (2008) Cas6 is an endoribonuclease that generates guide RNAs for invader defense in prokaryotes. Genes Dev 22:3489–3496

52. Haurwitz RE, Jinek M, Wiedenheft B, Zhou K, Doudna JA (2010) Sequence- and structure-specific RNA processing by a CRISPR endonuclease. Science 329:1355–1358

53. Cocozaki AI, Ramia NF, Shao Y, Hale CR, Terns RM, Terns MP, Li H (2012) Structure of the Cmr2 subunit of the CRISPR-Cas RNA silencing complex. Structure 20:545–553
54. Reeks J, Naismith JH, White MF (2013) CRISPR interference: a structural perspective. Biochem J 453:155–166
55. Reeks J, Graham S, Anderson L, Liu H, White MF, Naismith JH (2013) Structure of the archaeal Cascade subunit Csa5: relating the small subunits of CRISPR effector complexes. RNA Biol 10:762–769
56. Barrangou R, Fremaux C, Deveau H, Richards M, Boyaval P, Moineau S, Romero DA, Horvath P (2007) CRISPR provides acquired resistance against viruses in prokaryotes. Science 315:1709–1712
57. Garneau JE, Dupuis ME, Villion M, Romero DA, Barrangou R, Boyaval P, Fremaux C, Horvath P, Magadan AH, Moineau S (2010) The CRISPR/Cas bacterial immune system cleaves bacteriophage and plasmid DNA. Nature 468:67–71
58. Sapranauskas R, Gasiunas G, Fremaux C, Barrangou R, Horvath P, Siksnys V (2011) The *Streptococcus thermophilus* CRISPR/Cas system provides immunity in *Escherichia coli*. Nucleic Acids Res 39:9275–9282
59. Nishimasu H, Ran FA, Hsu PD, Konermann S, Shehata SI, Dohmae N, Ishitani R, Zhang F, Nureki O (2014) Crystal structure of Cas9 in complex with guide RNA and target DNA. Cell 156:935–949
60. Jinek M, Jiang F, Taylor DW, Sternberg SH, Kaya E, Ma E, Anders C, Hauer M, Zhou K, Lin S, Kaplan M, Iavarone AT, Charpentier E, Nogales E, Doudna JA (2014) Structures of Cas9 endonucleases reveal RNA-mediated conformational activation. Science 343:1247997
61. Mulepati S, Bailey S (2011) Structural and biochemical analysis of nuclease domain of clustered regularly interspaced short palindromic repeat (CRISPR)-associated protein 3 (Cas3). J Biol Chem 286:31896–31903
62. Beloglazova N, Petit P, Flick R, Brown G, Savchenko A, Yakunin AF (2011) Structure and activity of the Cas3 HD nuclease MJ0384, an effector enzyme of the CRISPR interference. EMBO J 30:4616–4627
63. Chakrabarti A, Desai P, Wickstrom E (2004) Transposon Tn7 protein TnsD binding to *Escherichia coli* attTn7 DNA and its eukaryotic orthologs. Biochemistry 43:2941–2946
64. Kholodii GY, Mindlin SZ, Bass IA, Yurieva OV, Minakhina SV, Nikiforov VG (1995) Four genes, two ends, and a res region are involved in transposition of Tn5053: a paradigm for a novel family of transposons carrying either a mer operon or an integron. Mol Microbiol 17:1189–1200
65. Jackson RN, Lavin M, Carter J, Wiedenheft B (2014) Fitting CRISPR-associated Cas3 into the helicase family tree. Curr Opin Struct Biol 24:106–114
66. Chylinski K, Le Rhun A, Charpentier E (2013) The tracrRNA and Cas9 families of type II CRISPR-Cas immunity systems. RNA Biol 10:726–737
67. Fonfara I, Le Rhun A, Chylinski K, Makarova KS, Lecrivain AL, Bzdrenga J, Koonin EV, Charpentier E (2014) Phylogeny of Cas9 determines functional exchangeability of dual-RNA and Cas9 among orthologous type II CRISPR-Cas systems. Nucleic Acids Res 42:2577–2590
68. Nam KH, Kurinov I, Ke A (2011) Crystal structure of clustered regularly interspaced short palindromic repeats (CRISPR)-associated Csn2 protein revealed Ca2+-dependent double-stranded DNA binding activity. J Biol Chem 286:30759–30768
69. Koo Y, Jung DK, Bae E (2012) Crystal structure of *Streptococcus pyogenes* Csn2 reveals calcium-dependent conformational changes in its tertiary and quaternary structure. PLoS One 7:e33401
70. Arslan Z, Wurm R, Brener O, Ellinger P, Nagel-Steger L, Oesterhelt F, Schmitt L, Willbold D, Wagner R, Gohlke H, Smits SH, Pul U (2013) Double-strand DNA end-binding and sliding of the toroidal CRISPR-associated protein Csn2. Nucleic Acids Res 41:6347–6359
71. Lee KH, Lee SG, Eun Lee K, Jeon H, Robinson H, Oh BH (2012) Identification, structural, and biochemical characterization of a group of large Csn2 proteins involved in CRISPR-mediated bacterial immunity. Proteins 80:2573–2582
72. Wei C, Liu J, Yu Z, Zhang B, Gao G, Jiao R (2013) TALEN or Cas9 - rapid, efficient and specific choices for genome modifications. J Genet Genomics 40:281–289
73. Pennisi E (2013) The CRISPR craze. Science 341:833–836
74. Anantharaman V, Iyer LM, Aravind L (2010) Presence of a classical RRM-fold palm domain in Thg1-type 3′-5′ nucleic acid polymerases and the origin of the GGDEF and CRISPR polymerase domains. Biol Direct 5:43
75. Pei J, Grishin NV (2001) GGDEF domain is homologous to adenylyl cyclase. Proteins 42:210–216
76. Makarova KS, Aravind L, Grishin NV, Rogozin IB, Koonin EV (2002) A DNA repair system specific for thermophilic Archaea and bacteria

predicted by genomic context analysis. Nucleic Acids Res 30:482–496

77. Zhu X, Ye K (2012) Crystal structure of Cmr2 suggests a nucleotide cyclase-related enzyme in type III CRISPR-Cas systems. FEBS Lett 586:939–945

78. Nickel L, Weidenbach K, Jager D, Backofen R, Lange SJ, Heidrich N, Schmitz RA (2013) Two CRISPR-Cas systems in Methanosarcina mazei strain Go1 display common processing features despite belonging to different types I and III. RNA Biol 10:779–791

79. Marraffini LA, Sontheimer EJ (2008) CRISPR interference limits horizontal gene transfer in staphylococci by targeting DNA. Science 322:1843–1845

80. White MF (2009) Structure, function and evolution of the XPD family of iron-sulfur-containing $5' \rightarrow 3'$ DNA helicases. Biochem Soc Trans 37:547–551

81. Makarova KS, Wolf YI, Forterre P, Prangishvili D, Krupovic M, Koonin EV (2014) Dark matter in archaeal genomes: a rich source of novel mobile elements, defense systems and secretory complexes. Extremophiles 18:877–893

82. Makarova KS, Wolf YI, Snir S, Koonin EV (2011) Defense islands in bacterial and archaeal genomes and prediction of novel defense systems. J Bacteriol 193:6039–6056

83. Datsenko KA, Pougach K, Tikhonov A, Wanner BL, Severinov K, Semenova E (2012) Molecular memory of prior infections activates the CRISPR/Cas adaptive bacterial immunity system. Nat Commun 3:945

84. Kim TY, Shin M, Huynh Thi Yen L, Kim JS (2013) Crystal structure of Cas1 from Archaeoglobus fulgidus and characterization of its nucleolytic activity. Biochem Biophys Res Commun 441:720–725

85. Krupovic M, Makarova KS, Forterre P, Prangishvili D, Koonin EV (2014) Casposons: a new superfamily of self-synthesizing DNA transposons at the origin of prokaryotic CRISPR-Cas immunity. BMC Biol 12:36

86. Makarova KS, Anantharaman V, Aravind L, Koonin EV (2012) Live virus-free or die: coupling of antivirus immunity and programmed suicide or dormancy in prokaryotes. Biol Direct 7:40

87. Lintner NG, Kerou M, Brumfield SK, Graham S, Liu H, Naismith JH, Sdano M, Peng N, She Q, Copie V, Young MJ, White MF, Lawrence CM (2011) Structural and functional characterization of an archaeal clustered regularly interspaced short palindromic repeat (CRISPR)-associated complex for antiviral defense (CASCADE). J Biol Chem 286:21643–21656

88. Shao Y, Cocozaki AI, Ramia NF, Terns RM, Terns MP, Li H (2013) Structure of the Cmr2-Cmr3 subcomplex of the Cmr RNA silencing complex. Structure 21:376–384

89. Jore MM, Brouns SJ, van der Oost J (2012) RNA in defense: CRISPRs protect prokaryotes against mobile genetic elements. Cold Spring Harb Perspect Biol 4:pii: a003657

90. Hrle A, Su AA, Ebert J, Benda C, Randau L, Conti E (2013) Structure and RNA-binding properties of the type III-A CRISPR-associated protein Csm3. RNA Biol 10:1670–1678

91. Osawa T, Inanaga H, Numata T (2013) Crystal structure of the Cmr2-Cmr3 subcomplex in the CRISPR-Cas RNA silencing effector complex. J Mol Biol 425:3811–3823

92. Quax TE, Wolf YI, Koehorst JJ, Wurtzel O, van der Oost R, Ran W, Blombach F, Makarova KS, Brouns SJ, Forster AC, Wagner EG, Sorek R, Koonin EV, van der Oost J (2013) Differential translation tunes uneven production of operon-encoded proteins. Cell Rep 4:938–944

93. Steitz TA (2004) The structural basis of the transition from initiation to elongation phases of transcription, as well as translocation and strand separation, by T7 RNA polymerase. Curr Opin Struct Biol 14:4–9

94. Anantharaman V, Makarova KS, Burroughs AM, Koonin EV, Aravind L (2013) Comprehensive analysis of the HEPN superfamily: identification of novel roles in intragenomic conflicts, defense, pathogenesis and RNA processing. Biol Direct 8:15

95. Penner M, Morad I, Snyder L, Kaufmann G (1995) Phage T4-coded Stp: double-edged effector of coupled DNA and tRNA-restriction systems. J Mol Biol 249:857–868

96. Wang C, Villion M, Semper C, Coros C, Moineau S, Zimmerly S (2011) A reverse transcriptase-related protein mediates phage resistance and polymerizes untemplated DNA in vitro. Nucleic Acids Res 39:7620–7629

97. Kim YK, Kim YG, Oh BH (2013) Crystal structure and nucleic acid-binding activity of the CRISPR-associated protein Csx1 of *Pyrococcus furiosus*. Proteins 81:261–270

98. Lintner NG, Frankel KA, Tsutakawa SE, Alsbury DL, Copie V, Young MJ, Tainer JA, Lawrence CM (2011) The structure of the CRISPR-associated protein Csa3 provides insight into the regulation of the CRISPR/Cas system. J Mol Biol 405:939–955

99. Hein S, Scholz I, Voss B, Hess WR (2013) Adaptation and modification of three CRISPR loci in two closely related cyanobacteria. RNA Biol 10:852–864

# Chapter 5

# Computational Detection of CRISPR/crRNA Targets

Ambarish Biswas, Peter C. Fineran, and Chris M. Brown

## Abstract

The CRISPR-Cas systems in bacteria and archaea provide protection by targeting foreign nucleic acids. The sequence of the "spacers" within CRISPR arrays specifically determines the targets in invader genomes. These spacers provide the short specific RNA nucleotide sequences within the guide crRNAs. In addition to complementarity in the spacer–target (protospacer) interaction, short flanking protospacer adjacent motifs (PAMs), or mismatching flanks have a discriminatory role in accurate target detection. Here, we describe a bioinformatic method, called CRISPRTarget, to use the sequence of a CRISPR array (e.g., predicted via CRISPRDetect/CRISPRDirection) to identify the foreign nucleic acids it targets.

**Key words** CRISPR-Cas, RNAi, Protospacer adjacent motifs, crRNA, Noncoding RNA

## 1 Introduction

CRISPR-Cas systems are adaptive bacterial immune systems that target invading nucleic acids such as bacteriophages and plasmids. CRISPR arrays consist of alternating direct repeats (DR) and spacers. It is the spacers within the array that determine the targeting specificity after the arrays are transcribed and processed into crRNAs [1, 2]. These crRNAs contain some or all of the spacer and parts of the flanking repeat(s). Chapter 4 has described the use of CRISPRFinder to identify CRISPR arrays [3] and classification of repeats into families can be performed using CRISPRMap [4]. We have also recently developed tools for the identification of CRISPR arrays (CRISPRDetect) and their direction (CRISPRDirection) [5].

Here we provide a method to answer the question of, given a predicted array, what are the targets of the spacers/crRNAs? In this method, we demonstrate how to use CRISPRTarget by utilizing a new worked example. Further examples are available in the original CRISPRTarget publication [6]. We have also recently used CRISPRTarget for the analysis of large numbers of new spacers incorporated in experimental acquisition assays in the *Escherichia coli* Type I-E and the *Pectobacterium atrosepticum*

Type I-F systems [7, 8]. The example utilized here is from the experimentally well-characterized system from *Streptococcus thermophilus* [6, 9, 10].

The CRISPR-Cas systems have also been adapted for use in precise genetic modification in many species [11] and for controlling gene expression [12]. For these purposes, targets need to be adjacent to PAM motifs and also be specific within the genome of interest (i.e., not match off-targets). This "off-target" problem is the complement of finding targets, but it is not specifically addressed by the method described here or by CRISPRTarget. There are several tools dedicated to "off-target" prediction [11, 13].

## 2 Materials

### 2.1 Prior Prediction of CRISPR Arrays

There are several methods available to predict arrays in genomic or metagenomic sequences. CRISPRFinder is an online tool, whereas others, such as CRT and PILER-CR, run on the command line [14–19]. We have recently developed a Web and online tool CRISPRDetect, which has advantages in CRISPR detection. The output of these different tools is a predicted array in different formats. For CRISPRTarget, PILER-CR/CRISPRDetect format is the best input, but other formats are supported.

Supported input formats:

The input files should be unedited as CRISPRTarget needs to extract specific information from the files.

#### 2.1.1 CRISPRDetect Format

Prediction using a web interface: To predict CRISPR array in your sequence using CRISPRDetect, you may use the web interface: http://brownlabtools.otago.ac.nz/CRISPRDetect/predict_crispr_array.html.

#### 2.1.2 PILER-CR Format

Prediction using PILER-CR by command line.

Program source: http://www.drive5.com/pilercr/.

Parameters used: While using PILER-CR, these parameters have been tested—otherwise the format may be changed and CRISPRTarget may not identify information correctly. However most parameters should work. An example of a PILER–CR output file is shown in Fig. 1.

Example command: "pilercr -noinfo -quiet -in ecoli.fna -out ecoli.out"

#### 2.1.3 CRT Format

Program source: CRT http://www.room220.com/crt/, MetaCRT http://omics.informatics.indiana.edu/mg/get.php?software=metaCRT.tz.

Sample command: "java -cp CRT1.2-CLI.jar crt ecoli.fna ecoli.out"

```
pilercr v1.02
By Robert C. Edgar
EF434469.fasta: 1 putative CRISPR arrays found.
DETAIL REPORT
Array 1
>gi|134103876|gb|EF434469.1| Streptococcus thermophilus strain DGCC7710 CRISPR1 locus genomic sequence

       Pos  Repeat    %id  Spacer  Left flank    Repeat                                    Spacer
==========  ======  =====  ======  ==========    ====================================      ======
        38      36  100.0      30  TTCATTTGAG    ....................................      TGTTTGACAGCAAATCAAGATTCGAATTGT
       104      36  100.0      30  TTCGAATTGT    ....................................      AATGACGAGGAGCTATTGGCACAACTTACA
       170      36  100.0      30  ACAACTTACA    ....................................      CGATTTGACAATCTGCTGACCACTGTTATC
       236      36  100.0      30  CACTGTTATC    ....................................      ACACTTGGCAGGCTTATTACTCAACAGCGA
       302      36  100.0      30  TCAACAGCGA    ....................................      CTGTTCCTTGTTCTTTTGTTGTATCTTTTC
       368      36  100.0      30  GTATCTTTTC    ....................................      TTCATTCTTCCGTTTTTGTTTGCGAATCCT
       434      36  100.0      30  TGCGAATCCT    ....................................      GCTGGCGAGGAAACGAACAAGGCCTCAACA
       500      36  100.0      30  GGCCTCAACA    ....................................      CATAGAGTGGAAAACTAGAAACAGATTCAA
       566      36  100.0      30  ACAGATTCAA    ....................................      ATAATGCCGTTGAATTACACGGCAAGGTCA
       632      36  100.0      30  GGCAAGGTCA    ....................................      GAGCGAGCTCGAAATAATCTTAATTACAAG
       698      36  100.0      30  TAATTACAAG    ....................................      GTTCGCTAGCGTCATGTGGTAACGTATTTA
       764      36  100.0      30  AACGTATTTA    ....................................      GGCGTCCCAATCCTGATTAATACTTACTCG
       830      36  100.0      30  TACTTACTCG    ....................................      AACACAGCAAGACAAGAGGATGATGCTATG
       896      36  100.0      29  TGATGCTATG    ....................................      CGACACAAGAACGTATGCAAGAGTTCAAG
       961      36  100.0      30  AGAGTTCAAG    ....................................      ACAATTCTTCATCCGGTAACTGCTCAAGTG
      1027      36  100.0      30  TGCTCAAGTG    ....................................      AATTAAGGGCATAGAAAGGGAGACAACATG
      1093      36  100.0      30  AGACAACATG    ....................................      CGATATTTAAAATCATTTTTCATAACTTCAT
      1159      36  100.0      30  ATAACTTCAT    ....................................      GCAGTATCAGCAAGCAAGCTGTTAGTTACT
      1225      36  100.0      30  GTTAGTTACT    ....................................      ATAAACTATGAAATTTTATAATTTTTAAGA
      1291      36  100.0      30  ATTTTTAAGA    ....................................      AATAATTTATGGTATAGCTTAATATCATTG
      1357      36  100.0      30  AATATCATTG    ....................................      TGCATCGAGCACGTTCGAGTTTACCGTTTC
      1423      36  100.0      30  TTACCGTTTC    ....................................      TCTATATCGAGGTCAACTAACAATTATGCT
      1489      36  100.0      30  CAATTATGCT    ....................................      AATCGTTCAAATTCTGTTTTAGGTACATTT
      1555      36  100.0      30  AGGTACATTT    ....................................      AATCAATACGACAAGAGTTAAAATGGTCTT
      1621      36  100.0      30  AAATGGTCTT    ....................................      GCTTAGCTGTCCAATCCACGAACGTGGATG
      1687      36  100.0      30  AACGTGGATG    ....................................      CAACCAACGGTAACAGCTACTTTTTTACAGT
      1753      36  100.0      30  TTTTTACAGT    ....................................      ATAACTGAAGGATAGGAGCTTGTAAAGTCT
      1819      36  100.0      30  TGTAAAGTCT    ....................................      TAATGCTACATCTCAAAGGATGATCCCAGA
      1885      36  100.0      30  TGATCCCAGA    ....................................      AAGTAGTTGATGACCTCTACAATGGTTTAT
      1951      36  100.0      30  AATGGTTTAT    ....................................      ACCTAGAAGCATTTGAGCGTATATTGATTG
      2017      36  100.0      30  ATATTGATTG    ....................................      AATTTTGCCCCTTCTTTGCCCCTTGACTAG
      2083      36  100.0          CCTTGACTAG    ....................................      ACCATTAGCA
==========  ======  =====  ======  ==========    ====================================      ======
        32      36              29                GTTTTTGTACTCTCAAGATTTAAGTAACTGTACAAC
SUMMARY BY SIMILARITY
Array          Sequence            Position      Length  # Copies  Repeat  Spacer  +  Consensus
=====  ==================      ==========  ==========  ========  ======  ======  =  =========
    1  gi|134103876|gb|                38        2081        32      36      29  +  GTTTTTGTACTCTCAAGATTTAAGTAACTGTACAAC

SUMMARY BY POSITION
>gi|134103876|gb|EF434469.1| Streptococcus thermophilus strain DGCC7710 CRISPR1 locus genomic sequence
Array          Sequence            Position      Length  # Copies  Repeat  Spacer  Distance  Consensus
=====  ==================      ==========  ==========  ========  ======  ======  ========  =========
    1  gi|134103876|gb|                38        2081        32      36      29            GTTTTTGTACTCTCAAGATTTAAGTAACTGTACAAC
```

**Fig. 1** Example PILER-CR input file. Prediction of an array from *S. thermophilus*. Arrays consist of near identical direct repeats (DR) with specific spacers. In this case, there are 32 identical repeats and 31 spacers. Both the spacer sequences (*right*) and the ends of the repeats (the handles) are used by CRISPRTarget. Extra blank lines have been removed for display here, but the program output should be unedited

## 2.1.4 CRISPRFinder Format

CRISPRFinder is a web application can be found at http://crispr.u-psud.fr/Server. How to obtain CRISPRFinder output:

To perform a CRISPR prediction and obtain the output file, follow these steps:

1. Upload or paste your genomic sequence in the corresponding text box and press submit.

2. The next page shows a table with headers "Confirmed CRISPRs" and "Questionable CRISPRs" along with links to the corresponding files. Clicking on a link will take you to the corresponding CRISPR arrays visualization.

3. Click on the button named "CRISPR Properties" will open the output file you need. Save the file, and you can use the file or its content as input in CRISPRTarget (*see* **Note 1**).

| | |
|---|---|
| **2.1.5 Upload Just the Spacers in FASTA or MultiFASTA Format** | Generate a FASTA or multiFASTA file of the spacers of interest (*see* **Note 2**). For example:<br><br>>identifier_1_1<br>GGGTTGGGGGTTTTA<br>>identifier_1_12<br>AACGGCGTTGGGGGTTATT |
| **2.1.6 Predicting Direction (Optional)** | With the exception of CRISPRDetect (which uses CRISPR Direction), all of the above programs do not predict direction of the CRISPR array. There is a command line tool, CRISPRDirection, that can be used to predict direction with ~94 % accuracy on a test set [5]. This step is optional, because if the direction is unknown then CRISPRTarget will analyze both strands (Subheading 3.2.1). The direction can be predicted with CRISPRDirection separately [5] or as part of the CRISPRDetect webserver. The CRISPRDirection output is an array in PILER-CR format in the most likely direction. For complete genomes there are precomputed arrays with direction available at the CRISPRDirection website (http://bioanalysis.otago.ac.nz/CRISPRDirection/). These could be used as input to CRISPRTarget. |
| **2.2 Databases of Targets**<br><br>*2.2.1 Genbank* | Genbank databases, released and updated bimonthly at CRISPRTarget (see the online help and news for the latest update http://brownlabtools.otago.ac.nz/CRISPRTarget/news_and_updates.html). |

1. Nonredundant nucleotide database: The nr/nt collection. This database contains "All GenBank + EMBL + DDBJ + PDB sequences (but no EST, STS, GSS, or phase 0, 1 or 2 HTGS sequences)." Size: ~59 billion bases, 4/6/2014.

2. Environmental database: env_nt. This contains "Sequences from environmental samples, such as uncultured bacterial samples isolated from soil or marine samples. The largest single source is Sargasso Sea project. This does not overlap with nucleotide nr." These sequences have no taxonomic classification other than metagenome. Size: 4.1 billion bases.

3. Phage division: This is one of the smallest Genbank divisions of 316 million bases, ~6,400 phage sequences.

*2.2.2 RefSeq Databases* — Several relevant divisions of the NCBI Reference Sequence databases are available, which contain better annotated (by NCBI) versions of GenBank sequences.

1. RefSeq-Plasmid. ~4,400 sequences, 336 million bases (Release 65, 12/5/2014).

2. RefSeq-Viral. ~4,500 sequences, 123 million bases.

3. RefSeq-Bacteria. ~5,500 complete microbial genomes, 7 billion bases.

| | |
|---|---|
| *2.2.3 CAMERA Database Sections* | We included viral parts of the CAMERA databases. 9,139,883 gene sequences, 1 billion bases (Files: CAM_PROJ_ReclaimedWaterVirues.read.fa, CAM_PROJ_MarineVirome.read.fa, CAM_P_0000909.read.fa, CAM_P_0000792.read.fa). ACLAME. 125,190 sequences, 96 million bases (V0.04, 8/2009, latest version at 6/2014). |
| *2.2.4 User Database* | Users can upload sequences of up to 50 Mb (if you wish to analyze larger databases, please contact the authors) (*see* **Note 3**). |

# 3 Methods

The following sections outline the procedure for analysis of CRISPR targets using CRISPRTarget.

*3.1 Input Section*

Figure 2 shows an example of a correctly formatted input with PILER-CR, but multiple input formats are accepted (*see* Subheading 2.1).

1. Upload or paste a file in one of the supported formats: The format needs to be unedited as the program needs to extract specific information from the files.

2. Remove redundant spacers: This default option is useful if your input has multiple spacers from a number of related species (e.g., all *E. coli* strains' spacers). Identical spacers will be removed and listed in a file. Identical reverse-complements of the spacers are not removed.

3. Upload the FASTA sequence file, which was used for generating the CRISPR: Uploading the source sequence of the CRISPR array is optional unless you want a longer handle region greater than the length of relative direct repeat(s). As most (except CRISPRFinder) CRISPR finding tools provide both the direct repeats as well as the spacers, the handle regions can be extracted from the adjacent direct repeats.

4. Select target databases: Hold down the Control key and click on any database from the list to select/unselect multiple or no databases. Relevant databases are provided by default in CRISPRTarget (*see* Subheading 2.2).

5. Parameters for the initial BLAST screen: The CRISPRTarget BLASTn parameters favor gapless matches but allow a number of mismatches at this screening stage by using a higher gap penalty of 10, rather than 5, which is the NCBI default (*see* **Note 4**). The initial CRISPRTarget defaults are Gap open −10, Gap extend −2, Match +1, Mismatch −1, Word size 7(11), and Expect (E): 1 (*see* **Notes 5** and **6**).

6. Changing BLAST parameters: Please note that only certain combinations of parameters produce valid statistics (others will not work; *see* **Notes 6–8**).

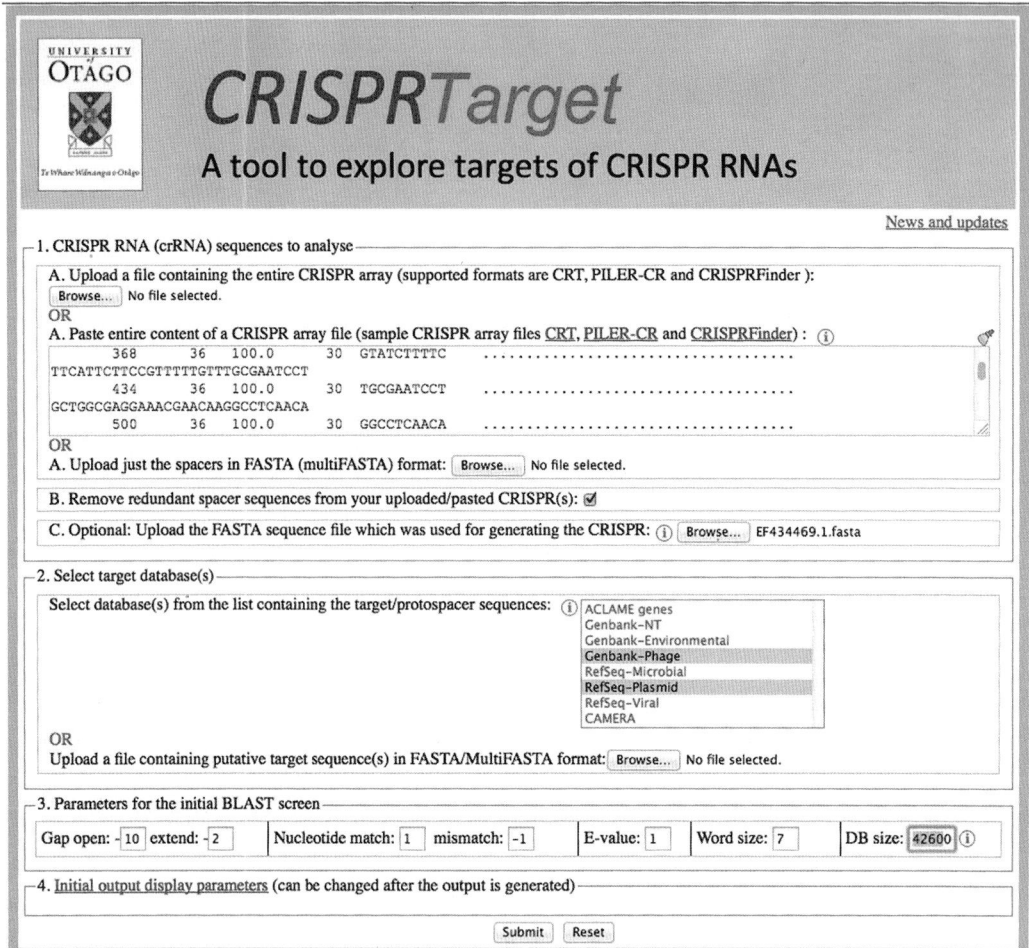

**Fig. 2** CRISPRTarget input screen. The input screen is shown with a PILER-CR input file, optional fasta file, GenBank-Phage and RefSeq-Plasmid selected (the defaults), and an (optional) DB size selected (*see* **Notes 7** and **8**)

7. Submit: Once the submit button is pressed, CRISPRTarget shows progress with links to intermediate files (Fig. 3). Typically, for a single CRISPR array (with relatively small number, e.g., 31 spacers in the above case), this takes just a few seconds. However, the total computational time depends on the number of databases selected as well as the total number of spacer sequences. After the BLAST and annotation steps are complete, you can save the "link" to check the output later. Closing the window will not terminate the job.

## 3.2 Interactively Assess the Output

1. An example output is shown in Fig. 4. Initially all the matches that pass the BLAST filters and CRISPRTarget score cutoff (e.g., default of 20) are shown. They can be reordered and scores can be recalculated.

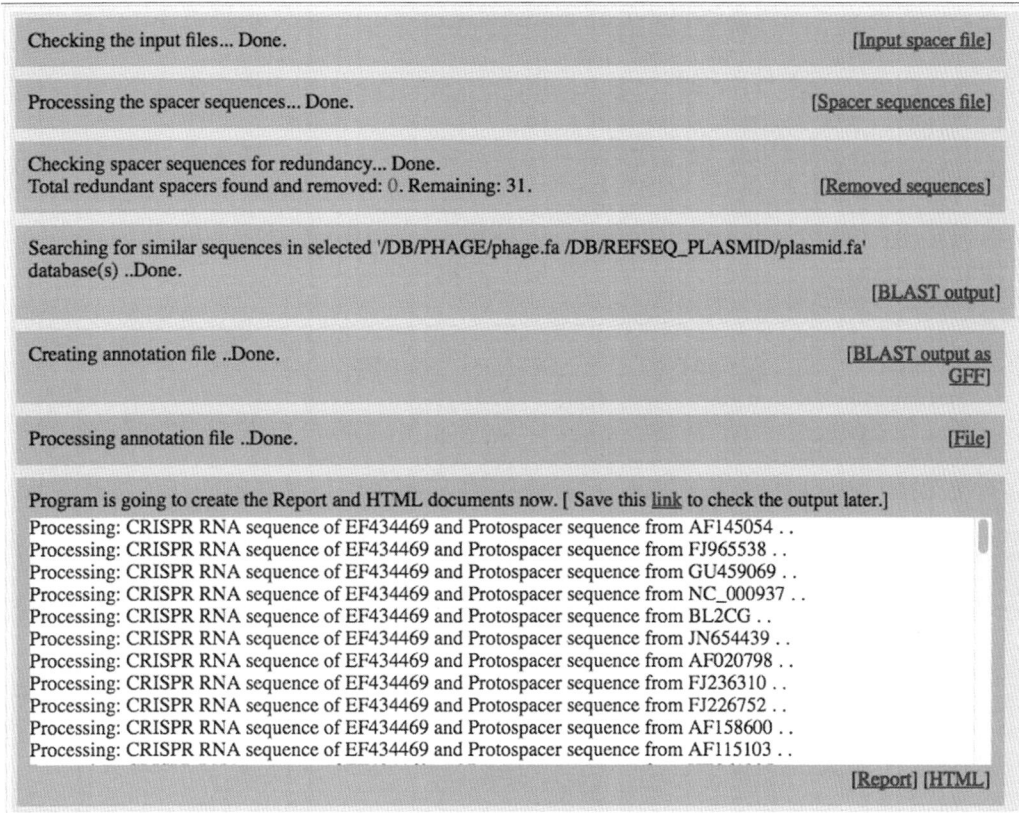

**Fig. 3** Progress log and intermediary files generated. The analysis should proceed rapidly though this progress screen. Intermediary files are provided as links from this screen

2. The protospacer target is extended automatically within CRISPRTarget by extracting the user-specified length of the 5′ or 3′ handle sequences from the BLAST database.

3. CRISPRTarget interactive scoring: All putative spacer/protospacer targets passing the BLAST screen are displayed in an interactive manner. An initial score is calculated by scoring matches (+1) and mismatches (−1) across the whole length of the spacer without gaps. Specific user-defined "seed" regions can be required to match at either or both ends of the protospacer. A match to predefined, or novel user defined, PAM sequences can increase the score.

4. In order to penalize self-matches that would match 100 % in both spacers and flanking handles (e.g., self-targeting to the original genomic array sequence), a score can be used that penalizes matches (e.g., −1) in the flanking handles. Mismatch penalties can also be used to identify targeting that is facilitated by mismatches in the handles (e.g., Type III-A systems).

5. Finally, a cutoff score can be applied to initially display only those matches with the best scores.

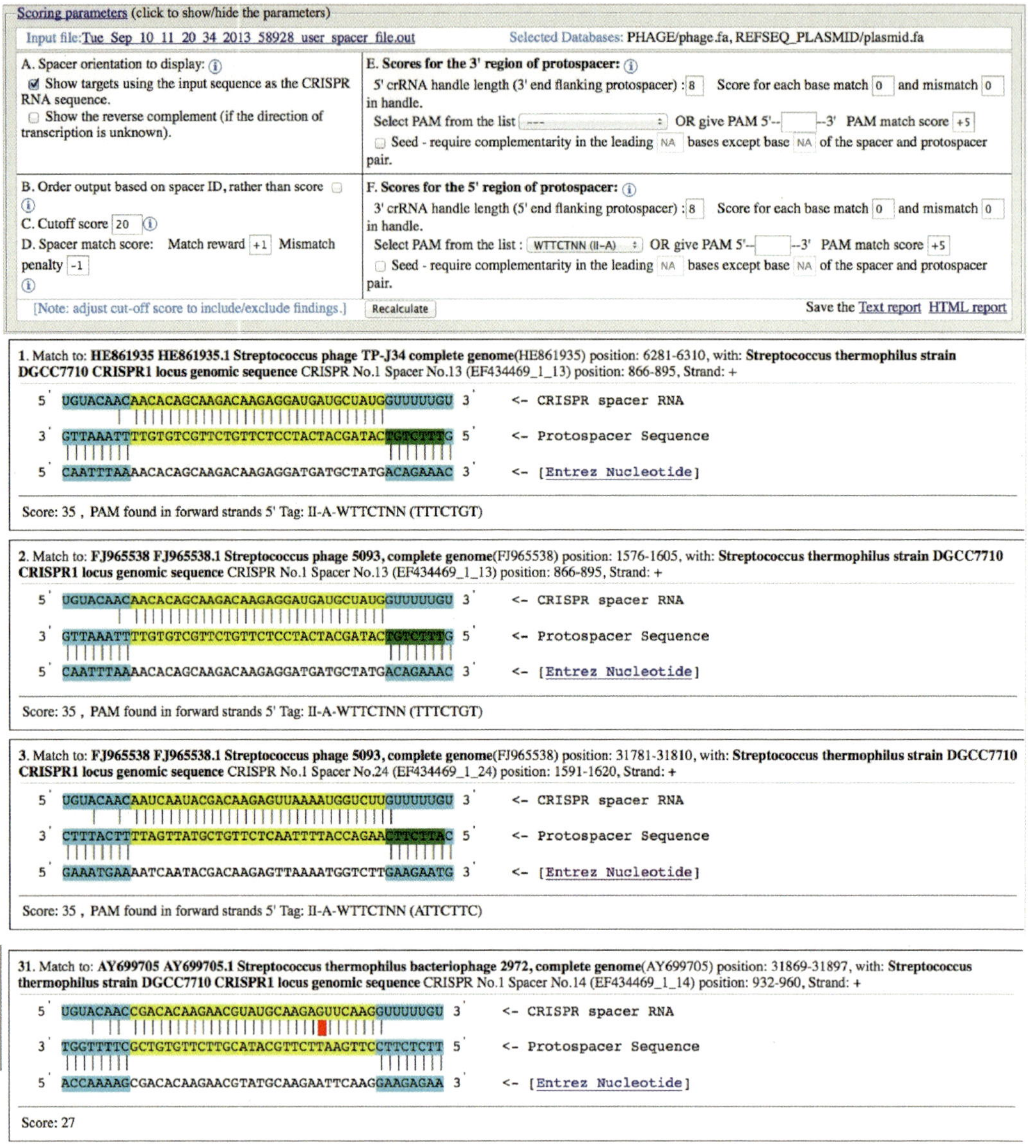

**Fig. 4** An example of a CRISPRTarget output. The top section (scoring parameters A-F) shows parameters that can be adjusted at this point in the process. The initial output has been interactively adjusted to reward a 5′ PAM as +5. The first match (1) is a perfect match, with consensus PAM to a phage from the same species from the 13th spacer in the array. The second match (2) is to a different named phage but has the same spacer and position of match, indicating a near identical phage with a different name. The third hit (3) is a different spacer (spacer 24) that matches a different region of the same phage. Finally, the 31st hit (31) matches a phage to which the bacterium is possibly susceptible, due to either the single mismatch or the lack of a consensus PAM. Note: this would score spuriously well in a simple BLAST search

### 3.2.1 Steps to Analyze the Output

1. Spacer orientation to display (Fig. 4 scoring parameter A): By default, the spacer sequence (top most in any set) is shown in a 5′ to 3′ orientation, and the protospacer sequence (the target sequence that base pairs with spacer sequence) is shown in 3′ to 5′ (*see* **Note 9**). The user can choose to display the other strand of the spacer sequence, which brings the other strand of the protospacer sequence to the middle. This option should be used when the orientation of the CRISPR array is not known/certain.

2. Order output based on spacer ID (Fig. 4 scoring parameter B): A spacer ID is represented with three elements, the sequence ID, CRISPR Index, and spacer Index, separated by underscore (e.g., EF434469_1_13). By default the output is sorted in descending order of the calculated score. However, if the user wants to show/arrange the output based on the spacer ID, selecting this option will achieve that. This option can be very useful in visually inspecting the output, as it maintains the order of the spacer for every CRISPR.

3. Cutoff score (Fig. 4 scoring parameter C): The cutoff score is used for filtering out the low scoring matches from the output. The default value is 20, but user can use any cutoff or no cutoff value to show/hide matches.

4. Spacer match score (Fig. 4 scoring parameter D): The default values for match and mismatch are +1 and −1, respectively. These values along with the cutoff score provide a way to push the matches with mismatches down the order or even omit them from display.

5. Scores for the 3′ region of protospacer (Fig. 4 scoring parameter E): This option can be used to increase the score of targets with particular PAM sequences and to disfavor self-targets with identical flanks (i.e., matches back to the original array). PAMs in the 3′ region of the protospacer typically belong to the Type I CRISPR-Cas systems. Predefined PAMs are I-A: NGG; I-B: NGG; I-E: CAT,CTT,CCT,CTC; I-C: GAA; I-F: GG, but users can also input specific user-defined sequences (*see* **steps 6–9**).

6. Scores for the 5′ region of protospacer (Fig. 4 scoring parameter F): All the parameters shown in Fig. 4 are for the 5′ region of the protospacer and its adjacent crRNA region. Each of the options are described in detail below (*see* **steps 7–10**) and are similar to those available for the 3′ region of the protospacer (Subheading 3.2.1, **step 5**).

7. 3′ crRNA handle length (Fig. 4 scoring parameter F): The default value used is 8, but the user can increase/decrease the length of the handle ranging from 0 to any number (e.g., 100). There is no upper limit/restriction, but if the source

sequence is not available, the length will be automatically adjusted. The handle sequence comes from the repeat sequences (unless the handle length is greater than repeat sequence length; see Subheading 3.1, **step 3**). Repeat lengths can be selected based on the biological knowledge of crRNA processing in the system being analyzed. For example, in the *E. coli* Type I-E system, crRNAs have a 5′ 8 nt handle and a 21 nt 3′ repeat flank. This corresponds to 3′ 8 nt and 21 nt 5′ regions of the protospacer.

8. Score for each base match and mismatch in the flanking region (Fig. 4 scoring parameter F): The default value is set to 0, but the user can alter the values to any positive or negative number (e.g., match: −1, mismatch: +1). If flanking sequence is present, these values can greatly help to identify and eliminate the self matches. For self matches, the handle sequences of spacers will match exactly. Therefore, penalizing base pairing in the handle region will send the self matches down the order or filter them out (using the cutoff score).

9. PAM scoring (Fig. 4 scoring parameter F): PAM (Protospacer Adjacent Motif) is an important indicator of true positive crRNA target matches. The PAM can be used to identify the targets of known CRISPR-Cas systems. The PAM types are II-A: WTTCTNN,TTTYRNNN; II-B:CNCCN,CCN. In Fig. 4, WTTCTNN is selected as this is the experimentally determined PAM for this system. Matches are shown in green highlighting on the output. The user can also give a PAM motif (e.g., CGT). CRISPRTarget supports any user given PAM that uses IUPAC codes as above. By default the PAM match score is +5, but this can be adjusted.

10. Seed sequence (Fig. 4 scoring parameter F): Specify the seed region if required. For example, for the *E. coli* Type I-E system, require a match to the first 8 bases, except base 6. "Seed—require complementarity in the leading 8 bases except base 6 of the spacer and protospacer pair." (*see* **Note 10**).

11. The results are shown in the order of decreasing scores. More information about the potential target in Genbank can be found by following the link to "Entrez Nucleotide" show the specific matches region of the target. This will be formatted in the NCBI graphical view (Fig. 5).

## 3.3 Saving the Output

From the progress screens intermediate files can be saved. The output file can be saved in text or html form. If saved in text form, this can be opened and edited easily in programs such as Excel. Bookmarked outputs will be retained on the server for at least a month.

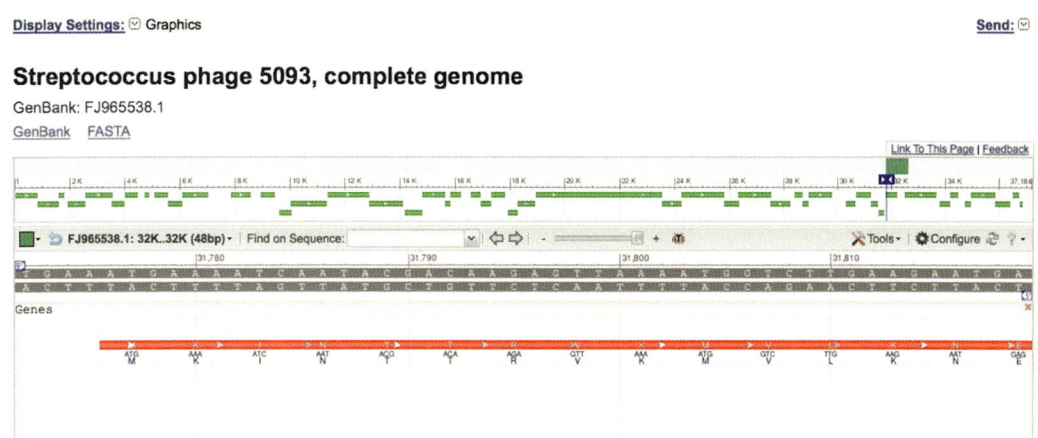

**Fig. 5** A CRISPRTarget match displayed in GenBank graphical view. The region indicated in Fig. 4 (hit 3) is shown on the virus genome at Genbank

## 4 Notes

1. You can concatenate all the predicted CRISPR Arrays (individual output files) in one file using a text editor (e.g., TextWrangler) and upload or paste it into CRISPRTarget.

2. If this option is used, the spacer sequence cannot be extended into the adjacent DR. However, the target will be extended to show the flanking handles.

3. If you are interested in analyzing SRA sequences, you need to download the specific sequences from SRA databases and convert them to FASTA format. The FASTA formatted sequences can be used as "User Database." For practical reasons, the redundancy of these data should be reduced by doing an assembly. More information on the SRA database is available here: http://www.ncbi.nlm.nih.gov/books/NBK47540/.

4. The default values used by NCBI BLASTN for short sequences <30 bases (defaults for long sequences are in brackets) are Gap open −5(−5), Gap extend −2(−2), Match +1(+1), Mismatch −3(−3), Word size 7(11), Expect (E): 1000 (10), Filter: No (Yes).

5. There is also no filter or masking for low complexity. BLAST calculates the scores over the length of the match and only shows this match. For example, a spacer of 32 bases that matches to a target in 17 of 20 bases would score 20−3 = 17, and 20 bases would be output. The expected (E) values of the match will be more likely to pass the filter if smaller databases are used (e.g., the default phg and plasmid).

6. Only certain combinations of BLAST parameters produce valid statistics. For +1, −1 an attempt to use some combinations

might fail. See the explanation of the following example that fails: $ blastn -db database -query myseq -gapopen 1 -gapextend 1 -reward 1 -penalty -1.

BLAST engine error: Error: Gap existence and extension values 1 and 1 are not supported for substitution scores 1 and –1.

3 and 2 are supported existence and extension values.

2 and 2 are supported existence and extension values.

1 and 2 are supported existence and extension values.

0 and 2 are supported existence and extension values.

4 and 1 are supported existence and extension values.

3 and 1 are supported existence and extension values.

2 and 1 are supported existence and extension values.

4 and 2 are supported existence and extension values.

Any values more stringent than 4 and 2 are supported (e.g., 10, 2).

Suggestion—Useful changes might be:

(a) Reducing the gap penalty to 4 or 5 if you have reason to believe that gaps (insertion/deletions) are tolerated in your system.

(b) Increasing the E to 10 or 100 in the unlikely event you are not getting hits.

(c) Increasing the mismatch penalty to –3 screens out mismatches.

7. Set the effective database size: This is optional. This should be the total size of the databases you search. BLAST calculates the E (Expect) value based on the size of the database searched. If one search against multiple databases is done the database need not be specified as BLAST does it internally. To compare the significance of matches in two or more consecutive searches of different databases, this value should be set as the sum of the two databases sizes (e.g., for RefSeq plasmid 309 Mb + Genbank phage 117 Mb = 426 Mb enter "426000000") (*see* **Note 8**).

8. The BLAST parameters need to be set here, optional parameters can be set here but can be changed when visualizing the output. If the BLAST parameters need to be changed, the analysis can be rerun.

9. The direction could have been predicted in advance with CRISPRDirection (*see* Subheading 2.1.5). Parameters that would guide likely orientation are described in Biswas et al. [6]. These include the presence of known repeats or ATTGAAAN at the end of the repeat. These features could be looked for in the input array to indicate which orientation to display.

10. If you want to exclude multiple bases of the spacer—protospacer match, then give them as comma separated. For the above

example, if you want to exclude base 3 and 5, then give the input as below: "Seed require complementarity in the leading 1–8 bases except base 3, 5 of the spacer and protospacer pair."

## Acknowledgements

This work was supported by a Rutherford Discovery Fellowship from the Royal Society of NZ to PCF, by a Human Frontier Science Program Grant to Ian Macara, Anne Spang and CMB. AB was a recipient of a University of Otago Postgraduate Scholarship and a Postgraduate Publishing Bursary.

## References

1. Richter C, Chang JT, Fineran PC (2012) Function and regulation of clustered regularly interspaced short palindromic repeats (CRISPR)/CRISPR associated (Cas) systems. Viruses 4:2291–2311
2. Westra ER, Buckling A, Fineran PC (2014) CRISPR-Cas systems: beyond adaptive immunity. Nat Rev Microbiol 12:317–326
3. Drevet C, Pourcel C (2012) How to identify CRISPRs in sequencing data. Methods Mol Biol 905:15–27
4. Lange SJ, Alkhnbashi OS, Rose D, Will S, Backofen R (2013) CRISPRmap: an automated classification of repeat conservation in prokaryotic adaptive immune systems. Nucleic Acids Res 41:8034–8044
5. Biswas A, Fineran PC, Brown CM (2014) Accurate computational prediction of the transcribed strand of CRISPR non-coding RNAs. Bioinformatics 30:1805–1813
6. Biswas A, Gagnon JN, Brouns SJ, Fineran PC, Brown CM (2013) CRISPRTarget: bioinformatic prediction and analysis of crRNA targets. RNA Biol 10:817–827
7. Fineran PC, Gerritzen MJ, Suarez-Diez M, Kunne T, Boekhorst J, van Hijum SA, Staals RH, Brouns SJ (2014) Degenerate target sites mediate rapid primed CRISPR adaptation. Proc Natl Acad Sci U S A 111:E1629–E1638
8. Richter C, Dy RL, McKenzie RE, Watson BN, Taylor C, Chang JT, McNeil M, Staals RHJ, Fineran PC (2014) Priming in the Type I-F CRISPR-Cas system triggers strand-independent spacer acquisition, bi-directionally from the primed protospacer. Nucleic Acids Res 42(13):8516–8526
9. Barrangou R, Fremaux C, Deveau H, Richards M, Boyaval P, Moineau S, Romero DA, Horvath P (2007) CRISPR provides acquired resistance against viruses in prokaryotes. Science 315:1709–1712
10. Deveau H, Barrangou R, Garneau JE, Labonte J, Fremaux C, Boyaval P, Romero DA, Horvath P, Moineau S (2008) Phage response to CRISPR-encoded resistance in *Streptococcus thermophilus*. J Bacteriol 190:1390–1400
11. Ran FA, Hsu PD, Wright J, Agarwala V, Scott DA, Zhang F (2013) Genome engineering using the CRISPR-Cas9 system. Nat Protoc 8:2281–2308
12. Fineran PC, Dy RL (2014) Gene regulation by engineered CRISPR-Cas systems. Curr Opin Microbiol 18:83–89
13. Yang L, Mali P, Kim-Kiselak C, Church G (2014) CRISPR-Cas-mediated targeted genome editing in human cells. Methods Mol Biol 1114:245–267
14. Bland C, Ramsey TL, Sabree F, Lowe M, Brown K, Kyrpides NC, Hugenholtz P (2007) CRISPR recognition tool (CRT): a tool for automatic detection of clustered regularly interspaced palindromic repeats. BMC Bioinformatics 8:209
15. Edgar RC (2007) PILER-CR: fast and accurate identification of CRISPR repeats. BMC Bioinformatics 8:18
16. Grissa I, Vergnaud G, Pourcel C (2007) CRISPRFinder: a web tool to identify clustered regularly interspaced short palindromic repeats. Nucleic Acids Res 35:W52–W57
17. Rousseau C, Gonnet M, Le Romancer M, Nicolas J (2009) CRISPI: a CRISPR interactive database. Bioinformatics 25:3317–3318
18. Skennerton CT, Imelfort M, Tyson GW (2013) Crass: identification and reconstruction of CRISPR from unassembled metagenomic data. Nucleic Acids Res 41:e105
19. Rho M, Wu YW, Tang H, Doak TG, Ye Y (2012) Diverse CRISPRs evolving in human microbiomes. PLoS Genet 8:e1002441

# Chapter 6

# High-Throughput CRISPR Typing of *Mycobacterium tuberculosis* Complex and *Salmonella enterica* Serotype Typhimurium

Christophe Sola, Edgar Abadia, Simon Le Hello, and François-Xavier Weill

## Abstract

Spoligotyping was developed almost 18 years ago and still remains a popular first-lane genotyping technique to identify and subtype *Mycobacterium tuberculosis* complex (MTC) clinical isolates at a phylogeographic level. For other pathogens, such as *Salmonella enterica*, recent studies suggest that specifically designed spoligotyping techniques could be interesting for public health purposes. Spoligotyping was in its original format a reverse line-blot hybridization method using capture probes designed on "spacers" and attached to a membrane's surface and a PCR product obtained from clustered regularly interspaced short palindromic repeats (CRISPRs). Cowan et al. and Fabre et al. were the first to propose a high-throughput Spoligotyping method based on microbeads for MTC and *S. enterica* serotype Typhimurium, respectively. The main advantages of the high-throughput Spoligotyping techniques we describe here are their low cost, their robustness, and the existence (at least for MTC) of very large databases that allow comparisons between spoligotypes from anywhere.

**Key words** Spoligotyping, CRISPR locus, Microbeads, High-throughput, Molecular epidemiology

## 1 Introduction

The discovery of a region within a *Mycobacterium bovis* BCG strain characterized by the presence of short repeats, each interspaced by unique sequences that were highly polymorphic allowed the invention of the Spoligotyping technique [1, 3, 4]. The name of this technique stands for *spacer oligonucleotide typing*, an acronym that was created by a research team in the National Institute of Health and Environment in Bilthoven, The Netherlands, who patented and standardized the technique for *Mycobacterium tuberculosis* complex (MTC). In 2002, the unique and peculiar genetic structure of this region was designated as CRISPR (clustered regularly interspaced short palindromic repeats) [5]. CRISPR loci were found to be present in nearly all archaea and in almost 50 % of bacteria [6, 7]. These structures represent at least in some species such as *Streptococcus thermophilus*, an adaptative immune system that allows the bacteria to defend against invader DNA or RNA [8]. The discovery of other physiological roles of this complex RNA-based interference mechanism of regulation is expanding [9]. The extreme molecular diversity of these CRISPR loci make them ideal to target bacterial strain diversity and perform subtyping, indirectly allowing clues on the natural history and evolutionary genetics of the underlying disease in the case of bacterial pathogens. Subtyping methods based on analyses of the spacers of CRISPR loci have since been developed for other bacteria of medical interest, such as *Yersinia pestis* and *Y. pseudotuberculosis* [10], *Corynebacterium diphtheriae* [11], *Salmonella enterica* [2, 12–14], *Legionella pneumophila* 1 [15], and *Streptococcus agalactiae* [16]. Hermans et al. revealed the presence of the DR region in MTC strains through sequencing [17]. The DR region consists of direct variable repeats (DVR), each made up of a constant and a variable part [3]; in MTC, the constant is represented by identical repeated sequences of 36 bp length (DR, direct repeats) interspaced by unique variable sequences (spacers) of 35–41 bp length that generate the polymorphism. The absence of some spacers may be the characteristic of a given subspecies or sublineages (e.g., the absence of spacer 3, 9, 16, and 39-43 in *M. bovis* BCG). Another example is the rare *M. canetti* subspecies that harbors specific spacers (69-104) or the signature of absence of the spacers 1-33 and presence of spacers 34-43 for the « Beijing » lineage. In 2000 van Embden et al. provided more knowledge on the genetic diversity of this locus on MTC strains and also some hypothesis about how the region may evolve [18]. Filliol et al. showed a good correlation between spoligotypes signatures and geographic regions, which in turn could be the result of MTC strains genomic changes and adaptation to their host [19]. It seems that the region evolves mainly losing spacers so the way particular spacers are being lost may represent phylogenetic signatures during their evolution. Gagneux et al. proposed that MTC lineages are adapted to particular human populations [20]. Indeed, strains from different lineages of MTC strongly associated with specific geographical regions and with patient country of origin [21, 22].

The locus's schematic view and the technique's principle are shown in Fig. 1, and a raw experiment and deduced pattern are shown in Fig. 2. Briefly, the power of the technique relies on the amplification of all the spacers that are present in the CRISPR region at once using one pair of primers that are complementary to each DR sequence. One of the primers needs to be biotin-labeled; thus biotin-labeled single-strand DNA of heterogeneous size will be produced. PCR products will be hybridized over a membrane to which a set of predefined complementary oligonucleotides

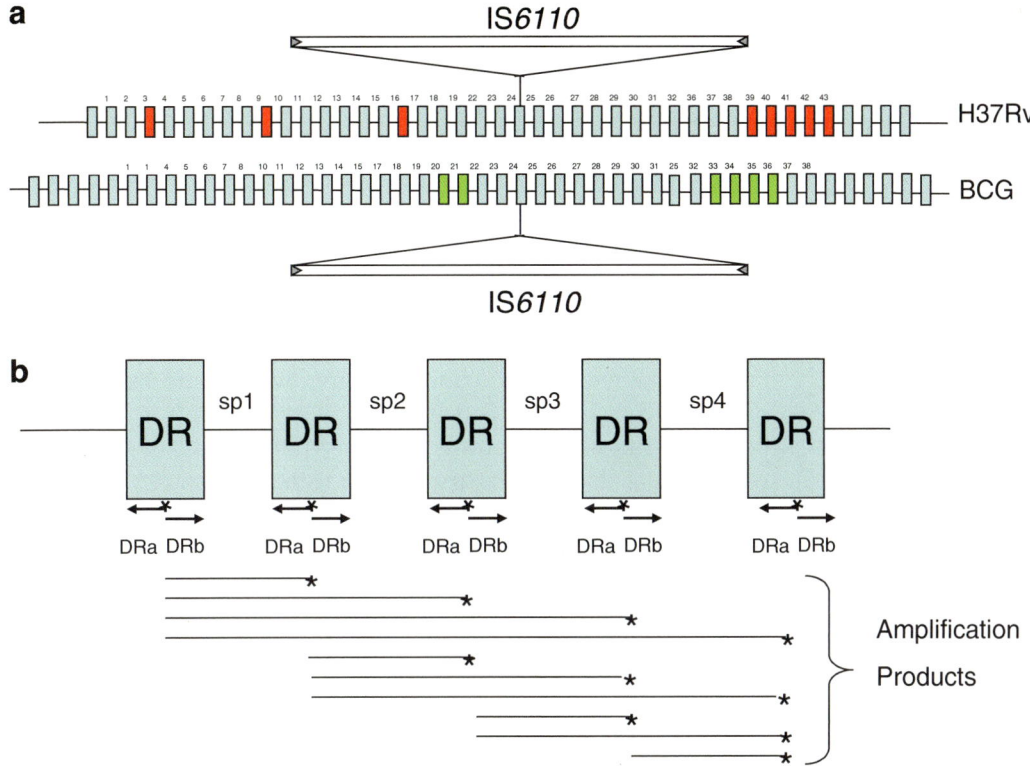

**Fig. 1** Principle of the MTC spoligotyping technique; a single couple of primers (Dra-Drb) allows to amplify by PCR a set of overlapping fragments that will be further detected by hybridization on a 2D or 3D device if one primer (Dra) is biotinylated and a detectable reporter is added (Streptavidin-Peroxydase using chemoluminescence, streptavidin-Phycoerythrin using microbead-based laser detection)

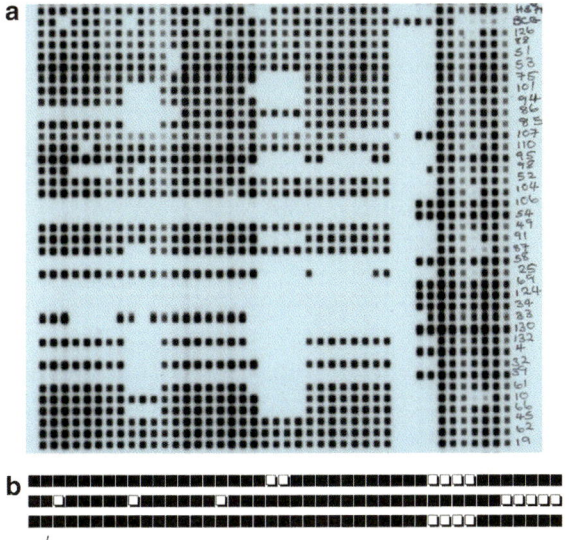

**Fig. 2** Example of an MTC membrane spoligotyping experiment (**a**) and a binary transcription of results done on Excel (**b**)

(capture probes) were previously chemically attached. The membrane is supported by a matricial device (miniblotter) in which the hybridization procedure takes place. After the hybridization, washing steps will be done and will allow to get rid of nonspecific hybridization events. The biotinylated-hybridized PCR fragments will be revealed after exposure to a streptavidin–peroxydase conjugate through a classical electrochemiluminescence autoradiogram. The result is a matrix of hybridized (black spots) or nonhybridized spots (no spots) depending on the presence or absence of the corresponding spacers in the original DNA sample (Fig. 2). Each DNA produces a unique pattern that will provide a first raw identity of a patient-specific clinical isolate. Some patterns are highly patient and strain-specific whereas others are highly common and poorly significant, requiring further typing. Patterns have to be compared to databases to reveal their informativeness [23].

The hybridizing/nonhybridizing spacers patterns transcription is done from the membrane to a spreadsheet of OpenOffice® or Excel® software. The order is strictly conserved from 1 to 43. For better display of results, an ultrametric font such as « Monotype Sort » or « Zapf Dingbats » size 10, using characters « n » for each spacer present and « o » for each spacer absent should be used. With such a display, a 43-character black/white pattern is easily recognized by human beings and even more easily by computers using machine-learning algorithms. Other spoligotype pattern display methods were developed according to the need of handling a code with fewer characters, like the octal code (15 digit) or the Hexadecimal code (12 characters) [24]. Just a single script (Excel's macro) can translate one code to another one (directly available from the authors upon request).

In *Salmonella*, there are two CRISPR loci, CRISPR1 and CRISPR2, separated by less than 20 kb. The CRISPR1 locus is located downstream from the *iap* gene, whereas CRISPR2 is located upstream from the *ygcF* gene. The ordered CRISPR-associated (*cas*) genes belonging to the *E. coli* subtype are located between the CRISPR loci. The DRs of both CRISPR loci were conserved. They were 29 bp long and have the consensus sequence 5′-CGGTTTATCCCCGCTGGCGCGGGGAACAC-3′. The CRISPR analysis by PCR and sequencing of 783 strains belonging to 130 serotypes revealed the presence of 3,800 spacers (mean size 32 bp) [2]. The spacer content was found correlated with both serotype and multilocus sequence type. Furthermore, spacer microevolution (duplication, triplication, loss or gain of spacers, presence of SNP variant spacers, or VNTR variant spacers) discriminated between subtypes within prevalent serotypes such as Typhimurium (STM), the most prevalent serotype worldwide. In eight genomes and 150 strains of serotype Typhimurium and its monophasic 1,4,(5),12:i:- variant, it was found 57 CRISPR1, 62

CRISPR2 alleles, and 83 CRISPR1-CRISPR2 combined alleles. Forty unique spacers (including four with variants, such as SNP or VNTR variants) were identified in CRISPR1. Thirty-nine unique spacers (including two with a SNP variant) were identified in CRISPR2. Particular well-characterized populations, such as multidrug-resistant DT104 isolates, African MDR ST313 isolates, and DT2 isolates from pigeons, each had typical CRISPR alleles. Based on this high polymorphism of the spacer contents, a microbead-based liquid hybridization assay, CRISPOL (for CRISPR polymorphism) has been developed for the serotype Typhimurium and its monophasic variant. This assay targets 72 of the 79 spacers identified previously as it is not possible, for the time-being, to distinguish between some of the remaining seven spacers by a Luminex approach. For example, STMB8var1 has a single SNP located in position 1 of the spacer compared with STMB8 or the four VNTR variants of STM18 only differ from each other by the number of an hexanucleotide repeat.

## 2 Materials

### 2.1 Consumables

- 1.5 mL microcentrifuge tubes.
- 1.5 mL Eppendorf Protein loBind microcentrifuge tubes (Fisher, VWR or other supplier).
- 2.0 mL screw-cap microcentrifuge tubes or ambered safe-lock tubes for storage.
- 0.2 mL PCR tubes-sterile.
- 10 µL, 250 µL, 1,000 µL pipette tip refills.
- 96 well microplate aluminium sealing tape.
- Thermowell 96-well P polycarbonate clear PCR Plates (Costar, Fisher or other supplier) (recommended for XYP heater block).
- Sealing Mat for 96-well Thermowell P Plates (Costar, Fisher or other supplier).
- Microseal "A" film (MJ Research).
- 1.2 µM PVDF filter microtiter plates (Millipore).
- 1 µM PTFE filter microtiter plates (Millipore).
- 1.2 µM Supor filter microtiter plates (Pall Life Sciences or VWR).
- 96-well black half-area flat bottom plates (Costar).
- 96-well half-area flat bottom plates (Costar) (white, non-treated, recommended for no-wash assays).
- 96-well round bottom polystyrene solid plates (Costar) (recommended for no-wash assays).

## 2.2 Chemicals

- MES (2(N-Morpholino) ethanesulfonic acid) (Euromedex, Sigma or other supplier).
- NaOH.
- TWEEN 20 (Polyoxyethylenesorbitan monolaurate).
- SDS (Sodium lauryl sulfate) powder or 10 % solution.
- EDTA powder or Tris-EDTA Buffer, pH 8.0, 100×.
- TMAC powder (MP Biomedicals, or Fisher) or 5 M TMAC—N-Lauroylsarcosine (sarkosyl) sodium salt 20 % solution (Sigma or other supplier).
- 1 M Tris–HCl, pH 8.0 and 0.5 M EDTA, pH 8.0 (prepared from powder or Tris-EDTA Buffer, pH 8.0, 100×, molecular biology grade).
- Molecular Biology grade water.
- SSPE, 20× (Phosphate buffer, pH 7.4, sodium chloride, EDTA) (Sigma or other supplier).
- Triton X-100 (MP Biomedicals or other supplier).
- Betain Chlorhydrate (VWR, Euromedex or other supplier).
- DMSO.
- EDC (1-ethyl-3-(3-dimethylaminopropyl)carbodiimide hydrochloride) (Perbio, Pierce or other supplier).

## 2.3 Buffers

- MES, pH 4.5 (Coupling buffer).
  For 250 mL: weight 4.88 g of MES, complete to <250 mL, adjust pH with NaOH (≅5 drops), complete to 250 mL, filter sterilize, and store at room temperature.
- 0.02 % Tween 20 (Washing buffer): pipet 50 μL of Tween 20, complete to 250 mL with water, filter sterilize, and store at room temperature.
- 0.1 % SDS (Washing buffer): pipet 2.5 mL of SDS 10 %, complete to 250 mL with water, filter sterilize and store at room temperature
- Tris-EDTA (TE) Buffer, pH 8.0, 1×: pipet 2.5 mL of TE 100×, complete to 250 mL with water; filter sterilize, and store at room temperature
- 1.5× TMAC Hybridization Solution (Microsphere diluent): pipet 225 mL of TMAC 5 M; add 1.88 mL of 20 % Sarkosyl solution; add 18.75 mL of 1 M Tris–HCl, pH 8.0; add 3 mL of 0.5 M EDTA, pH 8.0; complete with water to 250 mL (1.37 mL); and store at room temperature.
- X TMAC Hybridization Solution (Microsphere diluent): pipet 150 mL of TMAC 5 M; add 1.25 mL of 20 % Sarkosyl solution; add 12.5 mL of 1 M Tris–HCl, pH 8.0; add 2 mL of 0.5 M EDTA, pH 8.0; complete with water to 250 mL (84.25 mL); and store at room temperature.

| | |
|---|---|
| **2.4 Equipment** | – Centrifuge for 96-well plates. |
| | – Sonicator (mini) (Cole Parmer, Ultrasonic cleaner, VWR or other supplier). |
| | – Bench microcentrifuge. |
| | – Vortex Mixer. |
| | – DNA Engine PTC200 (Bio-Rad) or equivalent equipment. |
| | – Luminex 200®, BioPlex200®, or MagPix® (Luminex). |
| | – Hemacytometer (Sigma) or equivalent equipment, or Bio-Rad TC20 cell counter (Bio-Rad) or equivalent equipment. |
| | – Refrigerated microcentrifuge. |
| | – Pipettes of different volume range |
| **2.5 Oligonucleotides, Enzymes, PCR Reagents, Microbeads** | – 5′ amino-C12 linker oligonucleotides (IDT or Eurogentec); capture probes according to published sequences. |
| | – The specific list of probes for the TB-SPOL (43-Plex) and STM-CRISPOL (72-Plex) are found in Kamerbeek et al. modified by van Embden et al. for MTC and in Fabre et al. for STM [2, 4, 18]. |
| | – Standard and biotinylated oligonucleotides and PCR primers (IDT, Eurogentec, or other supplier). |

Specific Primers for MTC-Spoligotyping (TB-SPOL):

DRa (5′Biot-GGTTTTGGGTCTGACGAC-3′) and DRb (5′-CCGAGAGGGGACGGAAAC-3′).

Specific Primers for STM-CRISPOL:

DRSTMA (5′-CCGCTGGCGCGGGGAACA-3′) and DRSTMB (5′Biot-CGCCAGCGGGGATAAACC-3′)

– MicroPlex® Microspheres Regions 1 through 100, choose 43 regions for TB-SPOL, 72 regions for STM-CRISPOL among LC10001 to LC10100 references, or MagPlex® among MC10012 to MC10100 references (Luminex, Austin, TX).

– Or MagPlex® microspheres Regions 1 through 100 (to run only on MagPix®).

– Taq DNA Polymerase, deoxynucleotides, and buffers (Homemade or Promega).

– Streptavidin-R-Phycoerythrin Lumi Grade (Roche, Invitrogen, Interchim, or other supplier).

– Separately purchased Oligonucleotides and Luminex Microbeads can advantageously be replaced by directly available, quality controlled coupled-microspheres sold by Beamedex® (Université-Paris-Sud, Bât 400, Orsay, France) or by purchasing full reagent kits: TB-SPOL (43-Plex) and STM-CRISPOL

(72-Plex), that also include, buffers, positive controls, and Streptavidin-Phycoerythrin; please visit www.beamedex.com.

## 2.6 DNA

The quality of DNA may vary from crude to purified. For MTC, crude thermolysates, purified Cetyl-Trimethylammonium bromide (CTAB) extracted, or any commercial kit-extracted DNA can be used as templates. For STM, the following DNA extraction protocol can be followed. Take a 10 μl loop of bacteria and suspend it in 200 μl of molecular biology-grade water. The suspension is vortexed for 10 s, incubated at 95 °C for 10 min and then centrifuged for 5 min at $10,000 \times g$. The supernatant is transferred to a 1.5 mL microtube and stored at −20 °C until use.

# 3 Methods

## 3.1 Generalities

Since the advent of multiplexed analyzers, an alternative to membrane-based spoligotyping is high-throughput microbead-based spoligotyping [25]. The transfer from the membrane-based towards the microbead-based format was indeed achieved in USA by Cowan et al. at the CDC-Atlanta in 2004 and in 2009 by Zhang et al. at University of Paris-Sud [1, 26] and by the Institut Pasteur in 2011 on *S. enterica* serotype Typhimurium [2, 12, 13]. However, for cost reasons, in many laboratories worldwide the technique still relies on a membrane-based procedure run on the Immunetics miniblotter. Other alternative techniques, e.g., MALDI-TOF (Matrix-Assisted Laser Desorption/Ionization time-of-flight) mass spectrometry-based Spoligotyping were also recently developed and will not be described here [27].

Briefly the principle of the Luminex system relies on the use of polystyrene or magnetic colored microbeads of different types (up to 500 types in the latest FlexMap 3D® version, 100 types on Fig. 3a), which can be individually recognized by a laser (L1) in a microfluidic system (Fig. 3b, c). On each set of beads, it is possible to link a large variety of sensor-targets (antibodies, antigens, nucleic acids) that can thus be individually assessed. In our case, these markers are amino-linked oligonucleotides with a C12 linker. The second laser of the system (L2, Fig. 3b), combined with a second optical mean (in our case Streptavidin-Phycoerythrin or SA-PE), allows to detect the microbead-fixed ligands thus permitting the quantification of results on each microbead type. Alternatively to the use of the biotin/SA-PE detection principle, 5′-labeled oligonucleotides using Cyanine or Alexa Fluor markers can also be used for quantification by L2. As many as 500 analytes can theoretically be individually assayed in a unique sample. In Fig. 3c right, two microbead types are represented, type 1 and type n, each previously coupled with a specific oligonucleotide (DR1 to DRn capture sequence).

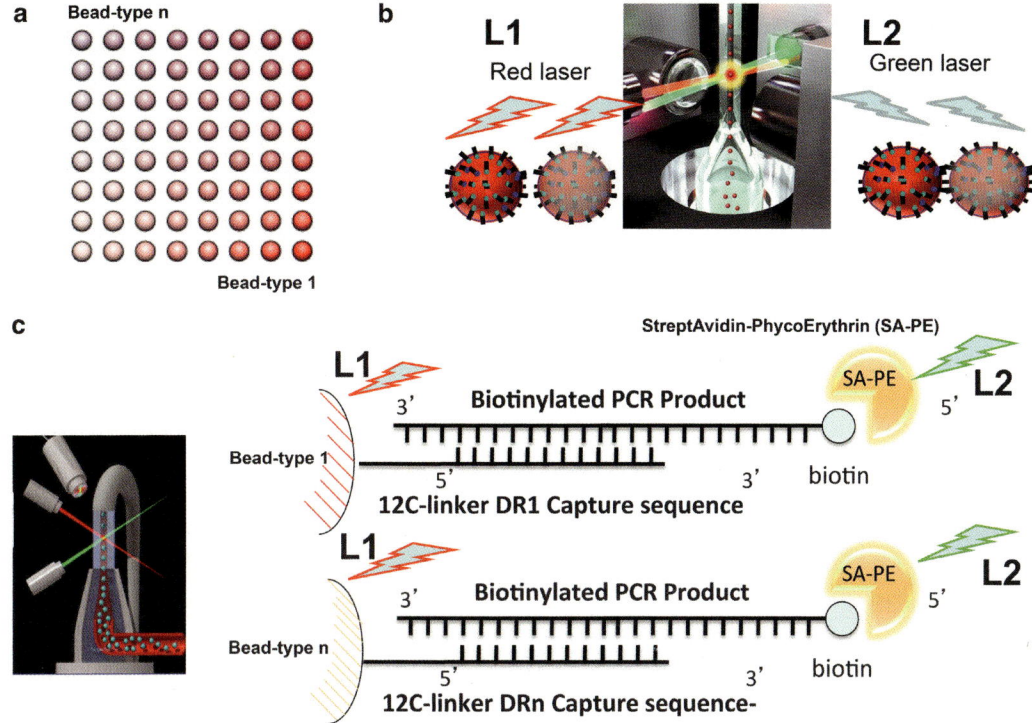

**Fig. 3** Principle of laser-based multiplexed experiment. (**a**) A set of 100 different commercial beads in the red region are available. (**b**) The L1 laser recognizes the beads and the L2 laser quantifies the signals that are present on « hairy » (oligonucleotide-coupled) and hybridized beads, by detection of SA-PE after previous hybridization with a single strand-biotinylated PCR amplified DNA fragment. (**c**) Artistic drawing of the analyzer showing the power of lasers combined to microfluidic systems

Users that have been previously producing their own spoligotyping membranes will easily be able to produce spoligotyping microbeads. Chemical constraints or precautions to link oligonucleotides to membranes are not much different. A simple list of requirements is to be followed: (1) always use low-binding Eppendorf tubes since polystyrene microbeads may adsorb to classical polypropylene tubes; (2) order 5′-amino oligonucleotides with a C12 amino-link instead of a C6 amino-link arm for membranes to increase gyration radius; (3) keep EDC powder frozen in aliquots at −20 °C and do not reuse freshly prepared solutions.

### 3.2 Protocols

#### 3.2.1 Coupling of Oligonucleotides to Microbeads

Step 1: microbead washing (*see* **Note 1**)

1. Let some fresh aliquot of EDC powder come back from −20 °C to laboratory temperature.
2. Dissolve amino-linked oligonucleotides ("probe" or "capture" oligo) to 1 mM (1 nmol/μL) in sterile water.
3. Resuspend the microbeads stock by vortexing (20 s) and sonication (20 s).

4. Transfer $5 \times 10^6$ (400 µL) of MicroPlex® or MagPlex® microbeads stocks into LowBind Eppendorf tubes.
5. Centrifuge microbeads at $8,000 \times g$ during 1–2 min.
6. Discard supernatant and resuspend the microbeads in 50 µL of 0.1 M MES (2-(*N*-morpholino)ethanesulfonic acid) buffer pH = 4.5 by vortexing and sonicating during 20 s. This washing step is done to eliminate microbeads conservation buffer including antimicrobial agents (*see* **Note 2**).

Step 2: chemical coupling of oligonucleotides to microbeads (*see* **Note 3**)

7. Prepare a 1/10 dilution of capture oligonucleotides in sterile water (0.1 nmole/µL; 0.1 mM or 100 µM).
8. Add 2 µl (0.2 nmol) of the 1/10 oligo-solution prepared above to the resuspended and vortexed microbeads.
9. Prepare a fresh EDC solution (10 mg/mL) in sterile water.
10. One by one, for each coupling reaction, add 2.5 µl of freshly prepared EDC solution to the microbeads, mix by vortexing.
11. Incubate during 30 min at room temperature in a dark room.
12. Prepare a new fresh EDC solution (10 mg/mL) in sterile water (*see* **Note 4**).
13. Again, one by one, for each coupling reaction, add 2.5 µl of freshly prepared EDC solution to the microbeads, mix by vortexing.
14. Again, incubate during 30 min at room temperature in a dark room.

Step 3: Washing of coupled microbeads

The microbeads are washed successively with Tween-20 and SDS to prevent microbeads aggregation and adsorption to Eppendorf tubes walls, as well as to block hydrophobic sites on the microbeads surfaces.

15. Add 1.0 mL of 0.02 % Tween-20 to the coupled microbeads (*see* **Note 5**).
16. Centrifuge the coupled microbeads at $8,000 \times g$ during 1–2 min.
17. discard the supernatant and resuspend the coupled microbeads within1 mL of 0.1 % SDS by vortexing (*see* **Note 6**).
18. Centrifuge the coupled microbeads by centrifuging at $8,000 \times g$ during 1–2 min.
19. Discard the supernatant and resuspend the coupled microbeads in 100 µl TE 1× pH = 8 by vortexing and sonicating during approximately 20 s (*see* **Note 7**).
20. Store coupled microbeads between 2 and 8 °C protected from light. Coupled beads can still be used after 6 months.

Step 4: counting microbeads on an hemacytometer

1. Dilute the coupled microbeads 1/100 in sterile water.
2. Load at the proper place 10 µL of microbeads dilution in an hemacytometer (cell counter).
3. Count all microbeads seen in one of the corner within a 4×4 grid as shown below (the model of classical hemacytometer may vary from country to country).
4. Compute the result by using the following formula

    total microbeads number = (Number read on 4×4 grids corner) × $(1 \times 10^4)$ × (dilution factor) × (volume of microbeads suspension in mL).

    The beads mix might alternatively be prepared and controlled on a TC20 Cell Counter (Bio-Rad, Hercules, CA), which provides the easiest way to check bead counts.

Step 5: Control of oligonucleotide-coupling. The objective of this step is to verify the right fabrication of oligonucleotides coupled MicroPlex or MagPlex microbeads reagents using complementary probes

1. Defrost 1 µL of biotinylated target stock oligonucleotides (antisense nucleotides).
2. Dilute the target oligonucleotides at 10 fmoles/µL in TE 1×, pH = 8.
3. Select the appropriate coupled microbeads and biotinylated target oligonucleotides sets to control.
4. Resuspend the microbeads by vortexing and sonication during approximately 20 s.
5. Prepare a « Microbeads working Mix » with 75 beads/µL in 1 mL of TMAC 1.5× hybridization Solution (add 1.5 µL of each coupled microbeads stock to 998,5 µL of TMAC 1.5× hybridization solution).
6. Mix the « Microbeads working Mix » by vortexing 20 s and sonication during 20 s.
7. In each assayed well of a 96-well plate, including the negative control add 33 µL of « Microbeads working Mix ».
8. in the negative control well(s), add 17 µL of TE 1×, pH = 8.
9. In each tested well, add the mixture of biotinylated complementary oligonucleotides (5–200 fmoles) and TE 1×, pH = 8 up to a total volume of 17 µL.
10. Smoothly mix the tested wells by pipeting up and down a couple of times.

11. Cover the 96-well plate to prevent evaporation and incubate at 95–100 °C during 3 min to break all oligonucleotides secondary structures.

12. Incubate the 96-well plate at the same temperature as the one used for the PCR-product hybridization assay (52 °C for TB-SPOL and 59 °C for STM-CRISPOL) during 20 min (*see* **Note 8**).

13. Centrifuge the plate during 1 min, eliminate as much as possible of the supernatant (25–35 μl) by pipeting carefully, replace with 25–35 μl of TMAC 1×.

14. Prepare a fresh « Reporter Mix » by adding 4 μL of Streptavidin-R-phycoerythrin in 996 μL of TMAC 1× hybridization solution to obtain a 4 μg/mL solution (4/1,000 of 1 mg/mL stock dilution).

15. Add 25 μL of « Reporter Mix » to each well and mix smoothly by pipeting up and down a few times.

16. Incubate the 96-well plate at 52 °C during 10 min.

17. Analyze 50 μL at 52 °C with the Luminex 200/BioPlex analyzer using user's manual (*see* **Note 9**).

### 3.2.2 High-Throughput MTC and STM Spoligotyping Protocols on Luminex® 200 or BioPlex®

Step 1: PCR protocols for MTC-Spoligotyping (DRa-DRb) and STM-CRISPOL (DRSTMA-DRSTMB)

A. PCR protocol for MTC-Spoligotyping (Dra-DRb)

| DNA (CTAB-extracted or thermolyzates) | 2 μL |
| --- | --- |
| dNTPs 2 μM | 2.5 μL |
| biotinylated-Dra 5 μM | 2.5 μL |
| Drb 5 μM | 2.5 μL |
| Betain 5×* | 5 μL |
| Q Buffer 10×* | 2.5 μL |
| H₂O | 8 μL |
| Home made or commercial Taq Pol*(1U) | 0.1 μL |
| Total volume | 25 μL |

*(*see* **Note 10**)

Cycling conditions:

96 °C, 3 min; (96 °C, 30s; 55 °C, 30s; 72 °C, 15 s); 72 °C, 5 min; repeat 20 cycles for CTAB, 25 cycles for thermolysates DNA (*see* **Note 11**)

B. PCR protocol for STM-CRISPOL

| | |
|---|---|
| DNA (thermolysate) | 1 μL |
| dNTPs | 200 μM |
| Biotinylated-DRSTMB | 50 pmol |
| DRSTMA | 50 pmol |
| MgCl2 | 1.5 mM |
| Go Taq Promega | 1.25 U |
| Go Taq Promega buffer 5× | 10 μL |
| $H_2O$ up to a total volume of | 50 μL |

Cycling conditions:

95 °C, 2 min; (95 °C, 1 min; 59 °C, 30s; 72 °C, 15 s) repeat 20 cycles (*see* **Note 12**)

Step 2: generic high-throughput hybridization protocol in 96 wells plates

1. Choose the appropriate set of oligonucleotides-coupled microbeads.
2. Resuspend the microbeads by vortexing 20 s and sonicating 20 s.
3. Prepare a « microbead working mix » at 75 microbeads/μL in a total volume of 1 mL of TMAC 1.5× (hybridization solution). Add 1.5 μL of each individual coupled-microbead stock (at 50,000 coupled microbeads/μl) within 1,000–(1.5×N) μL of TMAC 1.5× where N is the multiplexing level (*see* **Note 13**).
4. Mix the « microbead working mix» by vortexing 20 s and sonicating for 20 s.
5. Distribute 33 μL of the « microbead working mix » in each sample well and controls.
6. Dispense 17 μL of TE 1×, pH = 8 in the negative control well.
7. In the wells that contains samples, add biotinylated PCR-amplified DNA and TE 1×, pH = 8 in a total volume of 17 μL (*see* **Note 14**).
8. Mix gently by pipeting up and down a few times.
9. Place a lid on the reaction plate to prevent evaporation and incubate at 95–100 °C for 10 min to denature the amplified biotinylated DNA (PCR product).
10. Incubate at 52 °C (MTC-Spoligotyping) or 59 °C (STM-CRISPOL) for 20 min.
11. Centrifugation and supernatant elimination.

For TB-SPOL, Centrifuge the plate during 1 min, eliminate as much as possible of the supernatant (25–35 μl) by pipeting carefully. This step is done only when working with MicroPlex® (polystyrene) beads. This is used to lower the background and can be replaced by filtration if working with filter plates. Alternatively if using MagPlex® beads, you can use a magnet and simple upside down move of the plates. Filter plates and filtration can also be used.

For STM-CRISPOL, Centrifuge the plate during 3 min, eliminate as much as possible of the supernatant by pipeting carefully.

12. Prepare a fresh « reporter mix» by adding 4 μL (TB-SPOL) or 1.25 μL (STM-CRISPOL) of Streptavidin-R-phycoerythrin (1 mg/mL stock) to 996 μL (TB-SPOL) or 999 μL (STM-CRISPOL) of TMAC 1× Hybridization solution.

13. Detection step.

    For TB-SPOL, resuspend the beads by adding as much of TE 1× (*see* **Note 15**) than the TMAC quantity you removed at the **step 11**, in general 25–35 μl. Add 25 μl of this fresh « reporter mix » to each well and mix smoothly by pipeting up and down.

    *STM-CRISPOL.*

    Resuspend the beads by adding 90 μl of this fresh « reporter mix » to each well and mix smoothly by pipeting up and down.

Step 3: results interpretation; basic knowledge, advanced knowledge

The signals generated by the instrument are of two kinds: (1) real-time acquisition of data, which may allow direct control of the success/failure of the experiences run on the instrument and (2) final results data points files with quantitative MFI (mean fluorescence intensities) measures (Fig. 4a). The raw output is a .csv file (Fig. 4a) that can be easily transferred to .xls files, which are processed and analyzed using specifically designed macros, transforming the analogical signals (MFI) into digital values (positive/negative) after cut-off computation For CRISPR data analysis, numerical results are converted into binary states results (presence/absence) of a given sequence and translated into characters (white or empty squares, to create a string or « spoligotype », Figs. 2 and 4b). The distribution of negatives RFI compared to positive RFI shows a bimodal distribution (Fig. 4c), the full raw data file after cut-off calculation may be translated into a color-code (pink = presence, white = absence as shown in Fig. 4d).

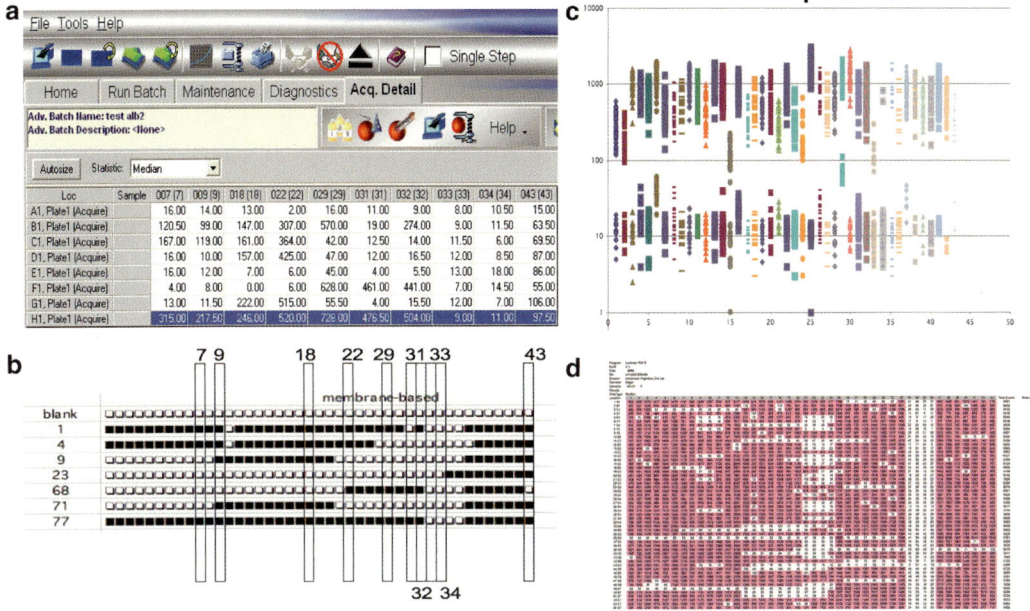

**Fig. 4** (**a**) Raw output results file of the Luminex 200 (preliminary trial experiment on 9 out of a total 43 targets) compared to (**b**) reference membrane-based spoligotypes. Raw values in (**a**) can intuitively be correlated with patterns shown on (**c**). After running experiments with enough samples with presence/absence of spacers it is shown that the experimental distribution of negative results does not overlap with the distribution of positive results (**d**). After introduction of automatized, cut-off based, interpretation of results, a colored pattern (pink/presence of the spacer sequence, white/absence of the spacer sequence) is obtained on Excel spreadsheet files. These files are further transcribed using Macros into final black/white patterns, as shown in Fig. 2. The same principles are applied to STM-CRISPOL results production

The computation of cut-offs will vary depending on instrument fine tuning and on techniques. Briefly, cut-offs are defined statistically using Mean Fluorescence Intensities (MFI) ± 2 standard errors when possible and ROC (Receiver Operating Characteristic) curves. If the experimental lowest reference positive results are always superior to the highest reference negative results, the technique is optimal. In some cases (e.g., 1–2 % of data points for the TB-SPOL technique) and only for suboptimal techniques, it may happen that a gray zone (zone for which the results may be positive or negative) has to be defined. In that case, rerunning the samples with doubtful results, and/or expert interpretation is required. This interpretation is often achieved, given (1) the level of signals obtained for other spacers that allow to provide indication on the quality/quantity of the starting material, (2) the neighborhood of the spacer, (3) available databases. Unique patterns should be re-run many times before inferring too rapid wrong conclusions.

For the sophisticated CRISPR-SNPs based method (not described in detail here), an adequate interpretation is also achieved thanks to the existence of internal wild-type/mutated alleles controls, since our probe design always uses two beads to check results, one with the wild-type nucleotide, one with the mutation assessed. It is out of the scope of this technical review to enter into all details and the reader will refer to our latest developments for further information [2, 28–32].

The full informativeness of spoligotyping profiles is then achieved through local or global database comparisons and mathematical similarity analysis. For MTC-Spoligotyping (TB-SPOL), updated SpolDB4.xls files are freely available from the author (CS). Alternatively the Institut Pasteur of Guadeloupe is managing a world-wide spoligotyping database that accepts scientific collaborations and new large data sets (www.pasteur-guadeloupe.fr:8081/SITVITDemo) and that, by the time of writing contains more than 110,000 patterns from more than 160 countries of origin (N. Rastogi, personal communication). For STM-CRISPOL, the CRISPOL database of the Pasteur Institute, Paris contains more than 7,000 strains resulting in more than 750 different CRISPOL types (CTs). This database will be soon made publicly available through a web tool. In the meantime, readers could contact the authors (SLH and FXW) to obtain the different validated CTs.

To conclude, we described in this chapter some of the technological, scientific and practical aspects of high-throughput CRISPR typing for *Mycobacterium tuberculosis* complex (TB-SPOL, 43-Plex method) and for *Salmonella enterica* serotype Typhimurium (STM-CRISPOL, 72-Plex method). Spoligotyping, even being a classic typing method was highly successful to describe the *Mycobacterium tuberculosis* genetic diversity, as well as within the molecular epidemiology research field. Molecular epidemiology is a moving field and Whole Genome Sequencing (WGS) will play an increasing role in the future [33–36]. However the advent of new technologies such as microbead-based suspension arrays, high-resolution melting or mass spectrometry will also provide complementary tools to give a second youth to spoligotyping and SNPs typing [26, 27, 37]. Moreover, the recently launched MagPix® new generation microbead-based assay with a decreasing cost and no need for air conditioning of the laboratory will undoubtedly create a growing environment of new Point Of Care diagnostic applications, similarly as what the iPhone has been for information technologies developers [38, 39]. Indeed, our increasing knowledge of CRISPR diversity on new bacterial targets could boost surveillance tools alternative to serotyping in combination with microbead-based hybridization systems [7].

## 4 Notes

1. Microbeads should be protected as much as possible from prolonged exposure to light throughout this procedure.

2. 0.1 M MES buffer provides adequate pH conditions for efficient carboxylic functions on microbeads- oligonucleotides- amino moiety coupling.

3. Addition of oligonucleotide and EDC. There are two rounds of EDC addition to maximize coupling efficiency. EDC has a very short half-life in solution; a single EDC addition is not sufficient to complete oligonucleotide coupling.

4. The EDC solution must not be reused.

5. 0.02 % Tween 20 is a washing buffer that allows to discard non coupled material. It also helps to recover microbeads from tube walls.

6. 0.1 % SDS is also a washing buffer that prevents microbeads aggregation.

7. TE 1× pH = 8 is a standard storage buffer of nucleic acids (TRIS 10 mM, EDTA 1 mM), prepared by extemporaneous dilution of stocks: TRIS 1 M pH = 8 (1/100), EDTA 0.5 M pH = 8 (1/500).

8. Steps 11 and 12 may be run in a thermocycler: Hold at 95 °C, 3 min; Hold at 52 °C or 59 °C (see above), 20 min.

9. Preparing a microbead working mix is possible, this mix can be kept at +4 °C for a couple of months (up to 6 months) if protected from light.

10. Go Taq Promega buffers and Polymerases can also be used efficiently as an alternative.

11. The final 72 °C—5 min elongation step is likely to be skipped in MTC since the products have a very short size, as this step was suppressed for the STM PCR Protocol.

12. The PCR products can be checked by electrophoresis in 1.2 % agarose gels (smears sizing 100–300 bp should be visible) and if not analyzed by the Luminex® platform on the same day, they might be stored at −20 °C for no more than three days. Negative (no PCR product, replaced with water) and Positive controls should always be used –H37Rv, *M. bovis* BCG P3 for MTC, SARA8, 81-784, 02-7015, 07-1777 (4 STM strains with a known spacer content covering all the spacers in the assay) for STM.

13. As an Example, if you run the 43-Plex assay (TB-SPOL), add 935.5 μl TMAC 1.5× to (43 × 1.5 = 64.5 μl of beads): if you run the 72-plex assay (STM-CRISPOL), add 892 μl TMAC 1.5× to 72 × 1.5 = 108 μl of beads.

14. We routinely use 2 μl (1–7 μL) of biotinylated PCR-amplified DNA for MTC-Spoligotyping and 7 μL for STM-CRISPOL.

15. The decrease in ionic strength is not a problem; at this step, you should normally add TMAC 1×, however evaporation during reading is much more intense with TMAC than with TE. We verified that TE substitutes without problem to TMAC 1× at this step.

## Acknowledgments

The authors would like to thank Jian Zhang, Marie Accou-Demartin, Lucile Sontag, Saïana de Romans, Catherine Lim and Laëtitia Fabre. M. François Topin and M. Jan van Gils from Luminex BV, The Netherlands, are also acknowledged for technical support.

## References

1. Cowan LS et al (2004) Transfer of a Mycobacterium tuberculosis genotyping method, spoligotyping, from a reverse line-blot hybridization, membrane-based assay to the Luminex multianalyte profiling system. J Clin Microbiol 42(1):474–477
2. Fabre L et al (2012) CRISPR typing and subtyping for improved laboratory surveillance of Salmonella infections. PLoS One 7(5):e36995
3. Groenen PM et al (1993) Nature of DNA polymorphism in the direct repeat cluster of Mycobacterium tuberculosis; application for strain differentiation by a novel typing method. Mol Microbiol 10(5):1057–1065
4. Kamerbeek J et al (1997) Simultaneous detection and strain differentiation of Mycobacterium tuberculosis for diagnosis and epidemiology. J Clin Microbiol 35(4):907–914
5. Jansen R et al (2002) Identification of a novel family of sequence repeats among prokaryotes. Genomics 6(1):23–33
6. Makarova KS et al (2006) A putative RNA-interference-based immune system in prokaryotes: computational analysis of the predicted enzymatic machinery, functional analogies with eukaryotic RNAi, and hypothetical mechanisms of action. Biol Direct 1:7
7. Sorek R, Kunin V, Hugenholtz P (2008) CRISPR—a widespread system that provides acquired resistance against phages in bacteria and archaea. Nat Rev Microbiol 6(3):181–186
8. Makarova KS et al (2011) Evolution and classification of the CRISPR-Cas systems. Nat Rev Microbiol 9(6):467–477
9. Viswanathan P et al (2007) Regulation of dev, an operon that includes genes essential for Myxococcus xanthus development and CRISPR-associated genes and repeats. J Bacteriol 189(10):3738–3750
10. Pourcel C, Salvignol G, Vergnaud G (2005) CRISPR elements in Yersinia pestis acquire new repeats by preferential uptake of bacteriophage DNA, and provide additional tools for evolutionary studies. Microbiology 151(Pt 3):653–663
11. Mokrousov I et al (2005) Efficient discrimination within a Corynebacterium diphtheriae epidemic clonal group by a novel macroarray-based method. J Clin Microbiol 43(4): 1662–1668
12. Weill FX, et al. 2009 Molecular typing and subtyping of Salmonella by identification of the variable nucleotide sequences of the CRISPR loci.https://data.epo.org/gpi/EP2255011A2-MOLECULAR-TYPING-AND-SUBTYPING-OF-SALMONELLA-BY-IDENTIFICATION-OF-THE-VARIABLE-NUCLEOTIDE-SEQUENCES-OF-THE-CRISPR-LOCI
13. Fabre L, et al. 2010 Improving laboratory surveillance of Salmonella infections by fast typing based on CRISPR polymorphisms. IS2, International Symposium on Salmonella
14. Liu F et al (2011) Subtyping Salmonella enterica serovar enteritidis isolates from different sources by using sequence typing based on virulence genes and clustered regularly interspaced short palindromic repeats (CRISPRs). Appl Environ Microbiol 77(13):4520–4526

15. Ginevra C et al (2012) Legionella pneumophila sequence type 1/Paris pulsotype subtyping by spoligotyping. J Clin Microbiol 50(3):696–701
16. Lopez-Sanchez MJ et al (2012) The highly dynamic CRISPR1 system of Streptococcus agalactiae controls the diversity of its mobilome. Mol Microbiol 85(6):1057–1071
17. Hermans PWM et al (1991) Insertion element IS987 from Mycobacterium bovis BCG is located in a hot-spot integration region for insertion elements in Mycobacterium tuberculosis complex strains. Infect Immun 59:2695–2705
18. van Embden JDA et al (2000) Genetic variation and evolutionary origin of the direct repeat locus of Mycobacterium tuberculosis complex bacteria. J Bacteriol 182:2393–2401
19. Filliol I et al (2003) Snapshot of moving and expanding clones of Mycobacterium tuberculosis and their global distribution assessed by spoligotyping in an international study. J Clin Microbiol 41(5):1963–1970
20. Gagneux S et al (2006) Variable host-pathogen compatibility in Mycobacterium tuberculosis. Proc Natl Acad Sci U S A 103(8): 2869–2873
21. Gagneux S, Small PM (2007) Global phylogeography of Mycobacterium tuberculosis and implications for tuberculosis product development. Lancet Infect Dis 7(5):328–337
22. Reed MB et al (2009) Major Mycobacterium tuberculosis lineages associate with patient country of origin. J Clin Microbiol 47(4): 1119–1128
23. Brudey K et al (2006) Mycobacterium tuberculosis complex genetic diversity: mining the fourth international spoligotyping database (SpolDB4) for classification, population genetics, and epidemiology. BMC Microbiol 6(6):23
24. Dale JW et al (2001) Spacer oligonucleotide typing of Mycobacterium tuberculosis: recommendations for standardized nomenclature. Int J Tuberc Lung Dis 5:216–219
25. Dunbar SA (2006) Applications of Luminex xMAP technology for rapid, high-throughput multiplexed nucleic acid detection. Clin Chim Acta 363(1–2):71–82
26. Zhang J et al (2010) Mycobacterium tuberculosis complex CRISPR genotyping: improving efficiency, throughput and discriminative power of "spoligotyping" with new spacers and a microbead-based hybridization assay. J Med Microbiol 59(Pt 3):285–294
27. Honisch C et al (2010) Replacing reverse line blot hybridization spoligotyping of the Mycobacterium tuberculosis complex. J Clin Microbiol 48(5):1520–1526
28. Zhang J et al (2011) A first assessment of the genetic diversity of Mycobacterium tuberculosis complex in Cambodia. BMC Infect Dis 11(1):42
29. Abadia E et al (2010) Resolving lineage assignation on Mycobacterium tuberculosis clinical isolates classified by spoligotyping with a new high-throughput 3R SNPs based method. Infect Genet Evol 10(7):1066–1074
30. Abadia E et al (2011) The use of microbead-based spoligotyping for Mycobacterium tuberculosis complex to evaluate the quality of the conventional method: providing guidelines for quality assurance when working on membranes. BMC Infect Dis 11:110
31. Gomgnimbou MK et al (2012) «Spoligoriftyping» a DPO-based direct-hybridization assay for TB control on a multi-analyte microbead-based hybridization system. J Clin Microbiol 50(10):3172–3179
32. Gomgnimbou MK et al (2013) "TB-SPRINT: TUBERCULOSIS-SPOLIGO-RIFAMPIN-ISONIAZID TYPING"; an all-in-one assay technique for surveillance and control of multi-drug resistant tuberculosis on Luminex® devices. J Clin Microbiol 51(11):3527–3534
33. Ioerger TR et al (2009) Genome analysis of multi- and extensively-drug-resistant tuberculosis from KwaZulu-Natal. PLoS One 4(11): e7778
34. Schurch AC et al (2010) High resolution typing by integration of genome sequencing data in a large tuberculosis cluster. J Clin Microbiol 48(9):3403–3406
35. Schurch AC et al (2011) Mutations in the regulatory network underlie the recent clonal expansion of a dominant subclone of the Mycobacterium tuberculosis Beijing genotype. Infect Genet Evol 11(3):587–597
36. Gardy JL et al (2011) Whole-genome sequencing and social-network analysis of a tuberculosis outbreak. N Engl J Med 364(8):730–739
37. Chen X et al (2011) Rapid detection of isoniazid, rifampin and ofloxacin resistance in Mycobacterium tuberculosis clinical isolates using high resolution melting analysis. J Clin Microbiol 49(10):3450–3457
38. Lin A et al (2011) Rapid O serogroup identification of the ten most clinically relevant STECs by Luminex microbead-based suspension array. J Microbiol Methods 87(1):105–110
39. Bergval I et al (2012) Combined species identification, genotyping, and drug resistance detection of mycobacterium tuberculosis cultures by MLPA on a bead-based array. PLoS One 7(8):e43240

# Chapter 7

# Spacer-Based Macroarrays for CRISPR Genotyping

Igor Mokrousov and Nalin Rastogi

## Abstract

Macroarray-based analysis is a powerful and economic format to study variations in "clustered regularly interspaced short palindromic repeat (CRISPR)" loci in bacteria. To date, it was used almost exclusively for *Mycobacterium tuberculosis* and was named spoligotyping (spacer oligonucleotides typing). Here, we describe the pipeline of this approach that includes search of loci and selection of spacers, preparation of the membrane with immobilized probes and spoligotyping itself (PCR and reverse hybridization).

**Key words** Spoligotyping, Reverse hybridization, Macroarrays

## 1 Introduction

The CRISPR ("clustered regularly interspaced short palindromic repeat") locus was first discovered in the genome of *Mycobacterium tuberculosis* and named Direct Cluster (accession number Z48304) since it contained multiple identical 36-bp direct repeat (DR) units interspersed by unique spacers 35–41 bp in length [1, 2]. Together with their nonrepetitive spacer sequences, the DR units constitute multiple direct variant repeats (DVRs) which show extensive polymorphism among *M. tuberculosis* clinical isolates. A macroarray-based method to study presence/absence of particular spacers named spoligotyping (spacer oligonucleotide typing) was developed to study the epidemiology of tuberculosis [3] (Fig. 1a, b). Spoligotyping has long been used for *M. tuberculosis* typing and Pubmed search for "spoligotyping OR spoligotype" yields 1117 articles (3 April 2015).

Because of its simplicity, binary result format and high reproducibility, spoligotyping is widely used for investigations on molecular epidemiology of *M. tuberculosis* [4]. The global database of *M. tuberculosis* spoligotypes created by Institut Pasteur de la Guadeloupe allows to have a global vision of *M. tuberculosis* genetic diversity based on standard 43-spacer spoligotyping format; its most recent published version named SITVITWEB [5] incorporates

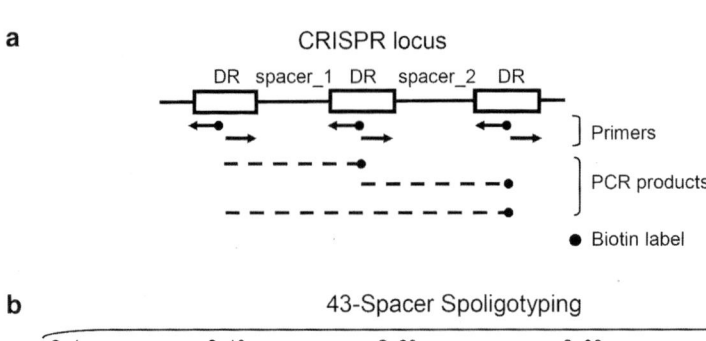

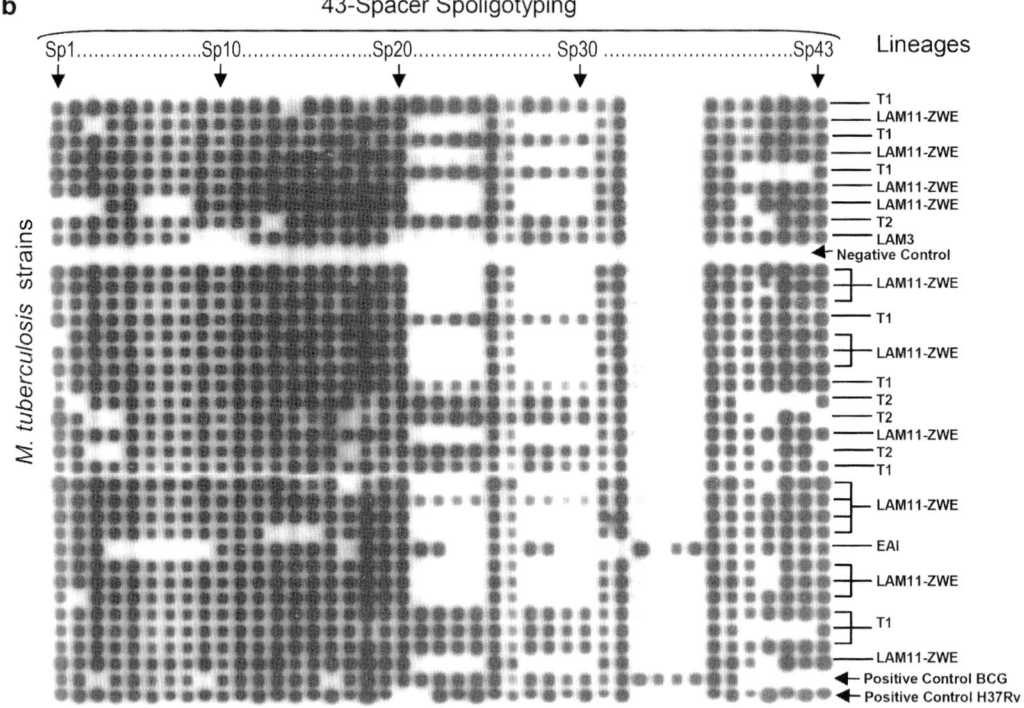

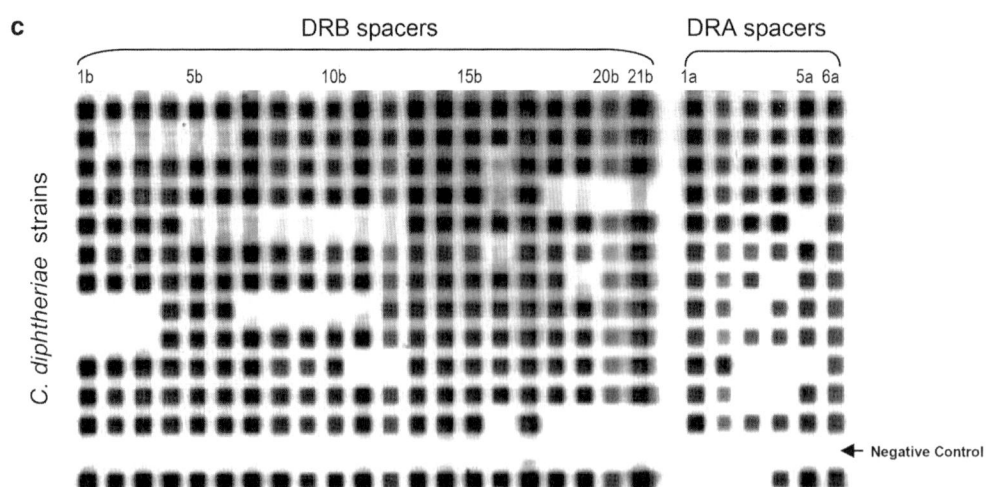

**Fig. 1** CRISPR-based spoligotyping in macroarrays format. (**a**) CRISPR locus structure; (**b**) spoligoprofiles of different *M. tuberculosis* strains; lineage designation according to SITVIT_WEB database (http://www.pasteur-guadeloupe.fr:8081/SITVIT_ONLINE/); (**c**) spoligoprofiles of different *C. diphtheriae* strains; reprinted from ref. [14] by permission of WILEY-VCH Verlag ©2007

spoligotyping data on 62,582 clinical isolates corresponding to 153 countries of patient origin (105 countries of isolation). A newer version of this database named SITVIT2 to be released in 2015 contains spoligotyping data on more than 112,000 *M. tuberculosis* complex clinical isolates from 170 countries of patient origin (N. Rastogi, personal communication).

As more spacers were discovered [6], the expanded format of *M. tuberculosis* spoligotyping based on 68 spacers was described [7], which requires use of additional membrane since the miniblotter MN45 apparatus permits up to 43 probes. Although subsequent papers reported on improved strain differentiation within the *M. tuberculosis* complex by using novel spacers [8–10], the commercial availability of spoligotyping kits and huge international databases with 43-spacer format largely favored its widespread use over extended spacer formats. Last but not least, 43-spacer spoligotyping is today almost exclusively used in conjunction with a second-line typing method for epidemiological investigations, preferentially MIRU-VNTR minisatellites [11]. The higher discrimination achieved by this "two PCR-based genotyping" has largely restricted the widespread use of extended spacer format spoligotyping.

The DR locus was long thought to be unique to *M. tuberculosis* but accumulation of new data in GenBank showed that this kind of locus is found in all archaea and 60 % of eubacteria [12]. While rarely used in the early 2000s for analysis of some bacteria using direct sequencing of CRISPR locus, the spoligotyping macroarray format was developed for analysis of two such loci found in *Corynebacterium diphtheriae* (Fig. 1c); *C. diphtheriae* spoligotyping demonstrated a very high discriminatory capacity of the novel method compared to all other available genotyping methods of this pathogen [13, 14]. As a whole, use of CRISPRs for genotyping of bacterial species is rapidly expanding [15].

## 2 Materials

### 2.1 Equipment

1. PCR machine.
2. Miniblotter MN45 (Immunetics, Boston, Massachusetts, USA, available exclusively through its distributors: http://www.immunetics.com/distributors.html).
3. Incubators: two, set at 60 °C and 42 °C (Mini Rotator oven with bottles; temperature range from ambient to 60 °C; bottle capacity 250 ml, speed 0–10 Rev/min).
4. Water bath, with shaker.
5. Shaker at room temperature.
6. Impulse sealer for plastic bags.

7. Washing boxes (size 15 × 15 cm at least, but not much larger to avoid overuse of reagents).

8. Boxes for development and fixing the film.

9. Exposure cassette.

10. Screen for exposure cassette.

11. Ice machine.

12. Forceps.

13. Scissors.

14. Alarm clock.

15. Dark room, red lamp.

## 2.2 Solutions and Disposables

*Reagents and solutions*

Most solutions may be prepared using ordinary distilled and autoclaved water (unless mentioned otherwise).

1. *Mycobacterium tuberculosis*: spoligotyping kit includes membrane with 43 immobilized oligonucleotides, forward primer and reverse 5′-biotinylated primer, 2 positive controls—DNAs of H37Rv and BCG (Ocimum Biosolutions: http://www3.ocimumbio.com/laboratory-sample-data-management/research-consumables/spoligotyping-kits/). Alternatively you may order reagents, probes and primers separately and prepare yourself as described below.

    Other bacteria: no kits are available and membrane and primers should be prepared in-house/ordered as described below.

2. Oligonucleotides (unlabeled forward primer, 5′-biotin labeled reverse primer, 5′-NH$_2$–labeled spacer-specific probes) are specific for CRISPR loci of different bacterial species (*see* **Note 1**).

3. EDAC (*N*-(3-dimethylaminopropyl)-*N*′-ethylcarbodiimide hydrochloride, Sigma-Aldrich).

4. NaHCO$_3$.

5. NaOH.

6. SDS (Sodium dodecyl sulfate or sodium lauryl sulfate). Stock solution: 10 % solution, dissolved at 80 °C (*see* **Note 2**).

7. EDTA disodium salt (ethylenediaminetetraacetic acid disodium salt dihydrate). Stock solution: 0.5 M EDTA, pH 8.0. (EDTA salt does not dissolve until pH is 8.0.)

8. 20× SSPE (0.2 M NaH$_2$PO$_4$ × H$_2$O, 3.6 M NaCl, 20 mM EDTA) is a stock solution.

    Dissolve 175.3 g NaCl, 27.6 g NaH$_2$PO$_4$ × H$_2$O in 800 ml H$_2$O, add 40 ml of 0.5 M EDTA solution, pH 8.0. Adjust pH to 7.4 with NaOH (~6.5 ml 10 N NaOH). Final volume: 1,000 ml.

9. Thermostable DNA Polymerase.
10. dNTP (deoxynucleotide triphosphates, mix of dATP, dGTP, dCTP, dTTP).
11. Streptavidine-POD (beta-peroxidase) conjugate, 500 U/ml (Roche Applied Science) (*see* **Note 3**).
12. Amersham ECL detection liquid (GE Healthcare).
13. India ink (alternatively called China Ink) (*see* **Note 4**).
14. Membrane preparation solutions.

    Prepare the following solutions at the day of membrane preparation.

    $NH_2$-labeled spacer-specific oligonucleotides diluted with dd$H_2$O to working concentration (5 pmol/µl).

    100 ml of 0.1 N NaOH (0.4 g in 100 ml distilled water).

    250 ml 0.1 % SDS/2× SSPE (25 ml 20× SSPE + 2.5 ml 10 % SDS + distilled water up to 250 ml).

    250 ml 20 mM EDTA (10 ml 0.5 M EDTA + distilled water up to 250 ml).

    India Ink solution: diluted 1:30 in 2× SSPE.

    $NaHCO_3$ 0.5 M, pH 8.4. Prepared the same or previous day by dissolving 4.2 g $NaHCO_3$ in 100 ml ultrapure sterilized water and adjusting pH to 8.4. The solution should be filtered through Millipore Sterivex™-GB 0.22 µm filter.

    EDAC (16 % w/v). Dissolve 1.92 g of EDAC in 12 ml of ultrapure sterilized water (*see* **Note 5**).

    NaOH 100 mM. Dissolve 1 ml NaOH 4 M in 39 ml ultrapure sterilized water.

15. Spoligotyping (hybridization and washing) solutions.

    The following buffers should be prepared using concentrated 10 % SDS and 20× SSPE and preheated in advance:

    150 ml 2× SSPE 0.1 % SDS.

    250 ml 2× SSPE 0.5 % SDS.

    250 ml 2× SSPE 0.5 % SDS.

    200 ml 2× SSPE.

(These volumes are used for 15 × 15 cm washing box.)

*Consumables*

   Membrane Biodyne® C 0.45 µm 20 × 20 cm (Pall Life Sciences).

   Sterivex™-GB 0.22 µm (Millipore).

Foam cushions PC200 for miniblotter MN45 (Immunetics, Boston, USA, available exclusively through its distributors (http://www.immunetics.com/distributors.html).

Amersham Hyperfilm ECL (GE Healthcare) (*see* **Note 6**).

Developer for films (e.g., Kodak XTOL Cat. 875 1752).

Fixer for films (e.g., Kodak Cat. 367 0346).

Plastic bag.

Plastic tubes 0.5 and 1.5 ml.

Plastic tubes 15 and 50 ml.

Filter Tips 20, 200, and 1,000 μl.

Disposable pipettes, 5 and 10 ml.

## 3 Methods

### 3.1 Search of Loci and Selection of Spacers. Optimization of Probe Concentrations

In case of new bacterial species you may use the CRISPR database (http://crispr.u-psud.fr/crispr/), other software such as CRISPR finder [16] or Tandem Repeat Finder [17], or published literature to find CRISPR loci.

The subsequent preparatory research work is not described in detail here but includes:

1. Computer analysis: search of CRISPR loci in published genome, identification of spacers, selection of the spacer-derived oligonucleotide probes (to be specific and be of the similar Tm, e.g., 60 °C). Selection of primers derived from the DR sequence, the reverse primer being 5′-biotin labeled; these primers will amplify all spacers located between direct repeats (Fig. 1a). *Since it is the reverse primer that is labeled with biotin, the probes should be selected from plus-strand sequence.*

2. Laboratory experiments: test four different concentrations of probes (5′-$NH_2$-labeled probes for membrane immobilization—see below) (e.g., from 10 to 100 pmol per slot). The PCR product (biotin-labeled) is hybridized to all immobilized probes and based on the signal strength, the optimal probe concentrations are chosen to avoid both false-positive and false-negative signals.

3. Search of new spacers in clinical isolates: amplify the entire locus, sequence it, find new spacers and study their diversity as described above.

It may be that there will be more than one such locus per genome so all spacers should be considered in initial analysis; the most relevant (discriminatory but nonredundant) may be used in the optimized typing scheme with a reduced number of spacers. As with other typing methods, the following should be considered: reproducibility, discriminatory power (compared to other typing methods used), stability of profiles. In case of more than 43 different spacers, other solutions to study their presence/absence are advised such as microarrays, or use more than one macroarrays membrane.

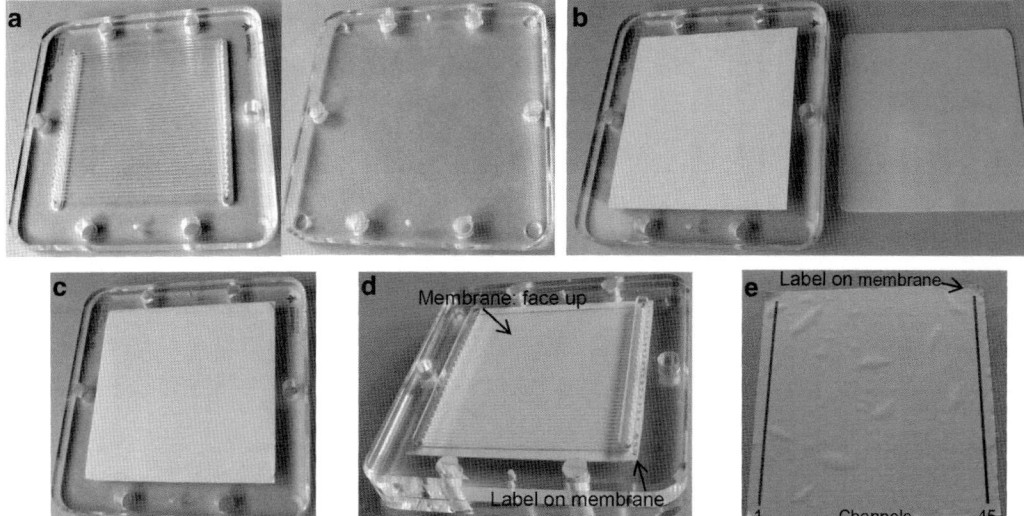

**Fig. 2** Manipulations at different steps of membrane preparation: (**a**) opened miniblotter, its cover part with channels up; (**b**) membrane placed onto the channels of miniblotter; and, separately foam cushion (right); (**c**) immediately after the previous step, a foam cushion is placed onto the membrane; (**d**) miniblotter with membrane and foam cushion covered with cover (remains to be screwed). India Ink will be loaded in the first and last channels, oligonucleotide solutions will be loaded in the other channels (empty channels to be filled in with NaHCO$_3$). (**e**) Spoligotyping membrane. *Black lines* show margins of the active area, *label* in the *upper right corner* shows orientation (last oligonucleotide)

## 3.2 Membrane Preparation

*Summary*

Oligonucleotides are 5′-NH$_2$ labeled. Biodyne® C membrane is negatively charged and thus not reacting with negatively charged DNA sodium salt. Membrane contains C=O groups that react with 5′-NH$_2$ group of oligonucleotide making a covalent bond C=N, and attaching oligonucleotides on the membrane.

The spacer-specific 5′-NH$_2$-labeled oligonucleotide probes are covalently bound to the membrane as follows. A membrane (Biodyne® C membrane) is activated by incubation with 16 % (wt/vol) EDAC (Sigma, USA) for 10 min. The oligonucleotides are diluted to the appropriate concentration in 0.5 M NaHCO$_3$, pH 8.4, and applied to the membrane in parallel slots (channels) by using the MN45 miniblotter apparatus. After 2–3 min incubation at room temperature, the probes are removed from channels and membrane is inactivated with 0.1 N NaOH for 10 min, washed twice in 2× SSPE/0.1 % SDS solution for 5 min at 60 °C (which corresponds to the temperature of hybridization in case of *M. tuberculosis* spoligotyping) and rinsed twice for 10 min in 20 mM EDTA at room temperature.

Membrane is stored with EDTA solution sealed in a plastic bag or in a box. Some essential manipulations with miniblotter, membrane and foam cushion are shown in Fig. 2.

**Table 1**
**Worksheet template for membrane preparation**

| Channel | Probe label | Working amount, pmol/150 μl solution | Probe (5 pmol/μl), μl | 0.5 M NaHCO$_3$ pH 8.4, μl |
|---|---|---|---|---|
| 1[a] | | | | |
| 2 | 1 | 35 | 7 | 143 |
| ............ | ............ | ............ | ............ | ............ |
| 44 | 43 | 100 | 20 | 130 |
| 45[a] | | | | |

[a]150 μl of India Ink (diluted) is added in channels 1 and 45

1. Prepare and print protocol with oligonucleotide names and amount of solutions (Table 1).

    If working with *M. tuberculosis* or *C. diphtheriae*, details on sequences of primers and probes and probe concentrations are available in references [3, 14].

2. Set water bath temperature to 60 °C (=hybridization and washing temperature), preheat 0.1 % SDS/2× SSPE solution in it. Make sure that rolling bottle is dry (see **step 7** below).

3. Always wear dry gloves cleaned with ethanol and water (to remove both fat and powder). Cut the membrane Biodyne C 14.5×14.5 cm. Label with pencil in the upper right corner, e.g., Mt_spol_150714.

4. Label 1.5 ml tubes from 1 to 43 (if 43 oligonucleotides will be used as in classical spoligotyping of *M. tuberculosis*).

5. According to the Table 1, distribute into each tube: NaHCO$_3$ and oligonucleotides (using working solutions, 5 pmol/μl), and close tubes. Preferably work on ice, or otherwise, place tubes into a fridge while doing the next step.

6. Weigh 1.92 g of EDAC. Add 12 ml of MilliQ water, close tube and dissolve EDAC by shaking.

7. Roll the membrane and place it into dry rolling bottle, add EDAC solution, close, make sure to remove all bubbles, and place bottle into rotating shaker.
    Incubate for 10 min at room temperature.
    Discard solution from bottle, add 20–30 ml of MilliQ water, incubate in rotating shaker 1 min at room temperature.
    Discard water.

8. Carefully take membrane out of bottle and place it (face down) onto the Miniblotter channels (so the label would appear in the right upper corner of the assembled miniblotter with membrane inside (Fig. 2b)).

Place foam cushion onto the membrane (Fig. 2c).
Place the Miniblotter cover (Fig. 2d), and tighten the screws.

9. Remove liquid from channels—using an automatic pipette or pump.
10. Fill the 1st and 45th channels with India Ink solution (diluted 1:30 in 2× SSPE).
11. Fill in all oligonucleotides in the respective channels (*see* **Note 7**).
12. Wait 2–3 min, and remove, first, ink from first and last channel, and next, oligonucleotide solutions from all channels (it is possible to use the same tip)—using automatic pipette or pump. If not using a pump, it is important to carefully remove all ink so you should wash these channels with 2× SSPE several times.
13. Pour 0.1 N NaOH solution into the box.

    Open miniblotter, carefully and rapidly remove the membrane and place it into 0.1 N NaOH (*see* **Note 8**)—for 10 min maximum, on a shaker (*see* **Note 9**).
14. Wash membrane in clean box in water bath at 60 °C with shaking—in preheated 125 ml, 0.1 % SDS/2× SSPE—for 5 min, two times (60 °C—is the temperature of hybridization).
15. Wash membrane in clean box with 125 ml of 20 mM EDTA on shaker for 15 min, two times, at room temperature.

    The membrane may be stored in 20 mM EDTA solution in a box, or ideally sealed in a plastic bag with a few milliliters of 20 mM EDTA and kept at +4 °C; do not fold or scratch (*see* **Note 10**).

    For hybridization (see below), the membrane will be placed into the miniblotter perpendicular to the oligonucleotides (and black ink lines that frame the active area of the membrane) so the label would be in the right lower corner (when all parts of miniblotter are assembled and tightened up (Fig. 3) (*see* **Note 11**).

## 3.3 PCR

General description of master mix:

| ddH$_2$O | qs |
|---|---|
| PCR buffer: | 1× final |
| dNTP | 200–250 µM final |
| MgCl$_2$ | 1.5 mM final (it may be up to 5.5 mM final and depends on polymerase used) |
| Forward primer | 15 pmol (*see* **Note 12**) |
| Reverse primer[a] | 15 pmol |
| *Tth/Taq* | 1–2.5 unit (*see* **Note 13**) |
| Total volume: | 30 µl (or 50 µl) |

[a]Reverse primer is labeled with biotin

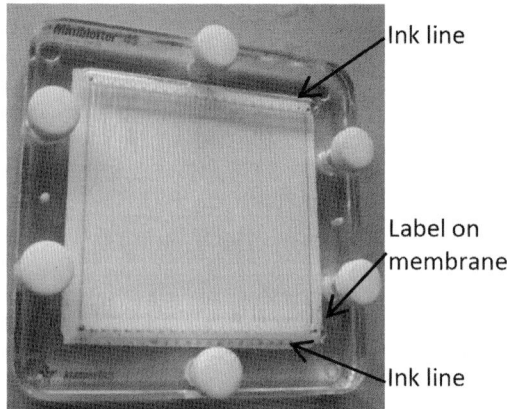

**Fig. 3** Assembled Miniblotter MN45 with membrane inside: ready for loading PCR products. Note that membrane (*first and last black lines*) is placed perpendicular to the channels of the miniblotter

Example of master mix for *M. tuberculosis* spoligotyping (1.5 mM $MgCl_2$, 200 µM dNTP, 1 U *Taq*, 15 pmol primers each):

|  | µl |
|---|---|
| dd$H_2O$ | 19 |
| PCR buffer, 10× | 3 |
| dNTP, 2 mM | 3 |
| $MgCl_2$, 25 mM | 1.8 |
| DRb primer, 10 µM | 1.5 |
| *DRa primer, 10 µM | 1.5 |
| Taq, 5 U/µl | 0.2 |
| Volume: | 30 µl |

Master Mix for all samples is prepared on ice in a single 1.5 ml tube and distributed into the PCR tubes.

DNA is added as follows: 0.1–1 µl DNA extracted from cultured cells (*see* **Note 14**) or up to 20 µl of cell lysate or DNA extracted from clinical sample (in such a case the total volume of PCR mix is 50 µl) (*see* **Note 15**).

*Example of cycling conditions (for M. tuberculosis spoligotyping)*
3 min, 95 °C; 95 °C, 50′, 55 °C, 50′, 72 °C, 1 min; 72 °C, 5 min.

25–35 cycles: for DNA extracted from cultured strains.

35–40 cycles: for DNA extracted from clinical samples or cell lysates (*see* **Note 16**).

You may check 10 µl of randomly selected PCR products in 1.5 % agarose gel. However, since many fragments are amplified

(Fig. 1a) you are likely to see a kind of smear-like ladder with a stronger band of ~70 bp (minimal PCR product of a single spacer plus length of two primers) (*see* **Note 17**).

25 μl of this PCR product is used for spoligotyping hybridization.

### 3.4 Hybridization and Detection

The biotin-labeled single-stranded PCR products (that theoretically contain all spacers present in the CRISPR locus of a given strain) are hybridized to the array of oligonucleotides immobilized on the membrane, at specific temperature conditions. After washing at the same temperature, the hybrids on the membrane are detected using an enzymatic reaction: the membrane is incubated with Streptavidin-Peroxidase (POD) conjugate at 37 °C, washed at the same temperature, and incubated with ECL detection liquid at room temperature (Fig. 4). ECL detection liquid consists of two reagents mixed immediately before use—luminal that is a substrate for peroxidase and an enhancer to maximize the signal. After this the membrane is sealed in a plastic bag and exposed to the light-sensitive film at 37 °C in dark room. After development, the patterns of black dots are seen on the film and make the spoligoprofiles of the strains. Up to 43 strains may be analyzed (this is defined by the design of this particular Miniblotter MN45 apparatus), including positive and negative control samples (*see* **Note 18**).

1. Add 25 μl of the PCR products to 125 μl 2× SSPE/0.1 % SDS preheated at 60 °C (*see* **Note 19**).

2. Heat-denature the diluted PCR product for 10 min at 99 °C (PCR machine) or in boiling water-bath (ensure that the tubes do not open) after which place the tubes in ice bath immediately, for 10 min. Spin tubes and place them back on ice.

3. During the previous step, incubate the membrane for 10–15 min (or twice 5 min) in 100 ml 2× SSPE/0.1 % SDS at 60 °C (60 °C is in case of *M. tuberculosis* spoligotyping).

4. Place the membrane on a support cushion into the miniblotter, in such a way that the slots are perpendicular to the line pattern of the applied oligonucleotides (Fig. 3).

5. Remove residual fluid from the slots of the miniblotter by aspiration.

6. Fill the slots with all amount (150 μl) of denatured PCR product (see **step 2** above) and avoid air bubbles. Do not use first and last slots. All empty slots should be filled with 2× SSPE/0.1 % SDS (*see* **Note 20**).

7. Hybridize for 60 min at 60 °C on a horizontal surface. Avoid cross flow to the neighboring slots (no shaking!) (*see* **Note 21**).

8. Remove the samples by aspiration (with a pipette or using a pump), open miniblotter, and remove the membrane from it using forceps.

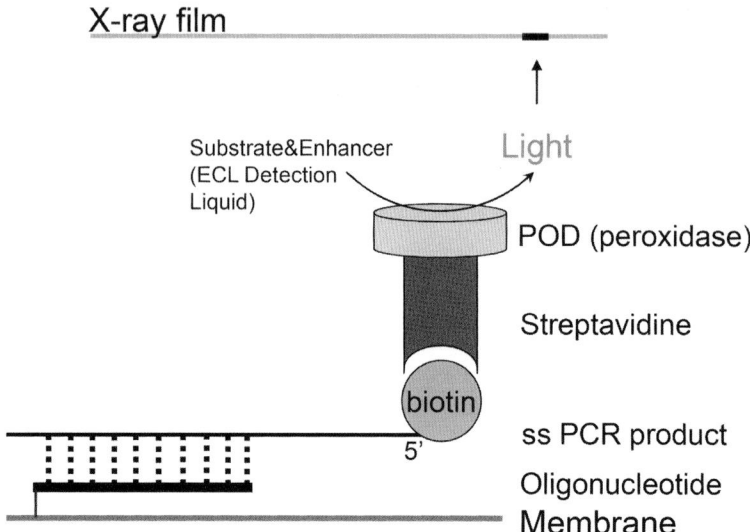

**Fig. 4** Detection of reverse hybridization result based on POD-catalyzed fluorescent reaction and exposure to light-sensitive film

9. Wash the membrane twice in a washing box in 125 ml 2× SSPE/0.5 % SDS (preheated to the hybridization temperature, 60 °C) for 10 min at 60 °C, in water bath with shaker.

10. Place the membrane in a rolling bottle and incubate with 10 ml 2× SSPE/0.5 % SDS containing 8 µl of 500 U/ml streptavidin-POD for 60–90 min at 37 °C (*see* **Note 22**).

11. Wash the membrane twice in a washing box in 125 ml of 2× SSPE/0.5 % SDS (preheated to 37 °C) for 10 min at 37 °C, in water bath with shaker (*see* **Note 23**).

12. Rinse the membrane twice in a box with 100 ml of 2× SSPE for 5 min at room temperature, on shaker (gentle agitation).

13. For chemoluminescent detection of hybridized DNA, incubate the membrane for 2 min in 20 ml ECL detection liquid in the ECL-incubation box (ECL detection mix prepared immediately before use, 10 ml of ECL1 plus 10 ml of ECL2) (*see* **Note 24**).

14. Seal membrane in plastic bag or using e.g., Saran Wrap, remove bubbles from inside the bag, remove all liquid from the surface of plastic bag (*see* **Note 25**).

15. *In dark room*. Expose a film (e.g., ECL Hyperfilm) to the face(DNA)-side of membrane for 10 min up to several hours or overnight in hypercassette (it is recommended to use hypercassette with hyperscreen) at 37 °C (*see* **Note 26**).

16. *In dark room*. Develop the film. If necessary the membrane can be directly used again to expose another film for a shorter or longer period (*see* **Note 27**).

## 3.5 Stripping of Membrane for Reuse

This step should be done after development of the film, the same or next day (next week latest).

Prior to reuse (up to 15 times), membrane must be stripped of hybridized PCR fragments (*see* **Note 28**).

1. Incubate twice in water bath with shaker in a box with 125 ml 1 % SDS, for 40 min at 80 °C.
2. Rinse twice on shaker in a box with 125 ml 20 mM EDTA at room temperature.
3. Seal in plastic bag and store at 4 °C.
4. Once a month, make sure the membrane is not dried up.

## 3.6 Treatment of Results

The generated results (presence or absence of the particular spacer) are by definition in binary format and make a 43-digit line (*see* **Note 29**).

For convenience, the octal format is also used [18] although it seems less convenient for direct visual comparison. For example, *M. tuberculosis* binary pattern 0000000000000000000000000 00000000111111111, which correspond to International Shared Type SIT1/Beijing lineage strain in the SITVITWEB database is encoded as 000000000003771 in 15-digit octal format.

The following kinds of analysis may be performed with spoligotyping binary data:

1. Manual entry to Excel file, use of automatic sorting function and comparison with previous results or database if available. In case of *M. tuberculosis*, it may be compared with international spoligotype database SITVITWEB (Institut Pasteur of Guadeloupe) at http://www.pasteur-guadeloupe.fr:8081/SITVIT_ONLINE/query, which allows interrogations using both binary and octal formats (*see* **Note 30**).

2. Minimum spanning tree (MST) can be constructed based on spoligotyping data alone or in association with another typing scheme such as VNTR or MLST-based (*see* **Note 31**). For phylogenetic analysis, BioNumerics (Applied Maths, Sint-Martens-Latem, Belgium) or other software may be used. The recent version of the BioNumerics software (version 6.6 and above) allows highlighting phylogenetical links between spoligotyping-defined lineages with different epidemiologic and demographic parameters such as sex, nationality and drug resistance.

3. SpolTools program [19, 20] available at http://www.emi.unsw.edu.au/spolTools/ may be used for analysis of binary spoligotyping data of *M. tuberculosis* and other microorganisms. As opposed to the MST, the Spoligoforest tree is directed and the patterns only evolve by loss of spacers. GraphViz software (http://www.graphviz.org) can be used to color the strains based on their lineages on the Spoligoforest trees [21, 22] (*see* **Note 32**).

4. Taxotron package [23] commercialized by Institute Pasteur, Paris (http://www.pasteur.fr/ip/easysite/pasteur/en/industrial-partnerships/products-and-services/taxotron-softwares) may be used to build UPGMA or NJ dendrograms of the binary data. Data should be entered in the Recogniser file and treated with further programs of the Taxotron package.

5. Other useful software is SpolPred, a Linux-operated program [24] that permits predicting spoligoprofile based on short reads data (SRA file).

## 3.7 More Precautions to Help Quality Controlled Laboratory Practice

1. Spoligotyping should be ideally performed by trained and qualified personnel and experiments should be made by the same person/staff.

2. Preceded by membrane preparation, Spoligotyping itself consists of three distinct steps: PCR, hybridization, and detection [3], in case of weak signals or lack of signals, all components of all steps (reagents and physical conditions) should be carefully checked.

3. The decay in signal control (by experiment) should be recorded and registered per membrane, i.e., each membrane has a specific reference number followed by the number of experiments done.

4. Stop using a membrane when spacers at the boundaries or elsewhere start to give lower signals.

5. Note that uneven intensity of spots within the same lane may be suggestive of mixed strain-population. Any doubtful results should be repeated at least twice before a final decision.

   Since mixed infections (dual infections of a single patient by distinct strains) are increasingly reported in high tuberculosis incidence areas, it is important to predict such cases before database entry so as not to attribute "false composite spoligotype pattern designations" since this misclassification would compromise database integrity. A computational approach was recently applied to illustrate how geographic context in specific cases may indicate that an observed pattern could arise due to potential cases of mixed infections; the use of MIRU-VNTRs is recommended to detect mixed infections in such cases [28].

# 4 Notes

1. Forward primer can be kept frozen at −20 °C for several years (5 or more). However 5′-biotin labeled reverse primer should be stored at +4 °C (do not freeze) and used within 1 year.

It may be that the oligonucleotides used as probe are not adequately labeled with $NH_2$ hence not immobilized on the membrane. In such a case, reclamation should be made to the company to repeat synthesis of such probes.

2. Quality of SDS is of critical importance. Do not boil. It should be prepared freshly and not stored for longer than 2 weeks.

3. Performance of Streptavidin-Peroxidase (POD)-conjugate decreases with age when older than 6 months (once dissolved in water, and until expiration date). POD can be frozen at –20 °C as aliquots for a single experiment each.

4. Pay attention to the quality of ink, it should be homogeneous and concentrated.

5. The EDAC solution should be prepared the day of experiment and used immediately (see protocol on Membrane preparation). EDAC powder is stored at –20 °C; close the bottle tightly after weighing.

6. Amersham Hyperfilm ECL (GE Healthcare) or medical X-ray film may be used.

7. When loading the $NH_2$-labeled oligonucleotide solutions into channels be careful not to contaminate: the covalent binding of $NH_2$-labeled probe to the C=O-labeled membrane is very fast and is assumed to take seconds.

8. Membrane should be covered with NaOH solution.

9. Be careful not to over-expose to NaOH: it is recommended to incubate with NaOH 10 min maximum but the acceptable range may be from 8 to 12 min.

10. Membrane should not be dried. However, if the membrane is dried it is possible to attempt its use after soaking for several hours in 20 mM EDTA.

11. A newly made membrane should be tested on its first use to see if all the 43 spacers throughout the membrane are functioning ideally. It can be done by plotting 3 previous amplification products (stored samples with known patterns) with profiles covering all the 43 spacers—on the top 3 lanes, middle 3 lanes, and bottom 3 lanes of the membrane. Verify that the intensity of the spots does not vary from top to bottom.

12. A number of steps and parameters involved may affect the quality of a PCR reaction, even in an experienced laboratory. Unfortunately, all the parameters are not clearly exposed to the user when the PCR reagents, reactives and enzymes are purchased through a manufacturer. For example, the synthesis of biotinylated primer for *M. tuberculosis* spoligotyping was changing over time reflecting differences in the length of the chain of biotine fixation to the oligonucleotide. This change was not mentioned by the manufacturer.

13. Each user should try different polymerases to compare their performance with previously characterized controls. An experience in one of our laboratories was that even the same company may produce (or, de facto, resell) different quality polymerase, and not always easy to identify, based on information provided. Different batches of Tth purchased did not bear enough information linking them to a single manufacturing batch, and the user was not automatically contacted through the retailer when a defect in the manufacturing process was discovered.

    Researchers did sometimes replace Tth with Taqstart for hot start PCR. However, no studies on the reproducibility of the spoligotyping using Tth, rTaq, or Taqstart have been performed. Further improvements may include the use of the TaqStart antibody for hot start PCR which is known to yield specific DNA fragments upon amplification when using high numbers of temperature cycles and a very low copy number of target DNA in a complex DNA background [25]. This may also eliminate or prevent the generation of nonspecific PCR templates that may be synthesized at ambient temperature prior to thermal cycling, rendering the technique more specific.

14. The amount of DNA used for PCR depends on concentration of your DNA, otherwise use similar amount as for standard PCR.

    Chromosomal DNA from cultured isolates may be best prepared using CTAB (cetyltrimethylammonium bromide) treatment, phenol–chloroform extraction, isopropanol precipitation; such DNA provides good results and may be transported dry at room temperature. DNA may also be prepared using available commercial kits and the amount used for PCR is the same as recommended for standard PCR. In case of too weak or too strong signals the quantity of DNA used for PCR may be changed.

15. Lysates from colonies or liquid culture medium are obtained by boiling, centrifugation and the pellet resuspension in 1×TE buffer or $H_2O$.

    DNA from clinical samples may be prepared using phenol extraction [26], or using commercial kits; the same amount of DNA as recommended for real-time PCR may be used for spoligotyping PCR.

    DNA from Ziehl–Neelsen stained slides and paraffin-embedded tissues may be prepared with Chelex resin based procedure [26, 27].

16. Generally, 30 cycles is optimum while 40 and more cycles may result in background signals.

17. In case of weak hybridization signals you may try increase number of PCR cycles or volume of PCR mix (up to 50 μl) all

of which to be used for hybridization. However, the signal does not necessarily increase twofold.

It may be helpful to keep PCR products from a previous experiment (with good results) for a later use as a positive control of PCR for hybridization (in case one encounters problems with weak signals). The PCR products are stable up to 6 months when stored at +4 °C (without loss of activity). However, despite a little though not significant loss of signal, the PCR products may be stored several years at –20 °C.

18. The spoligotyping signal detection step is mediated by chemiluminescence that is a sensitive procedure and may lead to false positive signals. False positive/negative profiles and uneven intensity of spots result from DNA contamination (at various steps) coupled to unnecessarily increased number of PCR cycles. For this reasons, appropriate controls should be included at different steps. These positive controls include strains with known profiles (in case of *M. tuberculosis* these are strains H37Rv and *M. bovis* BCG that may be supplemented by locally prevalent strains with characteristic profiles). Water may be used as negative control in PCR (Fig. 1b).

19. 60 °C is an example from *M. tuberculosis* spoligotyping. Temperature of this preincubation is the same as subsequent hybridization and washing temperature and partly depends on Tm of the probes. However, taking as example 43 probes used for *M. tuberculosis* spoligotyping, their Tm vary in range from 48 to 78°C. Still 60 °C is used for hybridization and washing and provides specific signals.

20. Overflow from one lane to an adjacent lane should be controlled and avoided. The incubator surface where you place the Miniblotter should be even.

21. Hybridization step (as well as POD incubation step) may be for 1 h but it is possible to extend for 2 h (in such a case, washing should be similarly increased).

22. Avoid very fast rotation during POD incubation step in rotating bottle.

23. Avoid very vigorous shaking when washing, especially when washing after POD incubation.

    Incubation with POD may be done in a flat box (easier to handle but less economic—more POD is needed to keep the concentration) or in a rolling bottle.

24. Performance of ECL can be controlled by mixing 100 µl of ECL reagent 1, 100 µl of ECL reagent 2, and 2 µl of POD. Instant production of light in the dark indicates good functioning of the ECL kit.

25. Plastic bag (in which you put membrane for exposure to light sensitive film) should not be too thick: it would reduce signal's strength. Make sure to completely seal the bag, to avoid that liquid leaks out—this would result in smears on the film after development.

26. Incubation with film may be at RT or at 37 °C—from several minutes to overnight (this step should be tested in each laboratory). Previously, temperature of incubation of the membrane with streptavidin-POD solution (and subsequent washing) was 42 °C; more recently, 37 °C was recommended.

27. Signal's strength is decreased after a few hours, but you can reincubate the membrane with ECL Detection liquid and repeat exposure to ECL Hyperfilm. However, the activity of peroxidase decreases after 10 h.

    Perform multiple exposures of a same blot, two ideal expositions being 10 min and 2 h for high intensity signals. However, low intensity signals are not rare over prolonged period of time in many laboratories due to multifactor variables (most often difficult to control). In this case, recommended ideal exposure times are 2–4 h and overnight.

28. Quality of membrane may influence strength of signals (commercial in case of *M. tuberculosis* vs. home-made). Membranes supplied with *M. tuberculosis* spoligotyping kit can be reused for at least 10 times. Home made membranes may be used even more if a careful monitoring of decay of results is done.

    If unexpected signals appear concomitantly to a strain signal (e.g., Beijing genotype) when reusing a membrane, one should verify the correct dehybridization (stripping) of the membrane.

29. For data capture (done manually), use a manual grid with double reading by two independent readers. Any discrepancies between the two readers are discussed, a final decision is made and a unique grid is transferred to an Excel file by one of these two persons.

30. It may be convenient to encode presence and absence of signal as 43-digit binary code (mention "1" or "n" for a positive spot, and 0 "zero" or letter "o" for negative spot), which allows representing a pattern as successive black and white symbols using Monotype sorts police (or Zapf dingbats). For a nicer visual presentations and publications (as equal size black and white boxes), one may use Wingdings2 font and encode presence as ¢ and absence as £ (still in Excel). From keyboard (Num lock switched on), a combination Alt+0162 gives ¢ and a combination Alt+0163 gives £.

    The data is recorded for each of the 43 spacers in individual excel cells separately (1 spacer/cell) using "1" or "n" for a positive spot, and 0 "zero" or letter "o" for negative spot. This allows visualization similar to the membrane image for easy visual control prior to conversion to 15-octal format.

Since it is difficult to compare membrane results visually with octal codes, the data capture is ideally controlled by an independent reviewer before octal conversion.

Once a report is obtained after interrogation of the international database, be extremely vigilant in case of an orphan spoligotype pattern. Verify that its orphan or unique nature is not due to any transcription or manipulation error(s).

31. MST is an undirected network in which all of the samples are linked together with the fewest possible linkages between nearest neighbors. The phylogenetic tree connects each pattern based on degree of changes required to go from one allele to another. The structure of the tree is represented by branches (continuous vs. dashed and dotted lines) and circles representing each individual pattern. The length of the branches represents the distance between patterns while the complexity of the lines (may be continuous, dashed, dotted etc.) denotes the number of allele/spacer changes between two patterns. Lastly, the size of the circle is proportional to the total number of isolates in a study, illustrating unique isolates (smaller nodes) versus clustered isolates (bigger nodes).

32. In spoligoforest graph, each spoligotyping pattern is represented by a node with area size being proportional to the total number of isolates with that specific pattern. Changes (loss of spacers) are represented by directed edges between nodes, with the arrowheads pointing to descendant spoligotypes. The heuristic used selects a single inbound edge with a maximum weight using a Zipf model as follows: solid black lines link patterns that are very similar, i.e., loss of one spacer only (maximum weigh being 1.0), while dashed lines represent links of weight comprised between 0.5 and 1, and dotted lines a weight less than 0.5. Two layout methods for Spoligoforests are provided in SpolTools: a hierarchical layout and a "burst" layout based on a Fruchterman–Reingold algorithm. The Hierarchical layout represents the changes between strains hierarchically; the more the strains evolve (lose spacers), the more they are present in the down layouts (note that in case of too many changes between two patterns, there are no links/edges linking them), while in the Fruchterman Reingold algorithm based tree the nodes containing the biggest number of strains are centered [22].

# Acknowledgements

We are grateful to Thierry Zozio, Elisabeth Streit and Julie Millet (Institut Pasteur de la Guadeloupe), Anna Vyazovaya and Olga Narvskaya (St. Petersburg Pasteur Institute) for helpful discussions. Igor Mokrousov gratefully acknowledges support from Russian Science Foundation (project 14-14-00292).

## References

1. Hermans PWM, van Soolingen D, Bik EM, de Haas PEW, Dale JW, van Embden JDA (1991) The insertion element IS987 from *Mycobacterium bovis* BCG is located in a hot-spot integration region for insertion elements in *Mycobacterium tuberculosis* complex strains. Infect Immun 59:2695–2705

2. Groenen PM, Bunschoten AE, van Soolingen D, van Embden JD (1995) Nature of DNA polymorphism in the direct repeat cluster of *Mycobacterium tuberculosis*; application for strain differentiation by a novel typing method. Mol Microbiol 10:1057–1065

3. Kamerbeek J, Schouls L, Kolk A, van Agterveld M, van Soolingen D, Kuijper S et al (1997) Simultaneous detection and strain differentiation of *Mycobacterium tuberculosis* for diagnosis and epidemiology. J Clin Microbiol 35:907–914

4. Kremer K, Bunschoten A, Schouls L, van Soolingen D, and van Embden J (2002) Spoligotyping—a PCR-based method to simultaneously detect and type *Mycobacterium tuberculosis* complex bacteria, version 4, 11/8/02. www.tuberculosis.rivm.nl/documents/ECDC/protocol%20spoligotyping.pdf

5. Demay C, Liens B, Burguière T, Hill V, Couvin D, Millet J et al (2012) SITVITWEB—a publicly available international multimarker database for studying Mycobacterium tuberculosis genetic diversity and molecular epidemiology. Infect Genet Evol 12:755–766

6. van Embden JDA, van Gorkom T, Kremer K, Jansen T, van der Zeijst BAM, Schouls LM (2000) Genetic variation and evolutionary origin of the direct repeat locus of *Mycobacterium tuberculosis* complex bacteria. J Bacteriol 182:2393–2401

7. van der Zanden AG, Kremer K, Schouls LM (2002) Improvement of differentiation and interpretability of spoligotyping for *Mycobacterium tuberculosis* complex isolates by introduction of new spacer oligonucleotides. J Clin Microbiol 40:4628–4639

8. Brudey K, Gutierrez MC, Vincent V, Parsons LM, Salfinger M, Rastogi N et al (2004) *Mycobacterium africanum* genotyping using novel spacer oligonucleotides in the direct repeat locus. J Clin Microbiol 42:5053–5057

9. Javed MT, Aranaz A, de Juan L, Bezos J, Romero B, Alvarez J et al (2007) Improvement of spoligotyping with additional spacer sequences for characterization of *Mycobacterium bovis* and *M. caprae* isolates from Spain. Tuberculosis 87:437–445

10. Zhang J, Abadia E, Refregier G, Tafaj S, Boschiroli ML, Guillard B, Andremont A et al (2010) *Mycobacterium tuberculosis* complex CRISPR genotyping: improving efficiency, throughput and discriminative power of "spoligotyping" with new spacers and a microbead-based hybridization assay. J Med Microbiol 59:285–294

11. Sola C, Filliol I, Legrand E, Lesjean S, Locht C, Supply P et al (2003) Genotyping of the *Mycobacterium tuberculosis* complex using MIRUs: association with VNTR and spoligotyping for molecular epidemiology and evolutionary genetics. Infect Genet Evol 3:125–133

12. Jansen R, Embden JD, Gaastra W, Schouls LM (2002) Identification of genes that are associated with DNA repeats in prokaryotes. Mol Microbiol 43:1565–1575

13. Mokrousov I, Narvskaya O, Limeschenko E, Vyazovaya A (2005) Efficient discrimination within a *Corynebacterium diphtheriae* epidemic clonal group by a novel macroarray-based method. J Clin Microbiol 43:1662–1668

14. Mokrousov I, Limeschenko E, Vyazovaya A, Narvskaya O (2007) Corynebacterium diphtheriae spoligotyping based on combined use of two CRISPR loci. Biotechnol J 2:901–906

15. Shariat N, Dudley EG (2014) CRISPRs: molecular signatures used for pathogen subtyping. Appl Environ Microbiol 80(2):430–439. doi:10.1128/AEM. 02790-13

16. Grissa I, Vergnaud G, Pourcel C (2007) CRISPRFinder: a web tool to identify clustered regularly interspaced short palindromic repeats. Nucleic Acids Res 35:W52–57

17. Benson G (1999) Tandem repeats finder: a program to analyze DNA sequences. Nucleic Acids Res 27:573–580

18. Dale JW, Brittain D, Cataldi AA, Cousins D, Crawford JT, Driscoll J et al (2001) Spacer oligonucleotide typing of bacteria of the Mycobacterium tuberculosis complex: recommendations for standardised nomenclature. Int J Tuberc Lung Dis 5:216–219

19. Tang C, Reyes JF, Luciani F, Francis AR, Tanaka MM (2008) spolTools: online utilities for analyzing spoligotypes of the *Mycobacterium tuberculosis* complex. Bioinformatics 24:2414–2415

20. Reyes JF, Francis AR, Tanaka MM (2008) Models of deletion for visualizing bacterial variation: an application to tuberculosis spoligotypes. BMC Bioinformatics 9:496

21. Ellson J, Gansner E, Koutsofios L, North SC, Woodhull G (2002) Graphviz—open source graph drawing tools (Mutzel P, Jünger M, Leipert S, eds.). Springer-Verlag Berlin, Heidelberg, pp 483–484
22. Groenheit R, Ghebremichael S, Pennhag A, Jonsson J, Hoffner S, Couvin D et al (2012) *Mycobacterium tuberculosis* strains potentially involved in the TB epidemic in Sweden a century ago. PLoS One 7:e46848
23. Grimont PAD (2000) Taxotron package. Taxolab, Institut Pasteur, Paris
24. Coll F, Mallard K, Preston MD, Bentley S, Parkhill J, McNerney R et al (2012) SpolPred: rapid and accurate prediction of *Mycobacterium tuberculosis* spoligotypes from short genomic sequences. Bioinformatics 28:2991–2993
25. Kellogg DE, Rybalkin I, Chen S, Mukhamedova N, Vlasik T, Siebert PD, Chencik A (1994) TaqStart antibody: "hot start" PCR facilitated by a neutralizing monoclonal antibody directed against *Taq* DNA polymerase. BioTechniques 16:1134–1137
26. Molhuizen HO, Bunschoten AE, Schouls LM, van Embden JD (1998) Rapid detection and simultaneous strain differentiation of *Mycobacterium tuberculosis* complex bacteria by spoligotyping. Methods Mol Biol 101: 381–394
27. Van Der Zanden AG, Te Koppele-Vije EM, Vijaya Bhanu N, Van Soolingen D, Schouls LM (2003) Use of DNA extracts from Ziehl-Neelsen-stained slides for molecular detection of rifampin resistance and spoligotyping of *Mycobacterium tuberculosis*. J Clin Microbiol 41:1101–1108
28. Lazzarini LC, Rosenfeld J, Huard RC, Hill V, Lapa e Silva JR, DeSalle R et al (2012) *Mycobacterium tuberculosis* spoligotypes that may derive from mixed strain infections are revealed by a novel computational approach. Infect Genet Evol 12:798–806

# Chapter 8

## Analysis of crRNA Using Liquid Chromatography Electrospray Ionization Mass Spectrometry (LC ESI MS)

**Sakharam P. Waghmare, Alison O. Nwokeoji, and Mark J. Dickman**

### Abstract

Mass spectrometry is a powerful tool for characterizing RNA. Here we describe a method for the identification and characterisation of crRNA using liquid chromatography interfaced with electrospray ionization mass spectrometry (LC ESI MS). The direct purification of crRNA from the Cascade-crRNA complex was performed using denaturing ion pair reverse phase chromatography. Following purification of the crRNA, the intact mass was determined by LC ESI MS. Using this approach, a significant reduction in metal ion adduct formation of the crRNA was observed. In addition, RNase mapping of the crRNA was performed using RNase digestion in conjunction with liquid chromatography tandem MS analysis. Using the intact mass of the crRNA, in conjunction with RNase mapping experiments enabled the identification and characterisation of the crRNA, providing further insight into crRNA processing in a number of type I CRISPR-Cas systems.

**Key words** crRNA processing, Liquid chromatography, Electrospray ionization mass spectrometry, RNase mapping

## 1 Introduction

Bacteria and archaea integrate short fragments of foreign nucleic acid into the host chromosome at one end of a repetitive element known as CRISPR (clustered regularly interspaced short palindromic repeat) [1, 2]. CRISPR loci are transcribed and the long primary transcript is processed into mature CRISPR RNAs (crRNAs) that each contains a sequence complementary to a previously encountered invading nucleic acid [3–6]. The silencing of invading nucleic acids is performed by ribonucleoprotein complexes that utilize crRNAs as guides for targeting and degradation of foreign nucleic acids [7–9]. The maturation of crRNAs is critical to the activation of all CRISPR/Cas immune systems and involves at least two distinct steps. CRISPR loci are initially transcribed as long precursor crRNAs (pre-crRNAs) from a promoter sequence in the leader. Subsequently these pre-crRNAs are processed into mature crRNA species by CRISPR-specific endoribonucleases.

In the *E. coli* system, Cas6e and its associated crRNA are essential components of a multisubunit macromolecular complex termed Cascade (CRISPR-associated complex for antiviral defense) [3, 4]. Mass spectrometry (MS) is a powerful tool for characterizing RNA [10, 11]. We have utilized mass spectrometry approaches to rapidly analyze crRNAs associated with a number of type I CRISPR-Cas systems. crRNAs were directly purified from the Cascade-crRNA complex using denaturing ion pair reverse phase chromatography (IP-RP-HPLC) [12, 13]. Following purification of the crRNA, the intact mass was determined using liquid chromatography interfaced with electrospray ionization mass spectrometry (LC ESI MS). Using this approach, a significant reduction in metal ion adduct formation of the crRNA was observed. In addition, RNase mapping of the crRNA was performed using RNase digestion in conjunction with liquid chromatography tandem MS analysis. Using the intact mass of the crRNA, in conjunction with RNase mapping experiments enabled the identification and characterisation of the crRNA, providing further insight into crRNA processing in a number of type I CRISPR-Cas systems [4, 14, 15].

## 2 Materials

Prepare all solutions for HPLC using HPLC grade solvents. Prepare all solutions for LC MS using LC MS grade solvents. All areas/benches where samples will be handled should be cleaned with 70 % ethanol. RNase-free solutions and RNase-free tubes should be used where appropriate. Minimize contact with potential sources of metal ions during sample preparation. Note that phenol–chloroform is hazardous and should be handled with care.

### 2.1 Ion Pair Reverse Phase High Performance Liquid Chromatography (IP-RP-HPLC)

1. HPLC with column oven.
2. HPLC Column: DNAsep (50 mm×4.6 mm ID) (Transgenomic, San Jose, CA).
3. HPLC buffer A: 0.1 M triethylammonium acetate (TEAA) (pH 7.0) (BioUltra grade Fluka). Store at 4 °C.
4. HPLC buffer B: 0.1 M triethylammonium acetate (TEAA) (pH 7.0) with 25 % (v/v). HPLC grade acetonitrile (Thermo Fisher, UK). Store at 4 °C.
5. HPLC buffer C: 0.1 M triethylammonium acetate (TEAA) (pH 7.0) with 25 mM EDTA (Sigma, UK). Store at 4 °C.
6. pUC18 HaeIII digest (Sigma-Aldrich, UK). Store at −20 °C.

### 2.2 RNA Mass Spectrometry

1. HPLC column for intact crRNA analysis: Monolithic (PS-DVB) capillary column (50 mm×0.2 mm ID, Thermo Fisher, UK).

2. HPLC column for oligoribonucleotide analysis: C-18 RP capillary column (150 mm × 0.3 mm ID).
3. HPLC grade water for the preparation of all reagents (Thermo Fisher).
4. TOF mass spectrometer.
5. HPLC cleaning buffer A: 50 % (v/v) acetonitrile, 0.1 M TEAA, 25 mM EDTA.
6. HPLC cleaning buffer B: 5 mM EDTA.
7. LC MS buffer A: 0.4 M 1,1,1,3,3,3,-Hexafluoro-2-propanol (HFIP, Sigma-Aldrich) with Triethylamine (TEA, Thermo Fisher, UK) to pH 7.0 and 0.1 mM TEAA.
8. LC MS buffer B: LC MS buffer A with 50 % methanol (v/v) (Thermo Fisher).
9. LC MS buffer C: 80 % (v/v) acetonitrile.
10. RNase A/T1 (Ambion).

## 2.3 Phenol–Chloroform Extraction

1. Sodium dodecyl sulfate (SDS) (Technical grade, 90 %, Epton Titration (VWR)).
2. Acid phenol–chloroform (pH 4.5), premixed with isoamyl alcohol (125:24:1 phenol–chloroform–isoamyl alcohol) (Life Technologies).
3. 2-propanol (ACS reagent, 99.5 % purity, Sigma-Aldrich).

## 2.4 Preparation of HPLC Buffers

1. Buffer A: To a 1 L volumetric flask add 100 ml of 1 M TEAA (from 100 ml volumetric flask), make up to 1 L mark using HPLC grade water (*see* **Note 1**).
2. Buffer B: To a 1 L volumetric flask add 100 ml of 1 M TEAA (from 100 ml volumetric flask), add 250 ml acetonitrile (from 250 ml volumetric flask) make up to 1 L using HPLC grade water.
3. Buffer C: To a 1 L volumetric flask add 100 ml of 1 M TEAA (from 100 ml volumetric flask), add 50 ml of 500 mM EDTA, make up to 1 L using HPLC grade water

## 2.5 Preparation of LC MS Buffers

1. LC MS Buffer A: Add 955.4 ml LC MS grade water to a clean 1 L volumetric flask, add 1 ml TEA and 1 ml TEAA and mix thoroughly. Finally add 42.6 ml HFIP (9.4 M) and mix. Take 3–4 ml aliquot in a 15 ml falcon tube and check the pH (~7.2) (*see* **Note 2**).
2. LC MS Buffer B: Add 228.3 ml LC MS grade water to a 0.5 L volumetric flask, add 250 ml LC MS grade methanol, add 0.125 ml TEA and 0.5 ml TEAA and mix thoroughly. Finally add 21.1 ml HFIP (9.4 M) and mix. Take 3–4 ml aliquot in a 15 ml falcon tube and check pH (~7.2).

3. LC MS Buffer C: To a 1 L volumetric flask add 800 ml LC MS grade acetonitrile, make up to 1 L using LC MS grade water (*see* **Note 3**).

## 3 Methods

### 3.1 Purification of crRNA Using Denaturing IP-RP-HPLC

1. Following preparation of HPLC buffers, purge the system, connect the HPLC column and equilibrate the system overnight at 90 % buffer A, 10 % buffer B at a flow rate of 50 μl/min (*see* **Note 4**).

2. Run the appropriate standard (pUC18 HaeIII digest) using the gradient below to ensure optimum performance of the IP RP HPLC. Starting with 65 % buffer A and 35 % buffer B extend the gradient to 50 % buffer B in 3 min, followed by a linear extension to 65 % buffer B over 12 min at a flow rate of 1.0 ml/min at temperature 50 °C.

3. Set the temperature to 70 °C and 95 % buffer C and 5 % buffer B at 1 ml/min flow rate. Inject 25 μl of the purified Cascade-crRNA protein complex ($\approx$1 μg/μl) (*see* **Note 5**).

4. After 5 min at 95 % buffer C and 5 % buffer B the gradient is switched to 85 % buffer A and 15 % buffer B and extended to 60 % buffer B in 12.5 min, followed by a linear extension to 100 % buffer B over 2 min at a flow rate of 1.0 ml/min at temperature 70 °C.

5. Fractionate the crRNA (typical retention time 10–12 min). The volume of fractions collected will depend on the peak width.

6. Concentrate the crRNA fraction to 10–20 μl on a vacuum concentrator.

The IP-RP-HPLC chromatogram following the direct injection of the Cascade-crRNA complex is shown in Fig. 1. The crRNA elutes after 10.0 min. To demonstrate the RNA has fully dissociated from the protein components of the complex, the fractions from the high acetonitrile wash at the end of the gradient can be analyzed using SDS PAGE (*see* inset Fig. 1). The SDS PAGE confirms the presence of the five protein components of the Cascade complex, demonstrating that under the chromatographic conditions employed, the protein components dissociate from the RNA-protein complex.

### 3.2 Purification of crRNA by Acid Phenol–Chloroform Extraction

An alternative approach to purify the crRNA from the Cascade-crRNA complex can also be used based on phenol–chloroform extraction procedure.

For a 100 μl final volume:

1. Add 10 μl of purified Cascade-crRNA complex (1 μg/μl) to 10 μl HPLC grade water.

2. Add 15 μl of 0.5 M EDTA and vortex for 5–10 s.

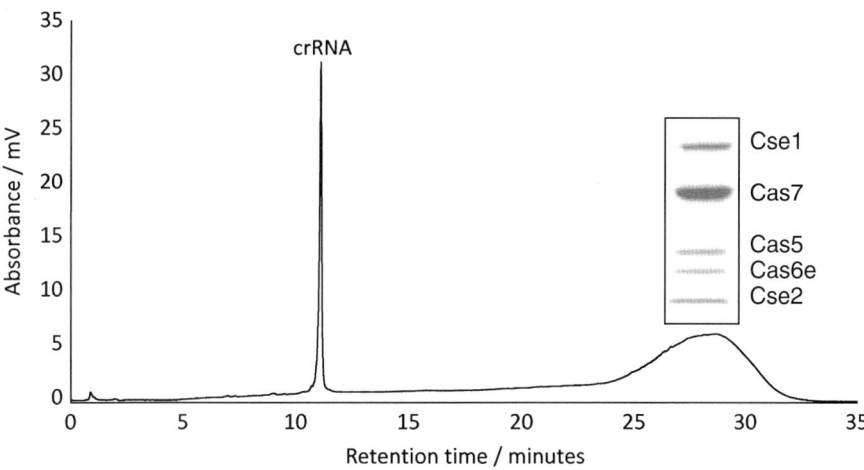

**Fig. 1** Direct purification of crRNA from the Cascade-crRNA complex using IP RP HPLC chromatography. 0.5 μg crRNA-Cascade complex was directly injected onto the stationary phase at 75 °C followed by a 25 mM EDTA wash for 5 min. At elevated temperature the complex dissociates separating the crRNA from the Cascade proteins under the chromatographic conditions. crRNA eluted at 10–12 min retention time was fractionated and the later eluting Cascade proteins were analyzed on SDS-PAGE (*see inset*)

3. Add 40 μl of 10 % SDS stock solution (4 % in final volume). Vortex for about 30 s. Make up to 100 μl with HPLC grade water.

4. Heat for 5 min at 75 °C.

5. Vortex for 1 min after heating.

6. Add 200 μl of HPLC grade water.

7. Add 300 μl of acid phenol–chloroform, vortex for 1 min and leave to stand for 3 min at room temperature.

8. Centrifuge for 5 min at 14,800 rpm (16,162 ×$g$).

9. Transfer the aqueous phase to new tube, careful not to disturb the bottom phenol–chloroform phase.

10. To the aqueous phase, add 1.2 ml of isopropanol. Vortex for 30 s and centrifuge for 15 min at 14,800 rpm (16,162 ×$g$) and remove the supernatant.

11. Resuspend the crRNA in RNase-free water.

## 3.3 Determination of RNA Concentration

The concentration of purified crRNA samples can be determined using UV spectrophotometry. Make appropriate dilutions of the sample and record the absorbance at 260 nm. Calculate the RNA concentration using molar extinction coefficient (0.025 (μg/ml)$^{-1}$cm$^{-1}$). A total 1–2 μg of crRNA sample is sufficient for the mass spectrometry analysis. Divide the sample in to three parts for downstream analysis including; (1) intact mass analysis (2) acid treatment for confirmation of chemical termini and (3) RNase A and RNase T1 digestion for oligonucleotide mass mapping.

### 3.4 Acid Treatment

To confirm the presence of cyclic phosphate at 3′-end the crRNA is treated with hydrochloric acid to hydrolyse 2′,3′-cyclic phosphate group to a 2′/3′-phosphate. The conversion of the 2′,3′-cyclic phosphate to 2′/3′-phosphate corresponds to increase in 18 Da.

1. Add HCl to HPLC purified crRNA to a final concentration of 0.1 M (*see* **Note 6**).
2. Incubate the reaction at 4 °C for 1 h.
3. The acid reaction can be stopped by addition of ammonium hydroxide to bring the pH up to ~7.0.

### 3.5 RNase Mapping

RNase mapping of crRNA is performed by generating a number of oligoribonucleotides using ribonucleases followed by LC-ESI MS/MS analysis. Ribonucleases cleave the intact RNA at specific positions to generate a sequence ladder. RNase T1 specifically cleaves the 3′-end of unpaired G and RNase A cleaves 3′-end of unpaired C and U residues.

1. Digest crRNA (200 ng) with 20 U RNase T1 and/or 20 U RNase A.
2. Incubate the reactions at 37 °C for 30 min.
3. Generate a list of theoretical monoisotopic masses of oligoribonucleotides using Mongo Oligo Mass Calculator (http://mods.rna.albany.edu/masspec/Mongo-Oligo). The output of this analysis produces a theoretical sequence ladder of oligoribonucleotides for the crRNA with all possible chemical termini including; 5′-OH, -phosphate, -cyclic phosphate and 3′-OH, -phosphate, -cyclic phosphate.
4. Mongo Oligo Mass Calculator can also be used to predict the CID fragment ions of the theoretical oligoribonucleotides.

### 3.6 LC ESI MS Analysis of crRNA

#### 3.6.1 Instrument Cleaning Prior to LC-ESI MS Analysis of crRNA

Prior to analysis of the RNA using LC ESI MS, it is important that the LC and MS instruments are clean and all contaminating metal ions are removed. The following protocol can be used.

1. Replace the column with a PEEK union.
2. Remove all pump inlet lines from the buffers and place in LC MS grade water.
3. Purge A, B, and C pumps with LC MS grade water.
4. Put the pump inlet lines in HPLC cleaning buffer A and purge A, B, and C pumps.
5. Close the purge valve and set the isocratic gradient with A 33 %, B 33 %, and C 34 %, set the flow rate 2 μl/min and leave the system running overnight.
6. Connect the column and post column capillary (connection to the mass spectrometer).

7. Set the flow rate 2 μl/min and leave the system running for ~1 h.
8. Stop the flow and put the pump inlet lines in HPLC cleaning buffer B.
9. Open the purge valve and purge A, B, and C pumps.
10. Close the purge valve and set the isocratic gradient with A 33 %, B 33 %, and C 34 %, set the flow rate 2 μl/min and leave the system running for 5–6 h.
11. Put the pump inlet lines in appropriate LC MS buffers.
12. Purge pump C followed by B and then A.
13. Close the purge valve.
14. Start flow 2 μl/min through 50 % A and 50 % B and run for 2–3 h.

*3.6.2 Cleaning of Syringe and ESI Spray Needle*

1. Clean the syringe and PEEK tubing (connected to MS) with acetonitrile (500 μl).
2. Clean the syringe with HPLC cleaning buffer A 2–3 times.
3. Connect the syringe to the ESI spray needle and pass the HPLC cleaning buffer A for 15 min at a flow rate 3 μl/min.
4. Fill the syringe with LC-MS buffer B and spray for 15 min at a flow rate of 3 μl/min.
5. Connect the ESI spray needle to the mass spectrometer.
6. Connect the HPLC outlet to the ESI Spray needle.
7. Interface the capillary liquid chromatography system to an ultrahigh-resolution time-of-flight (UHR-ToF) mass spectrometer.

## 3.7 LC ESI MS Analysis of Intact crRNA

### 3.7.1 UHR-ToF Mass Spectrometry Conditions

1. Set the ultrahigh-resolution time-of-flight (UHR-ToF) mass spectrometer to perform data acquisition in the negative mode with a selected mass range of 100–2,500 *m/z* (*see* **Note 7**).
2. Set the ionization voltage of –3,200 V to maintain capillary current between 30 and 50 nA. Set the temperature of nitrogen to 120 °C at a flow rate of 4.0 L/h and $N_2$ nebulizer gas pressure at 0.4 bar.
3. Inject 2–5 μl sample (50–200 μg) using capillary flow HPLC using the following gradient on a monolithic (PS-DVB) capillary column (50 mm × 0.2 mm ID): Starting with 80 % A and 20 % LC MS buffer B, extend the gradient to 40 % B in 5 min followed by a linear extension to 60 % B over 8 min at a flow rate of 2 μl/min.

### 3.7.2 Data Analysis

1. Data acquisition processing was performed using the following settings. The charge/mass deconvolution processing parameters were set to default except for the following. (a) Mass

list > Sum peak: peak width (FWHM) 30 points. (b) Charge deconvolution > Protein/large molecules (low mass = 5,000, high mass = 50,000). (c) Exclusion of the low intensity masses. 500–700 $m/z$ and 1,300–2,000 $m/z$ (see **Note 7**).

2. Compare the accurate molecular weights of acid treated and untreated crRNA. An 18 Da mass difference indicates the presence of cyclic phosphate, whereas no change in the mass confirms the absence of cyclic phosphate.

3. Using the accurate mass of the crRNA, in conjunction with the knowledge of the presence or absence of cyclic phosphate, predict the potential crRNA sequences, considering all possible cleavage points of the processing enzyme on the repeats of pre-crRNA. Generate a list of theoretical average masses using Mongo Oligo Mass Calculator (http://mods.rna.albany.edu/masspec/Mongo-Oligo).

Following purification of the crRNA using IP RP HPLC, the crRNA is analyzed using LC ESI MS. The results are shown in Fig. 2a. An enhanced view of the −21 charge state demonstrates the presence of minimal metal ion adducts (see Fig. 2b). Deconvolution is used to obtain an intact mass (19660.803 Da). The intact $M_W$ analysis of the crRNA predicted the presence of a 5′-hydroxyl and 2′,3′-cyclic phosphate termini. Further evidence of the presence of the 2′,3′-cyclic phosphate termini is obtained by acid treatment of the purified crRNA. An enhanced view of the 21-charge state of the crRNA ESI MS spectra before and after acid treatment is show in Fig. 2b. The corresponding mass shift (18 Da) is observed, demonstrating the conversion of the 2′,3′-cyclic phosphate to the 3′ (or 2′)-phosphate.

## 3.8 Analysis of crRNA Digests

### 3.8.1 LC ESI MS Analysis of crRNA Digests

1. Set the (UHR-ToF) mass spectrometer to perform data acquisition in the negative mode with a selected mass range of 100–2,500 $m/z$ (see **Note 7**).

2. Set the ionization voltage of −3,200 V to maintain capillary current between 30 and 50 nA. Set the temperature of nitrogen to 120 °C at a flow rate of 4.0 L/h and $N_2$ nebulizer gas pressure at 0.4 bar (see **Note 7**).

3. Set the collision energy to −10 eV and collision cell RF 1,200 Vpp with collision cell transfer time 199 μs. Set the precursors with charge −1 to −4 for the fragmentation and exclude the selected precursors that had been analyzed >2 times from analysis for 1 min (see **Note 7**).

4. Inject 2–5 μl sample (5–200 ng) using capillary flow HPLC (C18 RP capillary column (150 mm × 0.3 μm ID)), using the following gradient: Starting with 80 % LC MS buffer A and 20 % LC MS buffer B, extend the gradient 35 % B in 3 min followed by a linear extension to 70 % B over 40 min at a flow rate of 2 μl/min.

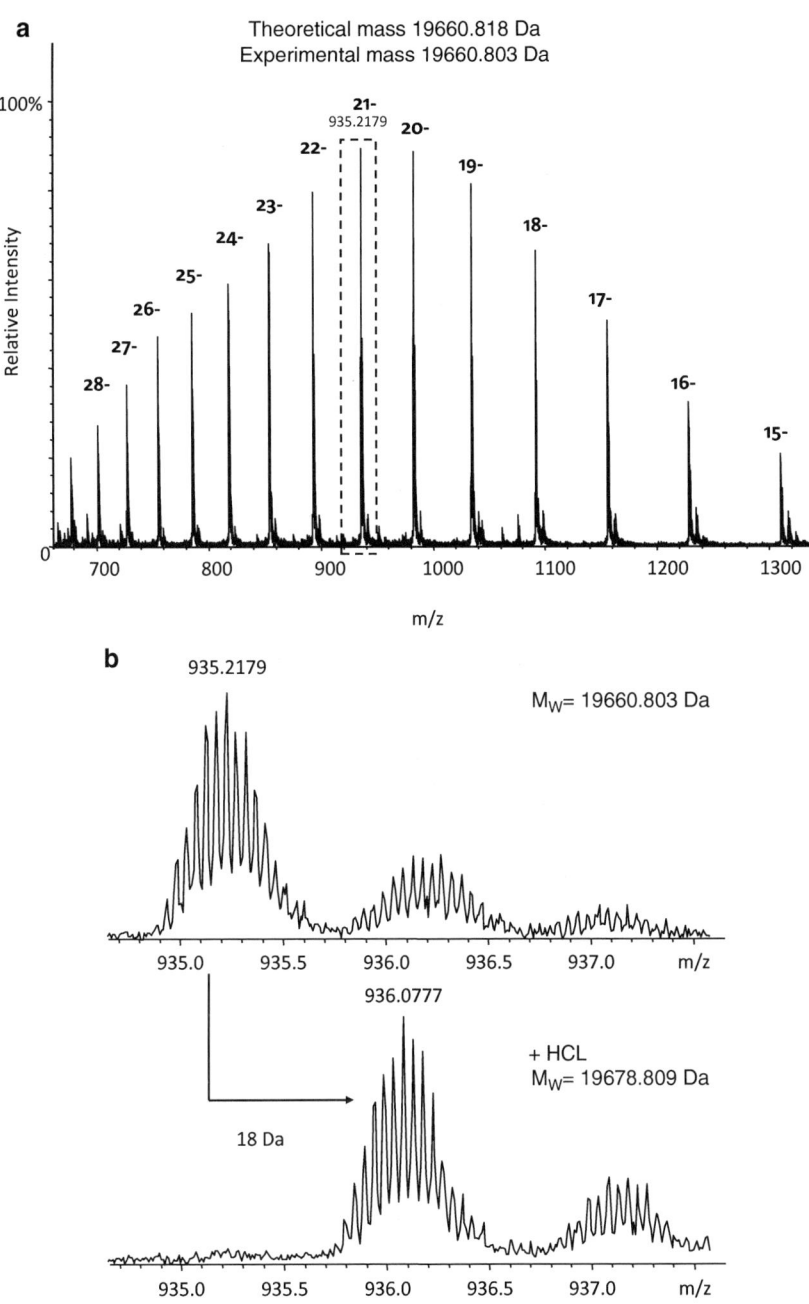

**Fig. 2** Mass spectrometry analysis of *E. coli* crRNA (**a**) LC ESI MS analysis of *E. coli* crRNA (50 ng). Following deconvolution the intact mass was obtained = 19660.803 Da consistent with a theoretical 61 nucleotide mature crRNA with a 5′-OH and 2′, 3′-cyclic phosphate termini resulting from a single Cas6e cleavage at the base of the hairpin loop in the pre-crRNA. (**b**) Enhanced view of the 21-charged species before (*upper panel*) and after acid treatment (*lower panel*) indicating the hydrolysis of the 2′,3′-cyclic phosphate to 2′/3′-phosphate with the corresponding 18 Da shift in molecular weight

3.8.2 *Data Analysis*

1. Internal calibration can be performed in –ve MS mode using hexakis-(1H,1H,3H-tetrafluoropentoxy)-phosphazene ($C_{18}H_{18}O_6N_3P_3F_{24}$, $m/z$ 1,033.988109) as calibrant.
2. From the processed data, identify the oligoribonucleotides by assigning the measured masses to the theoretical masses (*see* Table 1).
3. Use the CID fragmentation spectrum for each oligoribonucleotide for confirmation of the oligoribonucleotide sequence. The combined intact mass analysis and oligo mass mapping is used to provide details of the cleavage point on the pre-crRNA and chemical nature of mature crRNA termini.

Analysis of the mature crRNA using ESI MS/MS following RNase T1 and A digestion is shown in Fig. 3. A number of oligoribonucleotides generated from the T1 and RNase A digests can be assigned to the mature crRNA sequence (*see* Fig. 3 and Table 1). Assignment of the terminal oligonucleotide GGGGcp also confirmed the cleavage point which is present at the base of each hairpin loop of the repeat sequence on the pre-crRNA.

**Table 1**
**Summary of the identified oligoribonucleotides assigned to the mature *E. coli* crRNA from the RNase A digest**

| Position | Theoretical mass (Da) | Experimental mass (Da) | Sequence |
|---|---|---|---|
| A1:U2 | 653.393 | 653.1 | AUp |
| A3:C6 | 1,310.826 | 1,310.4 | AAACp |
| G8:C10 | 997.617 | 997.1 | GACp |
| G11:U13 | 1,014.602 | 1,014.2 | GGUp |
| A14:U15 | 653.393 | 653.1 | AUp |
| G17:U18 | 669.393 | 669.0 | GUp |
| A21:U24 | 1,327.811 | 1,327.3 | AGAUp |
| G28:C30 | 1,013.617 | 1,013.4 | GGCp |
| G33:C34 | 668.408 | 668.1 | GCp |
| A36:C38 | 981.617 | 981.2 | AACp |
| A39:U44 | 2,018.229 | 2,017.7 | AGGAGUp |
| G50:C51 | 668.408 | 668.1 | GCp |
| G52:C53 | 668.408 | 668.1 | GCp |
| A55:C57 | 997.617 | 997.1 | AGCp |
| G58:G61 | 1,398.851 | 1,398.4 | GGGGp |
| G58:G61 | 1,380.836 | 1,380.4 | GGGG^p |

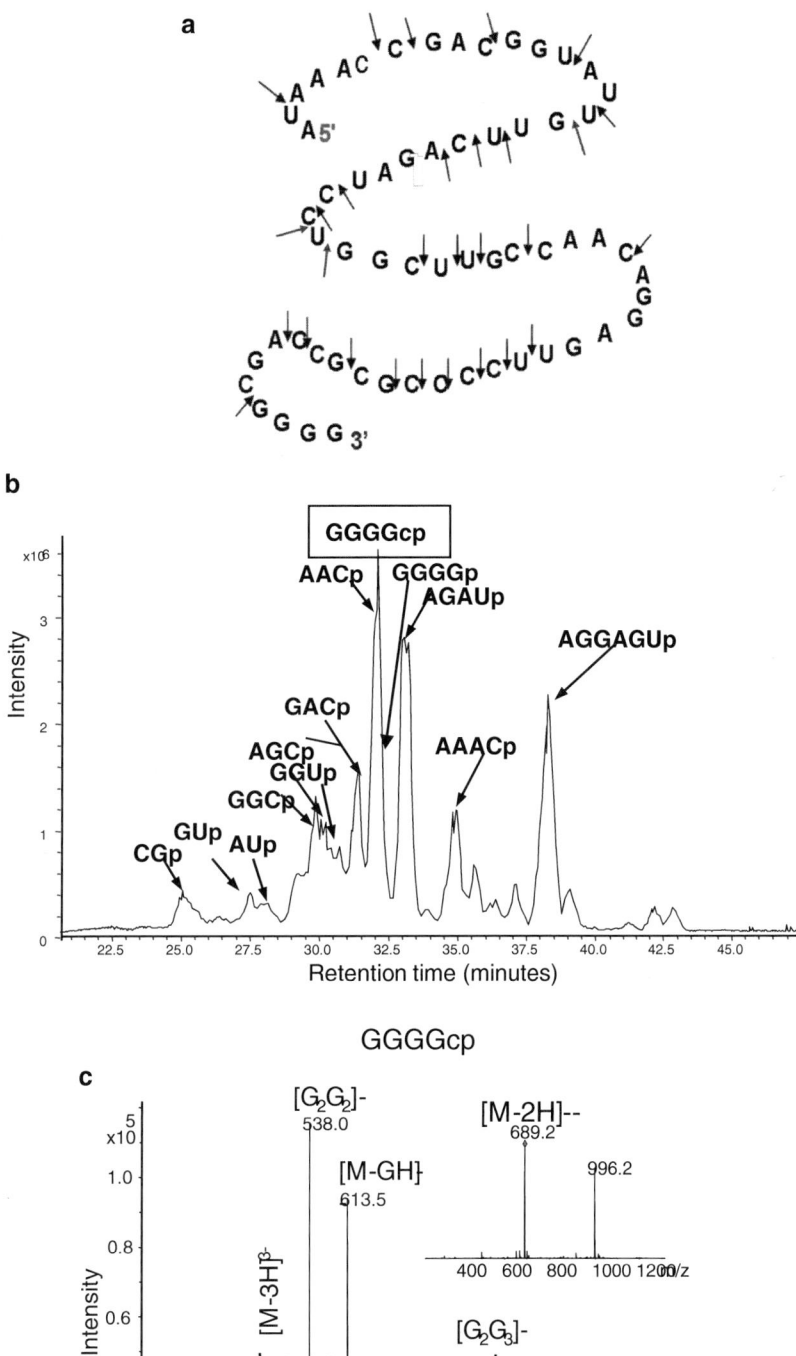

**Fig. 3** Characterisation of crRNA using LC ESI MS/MS. (**a**) Predicted RNase A cleavage sites in *E. coli* crRNA. (**b**) ESI MS analysis of the RNase A digest of crRNA. (**b**) Base peak chromatogram of the RNase A digest of mature crRNA. The predominant oligoribonucleotide peaks assigned to the mature crRNA are highlighted. cp indicates 2′,3′-cyclic phosphate. (**c**) Tandem mass spectrometry analysis of terminal oligoribonucleotide GGGGcp. The predominant fragment ions are highlighted

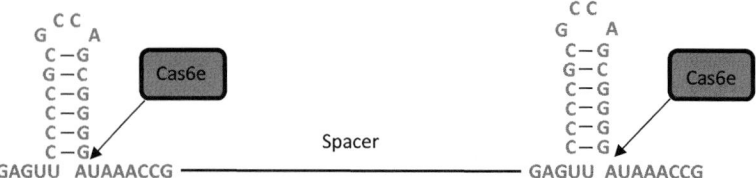

**Fig. 4** Cleavage position by Cas6e endoribonucleases in the precursor crRNA. Combination of the intact mass analysis in conjunction with the RNase mass mapping enabled the characterisation of the mature crRNAs from *E. coli* crRNA

Here, the assignment of the oligoribonucleotide at the 3′-terminus provides further evidence of the cleavage point and presence of cyclic phosphate or phosphate at the 3′-end. The combined intact mass analysis of the crRNA, in conjunction with the RNase digestion and MS/MS analysis enables the identification and characterisation of the crRNA. The MS analysis of the *E. coli* crRNA revealed a 61 nt RNA fragment, with a single Cas6e cleavage site present in each repeat (*see* Fig. 4).

# 4 Notes

1. Before preparation of HPLC and LC MS buffers make sure all glassware is clean and free of metal ions. Clean the glassware thoroughly with the Milli-Q water. Rinse with HPLC buffer C, followed by rinsing with Milli-Q water 3–4 times.

2. TEA is immiscible in 0.4 M HFIP therefore TEA should be first added to water.

3. HFIP is immiscible in acetonitrile therefore LC MS buffer A and LC MS buffer C should not be run together or immediately after each other. LC MS buffer B contains 50 % (v/v) methanol therefore is miscible with LC MS buffer C.

4. Prior to performing nucleic acid analysis using IP-RP-HPLC it is important to minimize all metal ions from the system to increase the chromatographic performance. Therefore, the HPLC system should be passivated by treating with 8 M nitric acid (35 %, w/w) and 200 mM EDTA (tetrasodium salt).

5. For the efficient column binding, the sample should be at the same pH as the starting buffer, and contain the same amount of TEAA (~100 mM). This can be either achieved by diluting the sample in the buffer A or by adding 1/10th volume of TEAA (from 1 M stock).

6. During the vacuum concentration of the HPLC purified crRNA, the TEAA does not completely evaporate. The residual TEAA effectively buffers this solution; therefore, it is important to check the pH. If not acidic then add more HCl to lower the pH.

7. Settings specified for UHR-ToF maXis mass spectrometer (Bruker Daltonics) or Bruker Compass DataAnalysis, other equipment/software may require different settings.

## Acknowledgments

MJD acknowledges funding support from the Engineering and Physical Sciences Research Council UK and the Biotechnology and Biological Sciences Research Council Research Council UK.

## References

1. Bolotin A, Quinquis B, Sorokin A, Ehrlich SD (2005) Clustered regularly interspaced short palindrome repeats (CRISPRs) have spacers of extrachromosomal origin. Microbiology 151:2551–2561
2. Mojica FJ, Díez-Villaseñor C, García-Martínez J, Soria E (2005) Intervening sequences of regularly spaced prokaryotic repeats derive from foreign genetic elements. J Mol Evol 60:174–182
3. Brouns SJ, Jore MM, Lundgren M, Westra ER, Slijkhuis RJ, Snijders AP, Dickman MJ, Makarova KS, Koonin EV, van der Oost J (2008) Small CRISPR RNAs guide antiviral defense in prokaryotes. Science 321:960–964
4. Jore MM, Lundgren M, van Duijn E, Bultema JB, Westra ER, Waghmare SP, Wiedenheft B, Pul U, Wurm R, Wagner R, Beijer MR, Barendregt A, Zhou K, Snijders AP, Dickman MJ, Doudna JA, Boekema EJ, Heck AJ, van der Oost J, Brouns SJ (2011) Structural basis for CRISPR RNA-guided DNA recognition by Cascade. Nat Struct Mol Biol 18:529–536
5. van der Oost J, Jore MM, Westra ER, Lundgren M, Brouns SJ (2009) CRISPR-based adaptive and heritable immunity in prokaryotes. Trends Biochem Sci 34:401–407
6. Sorek R, Kunin V, Hugenholtz P (2008) CRISPR–a widespread system that provides acquired resistance against phages in bacteria and archaea. Nat Rev Microbiol 6:181–186
7. Barrangou R, Fremaux C, Deveau H, Richards M, Boyaval P, Moineau S, Romero DA, Horvath P (2007) CRISPR provides acquired resistance against viruses in prokaryotes. Science 315:1709–1712
8. Wiedenheft B, Sternberg SH, Doudna JA (2012) RNA-guided genetic silencing systems in bacteria and archaea. Nature 482:331–338
9. Sorek R, Lawrence CM, Wiedenheft B (2013) CRISPR-mediated adaptive immune systems in bacteria and archaea. Annu Rev Biochem 82:237–266
10. Giessing AM, Kirpekar FJ (2012) Mass spectrometry in the biology of RNA and its modifications. J Proteomics 75:3434–3449
11. Waghmare SP, Dickman MJ (2011) Characterization and quantification of RNA post-transcriptional modifications using stable isotope labeling of RNA in conjunction with mass spectrometry analysis. Anal Chem 83:4894–4901
12. Dickman MJ, Hornby DP (2006) Enrichment and analysis of RNA centered on ion pair reverse phase methodology. RNA 12:691–696
13. Waghmare SP, Pousinis P, Hornby DP, Dickman MJ (2009) Studying the mechanism of RNA separations using RNA chromatography and its application in the analysis of ribosomal RNA and RNA:RNA interactions. J Chromatogr A 1216:1377–1382
14. Wiedenheft B, van Duijn E, Bultema JB, Waghmare SP, Zhou K, Barendregt A, Westphal W, Heck AJ, Boekema EJ, Dickman MJ, Doudna JA (2011) RNA-guided complex from a bacterial immune system enhances target recognition through seed sequence interactions. Proc Natl Acad Sci U S A 108:10092–10097
15. Sinkunas T, Gasiunas G, Waghmare SP, Dickman MJ, Barrangou R, Horvath P, Siksnys V (2013) In vitro reconstitution of Cascade-mediated CRISPR immunity in *Streptococcus thermophilus*. EMBO J 32:385–394

# Chapter 9

## Rapid Multiplex Creation of *Escherichia coli* Strains Capable of Interfering with Phage Infection Through CRISPR

Alexandra Strotksaya, Ekaterina Semenova, Ekaterina Savitskaya, and Konstantin Severinov

### Abstract

In *Escherichia coli*, acquisition of new spacers in the course of CRISPR-Cas adaptation is dramatically stimulated by preexisting partial matches between a bacterial CRISPR cassette spacer and a protospacer sequence in the DNA of the infecting bacteriophage or plasmid. This phenomenon, which we refer to as "priming," can be used for very simple and rapid construction of multiple *E. coli* strains capable of targeting, through CRISPR interference, any phage or plasmid of interest. Availability of such strains should allow rapid progress in the analysis of CRISPR-Cas system function against diverse mobile genetic elements.

**Key words** *Escherichia coli*, CRISPR, Cas proteins, Bacteriophage, Spacers, Strain construction

## 1 Introduction

In *Escherichia coli*, the CRISPR-Cas system is dormant, at least in laboratory conditions [1–3]. However, elevated expression of *cas* genes is sufficient to make cells resistant to transformation with plasmids or infection with some phages, provided that the CRISPR cassette contains a spacer that matches a protospacer in the plasmid being transformed or the phage used for infection [2, 4]. Increased *cas* gene expression can be achieved by deletion of a gene coding for global transcription repressor H-NS, which negatively controls *cas* gene transcription [1–3]. Another strategy is to co-overexpress *cas* genes from a plasmid. Compatible plasmids expressing the *cas3* gene, and the *casABCDE12* genes have been described [4]. We developed a series of *E. coli* K12-based strains with chromosomal *cas* genes fused to inducible promoters (*see* ref. 5, and Fig. 1).

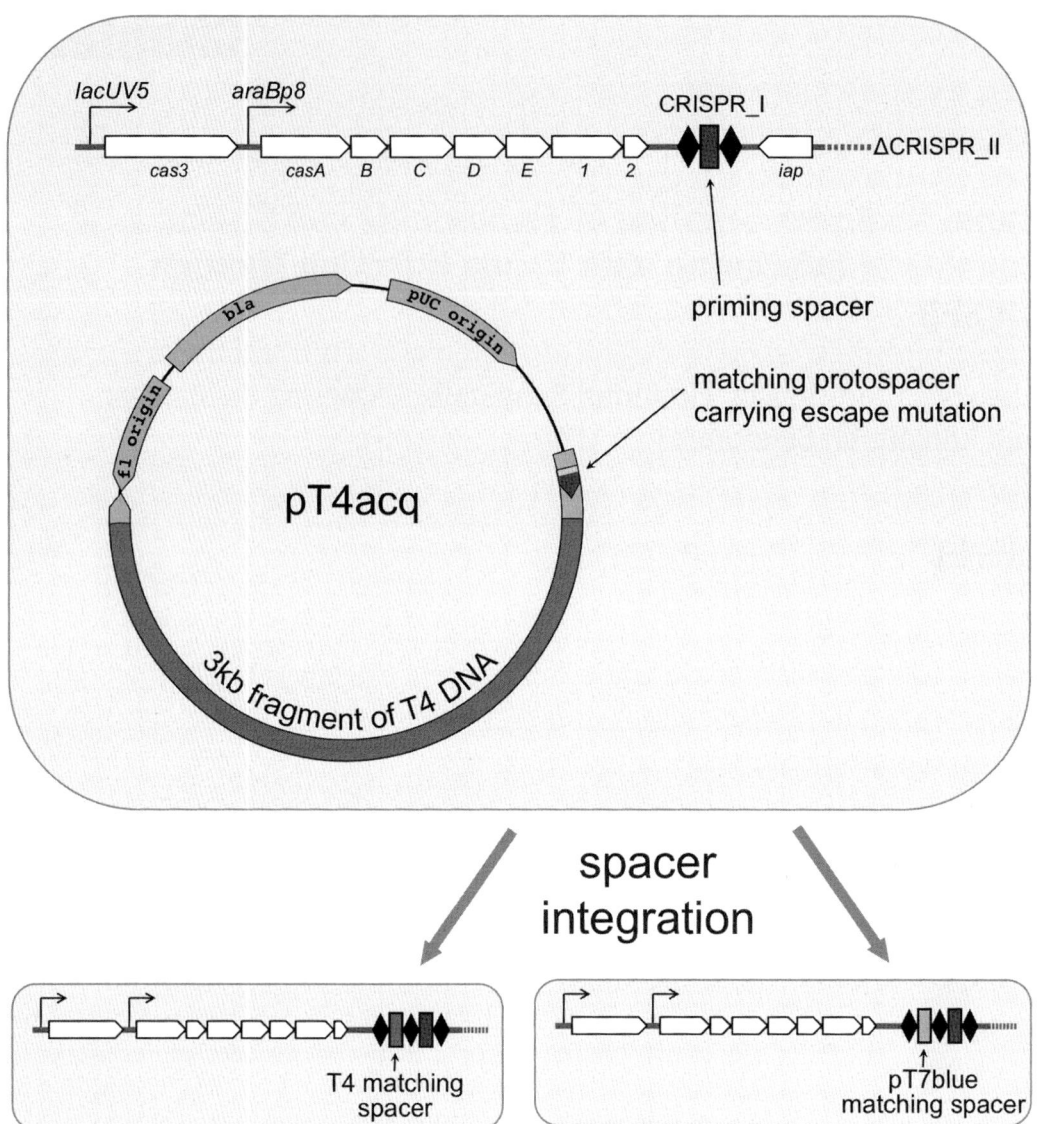

**Fig. 1** Schematic representation of *E. coli* strain KD263 transformed with the T4 capture plasmid pT4acq. The strain contains *cas* genes driven by the *lac*UV5 (*cas3*) and *ara*Bp8 (*casABCDE12*) promoters. The CRISPR II cassette has been deleted, while the CRISPR I cassette has been reduced and contains two repeats separated by a g8 spacer (*blue*) derived from the M13 phage [6]. pT4acq is based on the pT7blue vector and contains a fragment corresponding to g8 protospacer with escape substitution ([6], *blue*) and a ~3 kbp insert of bacteriophage T4 genomic DNA (*red*). Primed spacer acquisition leads to an expanded CRISPR I cassette carrying phage insert of plasmid vector derived spacer (below) (Color figure online)

These strains appear to be preferable to plasmid-borne *cas* co-overexpression since there is no gross overproduction of Cas proteins in induced cells. On the other hand, the level of expression is sufficiently high to lead to a much more prominent

CRISPR interference/adaptation response than that observed in the *hns* mutant. The latter strain also does not allow one to control the *cas* gene expression level and has various pleiotropic effects due to derepression of multiple genes that are normally repressed by H-NS.

To study the CRISPR interference response directed to a specific mobile genetic element, one needs, in addition to sufficient levels of *cas* gene expression, a CRISPR cassette with a spacer targeting a protospacer in foreign DNA. In practice, one has to test multiple different spacers to observe strong interference, for the magnitude of CRISPR interference appears to vary from one spacer to another. The reasons for this spacer to spacer variation are not yet understood, and may involve not just the strength of crRNA-Cascade interaction with target DNA but also the location of the protospacer in the phage or plasmid genome. Until recently, specific "targeting" *E. coli* strains had to be created by genetic engineering of genomic CRISPR cassettes using a modified version of the recombineering technique [5]. For various reasons (mostly having to do with the presence of multiple identical repeats in the cassette) the procedure for engineering expanded CRISPR cassettes has been technically challenging, time-consuming, and did not lend itself to multiplexing.

While studying the CRISPR adaptation process in *E. coli* we discovered a curious phenomenon that we refer to as "priming." We observed that if foreign DNA contained a mutated protospacer that rendered CRISPR interference inactive (due to a single-nucleotide substitution in the PAM or the seed region, *see* ref. 6), very efficient acquisition of additional spacers derived from DNA located *in cis* with respect to the mutated protospacer is observed [5]. The molecular mechanism of the priming phenomenon is not understood yet; however, it is clear that it must be biologically significant, for it provides molecular "memory" of prior encounters with foreign DNA that allows specific adaptive response to targets that managed to escape initial interference. The priming phenomenon is very useful for facile construction of new strains targeting various regions of phage and/or plasmid DNA. Such strains can be used to study the molecular details of CRISPR-Cas function directed against various mobile genetic elements. A general procedure to use priming-mediated construction of targeting *E. coli* strains is described in this chapter.

Priming-mediated strain construction involves creation of a plasmid containing an escape protospacer targeted by a resident CRISPR spacer and a fragment of DNA that needs to be targeted. Following transformation and induction of CRISPR adaptation, clones that have lost the plasmid and expanded their CRISPR cassettes are identified. Upon sequencing, clones that acquired spacers from the DNA of interest are identified and used for downstream applications.

## 2 Materials

### 2.1 Strains and Plasmids

1. The KD263 *E. coli* strain [7] contains the *cas3* gene under the control of *lac*UV5 promoter and the *casABCDE12* operon under the control of *ara*Bp8 promoter (Fig. 1). The strain has been additionally modified to remove the CRISPR II cassette and the CRISPR I cassette has been reduced to a minimum, with just a single repeat-spacer-repeat unit. The spacer, referred to as g8, is derived from bacteriophage M13 (*see* ref. 6). Similar strains containing both *E. coli* K12 CRISPR cassettes with CRISPR I expanded (or not) by a single repeat-spacer unit containing the g8 spacer are also available (BW40119 and BW40114, respectively, ref. 5). The strains are available from the authors upon request.

2. The starting plasmid is a pT7blue vector (EMD Millipore, USA) with a DNA fragment corresponding to one of the resident CRISPR spacers cloned into the *Eco*RV site. The plasmid-borne protospacer should either carry a single-nucleotide substitution in the PAM or in the seed (as in the example below) to render CRISPR-Cas interference inactive. When using strains other than KD263 or BW40119, for example, BW40114, which contains inducible *cas* genes and natural CRISPR I and CRISPR II cassettes [5], *hns* mutants [6], or cells expressing *cas* genes and pre-crRNA from plasmids [4], plasmids harboring a priming protospacer matching one of the spacers in genomic CRISPR cassette have to be created. The protocol presented below, used for construction of pT7blue plasmid with a g8 protospacer insert, works well.

### 2.2 Materials for Generating a Priming Protospacer Plasmid

1. Primed protospacer oligonucleotides: pg8_F: 5'-<u>ATG</u>**T**TGTC TTTCGCTGCTGAGGGTGACGATCCCGC-3' and pg8_R: 5'-GCGGGATCGTCACCCTCAGCAGCGAAA GACAA<u>CAT</u>-3' (the substitution at the seed region (C1T, highlighted in bold typeface) and functional ATG PAM sequence (underlined) are shown).

2. Oligonucleotide buffer: 10 mM Tris–HCl pH 7.5, 100 mM NaCl.

3. Heat block or water bath.

4. Nuclease-free water.

5. End Conversion Mix (Perfectly Blunt® Cloning Kit from the EMD Millipore) or T4 Polynucleotide Kinase (PNK from New England BioLabs® Inc.)

6. T4 DNA ligase and ligase buffer (New England BioLabs® Inc.).

7. pT7blue blunt-end vector (Perfectly Blunt® Cloning Kit).

8. Competent *E. coli* cells (e.g., NovaBlue Singles™ from EMD Millipore).

9. Ice-cold 0.1 M CaCl$_2$ solution.
10. Liquid nitrogen.
11. LB plates containing 50 μg/mL carbenicillin or 50 μg/mL ampicillin, 80 μM Isopropyl-β-D-thiogalactopyranoside (IPTG) and 70 μg/mL 5-bromo-4-chloro-3-indolyl-β-D-galactopyranoside (X-gal).
12. Commercial plasmid purification kit.

### 2.3 Materials for Generating a Capture Plasmid Harboring an Escape Protospacer and a Fragment of Phage DNA

1. Oligonucleotides for T4 phage insert: T4B_7-F: 5′-TTTTT<u>GGATCC</u>GCGACTTTACCAGCGAATG-3′ and T4B_7-R: 5′-TTTTT<u>GAGCTC</u>GGTAATGCAGCTTCAGGAAAA-3′ with *Bam*HI and *Sac*I restriction sites (underlined).
2. Oligonucleotide buffer: 10 mM Tris–HCl pH 7.5, 100 mM NaCl.
3. Heat block or water bath.
4. Nuclease-free water.
5. T4 DNA ligase and ligase buffer (New England BioLabs® Inc.).
6. pT7blue_G8 plasmid containing the g8 protospacer with substitution at the seed region with sticky-ends.
7. Competent *E. coli* cells (e.g., NovaBlue Singles™ from EMD Millipore).
8. Ice-cold 0.1 M CaCl$_2$ solution.
9. Liquid nitrogen.
10. LB plates containing 50 μg/mL carbenicillin or 50 μg/mL ampicillin (*see* **Note 4**).
11. Commercial plasmid purification and PCR purification/gel extraction kits.

### 2.4 Materials for Primed Adaptation

1. Chemically competent cells of the targeting strain KD263 (K-12 F+, *lac*UV5-*cas3 ara*Bp8-*casA*, CRISPR I: repeat-spacer g8-repeat, CRISPR II deleted; sequence of the g8 protospacer: 5′-CTGTCTTTCGCTGCTGAGGGTGACGATCCCGC-3′).
2. Heat block or water bath.
3. Nuclease-free water.
4. Ice-cold 0.1 M CaCl$_2$ solution.
5. Liquid nitrogen.
6. LB plates containing 50 μg/mL carbenicillin or 50 μg/mL ampicillin (*see* **Note 4**).
7. Sterile Difco Luria broth (LB).
8. Overnight 5 mL broth culture of the transformed bacterium grown in LB medium supplemented with 50 μg/mL carbenicillin or 50 μg/mL ampicillin.

9. 1 M IPTG.
10. 10 % (0.67 M) arabinose.
11. Overnight 5 mL broth culture of bacterium grown in LB medium supplemented with 1 mM IPTG, 1 mM arabinose.
12. Oligonucleotides amplifying promoter–g8_spacer for check CRISPR expansion: Ec_LDR-F: 5′-AAGGTTGGTGGGTT GTTTTTATGG-3′, M13_g8: 5′-GGATCGTCACCCTCAG CAGCG-3′.

## 3 Methods

### 3.1 Creating a Plasmid Harboring a Priming Protospacer

A plasmid carrying a mutant protospacer is constructed by cloning a synthetic oligonucleotide duplex in the pT7blue vector. In the example provided below, two complementary oligonucleotides (pg8_F and pg8_R) whose sequences correspond to the g8 spacer [6] with a substitution at the seed region (C1T, highlighted in bold typeface) and functional ATG PAM sequence (underlined) are used (*see* Subheading 2.2, **item 1**).

#### 3.1.1 Preparation of Synthetic Priming Protospacer Duplex and Its Ligation into a Plasmid

1. Combine equimolar amounts of each oligonucleotide (100 pmol) in 10 µl of oligonucleotide buffer, incubate for 3 min at 95 °C in a heat block or water bath and then transfer to room temperature and leave for 1 h to anneal.

2. Dilute annealed oligonucleotides 100-fold with nuclease-free water and mix 1 µL of the duplex (0.1 pmol) with 4 µL of water and 5 µL of End Conversion Mix. Incubate the mixture at 22 °C for 15 min. Incubate reactions at 75 °C for 5 min and briefly cool on ice for 2 min. Alternatively, the duplex can be phosphorylated using T4 PNK. In this case, perform the reaction in 10 µL of 1× ligase buffer, at 37 °C for 30 min and inactivate the enzyme by heating at 65 °C for 20 min (*see* **Note 1**).

3. For ligation, add 1 µL of the pT7blue blunt-end vector and 1 µL of T4 DNA ligase directly to 10 µL of phosphorylated duplex obtained in the previous step. Incubate at 22 °C for 30 min. An insert–vector molar ratio under these conditions is 3:1 and gives a good yield of recombinant plasmids containing single-copy inserts. The blunt-end vector can also be prepared in house from intact pT7blue plasmid using appropriate standard procedures (digestion with EcoRV and dephosphorylation).

4. Transform the ligation mixture into competent *E. coli*. Ready-to-use competent *E. coli* cells can be obtained from various suppliers. Competent *E. coli* cells can also be prepared in house (*see* Subheading 3.1.2) and transformation is performed as described in Subheading 3.1.3.

### 3.1.2 Preparation of Chemically Competent E. coli Cells

1. Streak an appropriate *E. coli* strain (such as DH5α or similar) on an LB agar plate. Grow the plate overnight at 37 °C.
2. Inoculate a single colony from the plate into 3 mL of liquid LB medium. Grow cells at 37 °C with agitation overnight.
3. Add 1 mL of overnight culture to 100 mL liquid LB and incubate at 37 °C with agitation until $OD_{600}$ reaches 0.5. Chill the culture on ice for 10 min. It is important to keep the cells at 4 °C for the rest of the procedure. Collect the cells by centrifuging at $3,000 \times g$ for 10 min at 4 °C. Pour off the supernatant and gently resuspend the cell pellet in 50 mL of ice-cold 0.1 M $CaCl_2$ solution. Keep cells on ice for 30 min.
4. Centrifuge at $3,000 \times g$ for 10 min at 4 °C and resuspend the cell pellet in 10 mL of ice-cold 0.1 M $CaCl_2$. Incubate on ice for 30 min.
5. Repeat the centrifugation step and resuspend cells in 0.5 mL of ice-cold 0.1 M $CaCl_2$. Incubate on ice for 60 min.
6. Distribute 50 μL aliquots into sterile prechilled 1.5 mL tubes, and flash-freeze in liquid nitrogen. Store at –80 °C.

### 3.1.3 Transformation of Competent E. coli

When using commercial competent cells, follow the protocol provided by the supplier. When using in-house cells:

1. Add 1–2 μL of ligation reaction (above) or solution containing ~1 ng of pure plasmid to a 50 μL aliquot of thawed ice-cold competent cells.
2. Incubate the mixture on ice for 5 min, heat-shock at 42 °C for 30 s, and return immediately on ice for 2 min.
3. Add 250 μL of pre-warmed (37 °C) liquid LB and incubate for 30–60 min at 37 °C with agitation (*see* **Note 2**).
4. Plate 10 and 100 μL on two separate LB agar plates containing 50 μg/mL carbenicillin or 50 μg/mL ampicillin (*see* **Note 3**), 80 μM IPTG and 70 μg/mL X-gal (*see* **Note 4**). Incubate the plates overnight at 37 °C.
5. Pick up several white colonies that are likely to contain an insert using a sterile toothpick, inoculate 3 mL of liquid LB containing 50 μg/mL ampicillin, grow overnight and prepare plasmid DNA using a suitable commercial plasmid purification kit.
6. Confirm the presence of insert by sequencing.

A plasmid created in this way using the g8 protospacer oligonucleotides has been named pT7blue_G8. The orientation of the protospacer (defined by the location of the ATG PAM) matches the direction of the *lacZα* peptide transcription (Fig. 1). However, plasmids containing either orientation of the protospacer insert can be used.

## 3.2 Creating a Capture Plasmid Harboring an Escape Protospacer and a Fragment of Phage DNA

Once a plasmid containing a priming protospacer is created, it is used to construct a plasmid containing a fragment of DNA from a phage or another plasmid one is interested in. A fragment of phage DNA is amplified using appropriate primers and cloned into the priming protospacer plasmid using standard procedures. The procedure presented below was used to create a pT4acq plasmid containing a fragment of bacteriophage T4 DNA.

### 3.2.1 Preparation of Phage Genomic DNA

While in most cases PCR amplification directly from cell lysates obtained after phage infection can by successfully used, we find that longer fragments of phage DNA (5 kbp or more) are best amplified from purified phage genomic DNA. A method described in Lee and Rasheed [8] developed for plasmid purification works well for most phage lysates.

1. Take 450 µL of phage lysate (a titer of $10^9$ plaque forming units/mL or higher), add 50 µL of 0.5 M EDTA, and 400 µL buffer-saturated phenol pH 8. Vortex for 10 s at the highest vortex settings. Wear gloves as phenol may leak.

2. Centrifuge on a microfuge at highest speed for 10 min at room temperature and transfer the upper aqueous phase into a fresh 1.5 mL microcentrifuge tube.

3. Add 400 µL of phenol-chloroform (50:50) solution and vortex/centrifuge as above.

4. Transfer the aqueous phase to a fresh microcentrifuge tube and extract with 400 µL chloroform as above.

5. Add 1 mL of ice-cold 95 % ethanol and precipitate phage DNA by incubating at −20 °C for 15 min.

6. Centrifuge on a microfuge at highest speed for 10 min at room temperature, discard the supernatant.

7. Wash the precipitate, which contains phage DNA with ice-cold 70 % ethanol.

8. Air-dry the pellet and dissolve in 100 µL of distilled water.

### 3.2.2 Amplification of Phage DNA Fragment for Cloning

1. Design a primer pair to amplify a 1–5 kbp fragment of phage DNA. In the example given here primers T4B_7–F (5′-TTTTTGGATCCGCGACTTTACCAGCGAATG-3′), and T4S_7–R (5′-TTTTTGAGCTCGGTAATGCAGCTTCAGGAAAA-3′) were used to amplify a 2,991 bp fragment of bacteriophage T4 DNA between genomic positions 83,951 and 86,941. The amplified fragment contains the entire T4 gene *7* and part of gene *8* (both genes encode T4 virion proteins) (*see* **Note 5**). T4B_7–F and T4S_7–R primers contain, correspondingly, engineered *Bam*HI and *Sac*I recognition sites (underlined). The amplified T4 fragment does not contain the recognition sites for these restriction endonucleases. When designing your primer pairs, different restriction endonuclease

sites can be introduced into oligonucleotide primers; however, the amplified phage DNA fragment shall not contain the recognition sites for these enzymes. After PCR amplification, the amplified phage DNA fragment was purified from an agarose gel, treated with *Bam*HI and *Sac*I and then repurified using PCR purification kit. The T4 fragment was cloned between the *Bam*HI and *Sac*I sites located in the polylinker of pT7blue_G8 plasmid that has been digested with *Bam*HI and *Sac*I. The resulting plasmid was named pT4acq. It allows one to capture T4-insert derived spacers as described below.

### 3.3 Primed Adaptation

Generally, the chance of obtaining an *E. coli* strain with a CRISPR spacer derived by primed adaption from phage DNA segment of a spacer capture plasmid increases together with the increase of the size of phage insert. We routinely use fragments of 1–5 kbp. Since the size of pT7blue is 2,922 bp, screening of strains that acquired spacers from phage becomes very easy. The procedure presented below is based on an example with the pT4acq plasmid and the KD263 cells.

#### 3.3.1 The Adaptation Experiment

1. Transform competent KD263 cells with spacer capture plasmid pT4acq and select transformants on LB agar plates containing 50 µg/mL carbenicillin or ampicillin.

2. Pick up an individual colony and grow overnight in 5 mL of liquid LB in the presence of carbenicillin or ampicillin at 37 °C.

3. Transfer 0.1 mL of overnight culture into a 5 mL of liquid LB and after 1 h of growth in incubator shaker at 37 °C induce the culture by the addition of 1 mM arabinose and 1 mM IPTG. Allow the culture to grow for 8 h or overnight.

4. Plate aliquots of serial dilutions of induced culture on LB agar plates to obtain individual colonies.

5. Select a dozen individual colonies and plate each using toothpicks on LB agar plates with and without ampicillin. Be sure to properly label the plates such that matching clones on both plates can be identified.

6. After overnight growth at 37 °C, identify clones that have lost ampicillin resistance (usually, less than 10 % of colonies remain ampicillin-resistant, and, therefore, harbor the plasmid after growth in the presence of inducers).

#### 3.3.2 Screening for CRISPR Cassette Expansion

1. Use colony PCR to screen ampicillin-sensitive colonies for CRISPR cassette expansion with Ec_LDR-F (5′-AAGGTTGG TGGGTTGTTTTTATGG-3′) and M13_g8 (5′-GGATCGTC ACCCTCAGCAGCG-3′) primers. The latter primer anneals to the g8 spacer. A different primer has to be designed when looking for expansion of CRISPR cassettes using capture

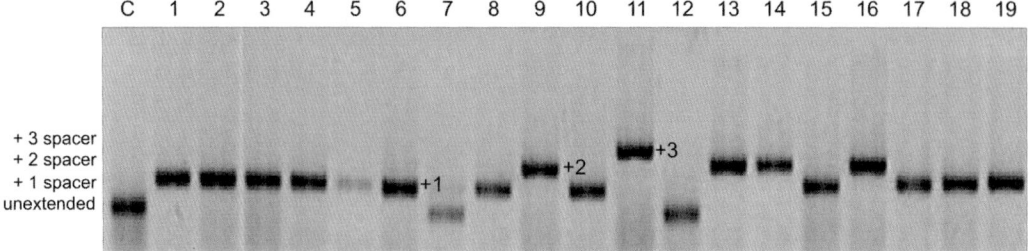

**Fig. 2** Screening for CRISPR cassette expansion. An agarose gel of the results of analysis of 18 clones obtained from an induced culture of pT4acq-containing KD263 *E. coli* that lost ampicillin-resistance is shown. Lane labeled "C" is a control lane (amplification of CRISPR cassette from KD263 *E. coli* that has not been transformed with any plasmid). Clones that have undergone cassette expansion by 1, 2, or 3 repeat-spacer units are indicated. Clones containing a single new spacer were sequenced and are summarized in Table 1

plasmids harboring priming protospacers corresponding to different CRISPR cassette spacers. Primer Ec_LDR-F anneals to the leader sequence of *E. coli* CRISPR I cassette. A different primer is needed when looking for expansion of CRISPR II cassette (recall that this cassette is absent in KD263). A very good protocol for colony PCR can be found in the pET system manual (http://lifeserv.bgu.ac.il/wb/zarivach/media/protocols/Novagen%20pET%20system%20manual.pdf). Results of a typical experiment obtained with cells that have lost pT4acq are shown in Fig. 2.

2. Identify clones containing CRISPR cassettes expanded by a single repeat-spacer unit (Fig. 2) and subject several amplified expanded cassette DNA fragments to sequencing to identify phage-derived spacers (*see* **Note 6**) A typical result is presented in Table 1.

### 3.3.3 Phage Sensitivity Test

The sensitivity of cells that acquired distinct phage-derived spacers to phage infection is easily determined by a spot test method.

1. Grow selected clones that acquired phage-derived spacers in LB supplemented with 1 mM arabinose and 1 mM IPTG at 37 °C until $OD_{600}$ reaches 0.5–0.7.

2. Overlay the LB agar plate containing 1 mM arabinose and 1 mM IPTG with 5 mL of melted soft (0.75 %) agar containing 100 μL of cell culture. The soft agar temperature should be no higher than 45 °C when the cells are added.

3. After solidification for 5 min, deposit 5 μL drops of 100-fold serial dilutions of phage lysate on the soft agar surface.

4. The serial dilutions of phage lysate are created in sterile 1.5 mL microcentrifuge tubes clearly marked ("$10^{-2}$", "$10^{-4}$" etc.) by adding 10 μL of stock phage lysate of known titer to 990 μL of liquid LB, mixing, then withdrawing 10 μL and placing that

## Table 1
### Sequence analysis of spacers acquired during primed acquisition experiment with the pT4acq plasmid

| No. | Protospacer sequence (5′–3′) | PAM sequence | Phage/Plasmid | Strand | Protospacer position |
|---|---|---|---|---|---|
| 1 | CCATACCAAACGACGAGCGTGACACCACGATG | AAG | pT7blue vector | nP | 1378 |
| 2 | TATATATGAGTAAACTTGGTCTGACAGTTACC | AAG | pT7blue vector | P | 1759 |
| 3 | TTGGCCGCAGTGTTATCACTCATGGTTATGGC | AAG | pT7blue vector | P | 1272 |
| 4 | TCATTCTGAGAATAGTGTATGCGGCGACCGAG | AAG | pT7blue vector | P | 1167 |
| 5 | TGCTCATCATTGGAAAACGTTCTTCGGGGCGA | AAG | pT7blue vector | P | 1076 |
| 6 | AAAGAAGACGTATTCAACCCGGATATGCGAAT | AAG | T4 | nP | 84327 |
| 8 | ACCCGACTAGATGGGGATATGAAGATAATCTC | AAG | T4 | nP | 86806 |
| 10 | GAACCACGATATATTCATTCGTGCATCTATTT | AAG | T4 | P | 86781 |
| 15 | ATGCTATTGAACACATTCCGGTATCAGGAACA | AAG | T4 | P | 86586 |
| 17 | GGAGCTGAGTTACACACTACAATATCGTTAAT | TAA | T4 | P | 86503 |
| 18 | CAAATCCTTTCCTTTAACCCCACGAATAATTT | AAG | T4 | P | 85600 |
| 19 | ATAACACTTGAATCATTCATCTATTTTAACCT | TAG | T4 | P | 86170 |

The table presents the results of analysis of 12 clones containing CRISPR cassettes expanded by a single repeat-spacer unit (*see* Fig. 2). For each spacer, the sequence, the PAM of the corresponding protospacer (shown 5′–3′ on non-target strand), the source of the spacer (plasmid or T4 insert part of the pT4acq plasmid) the strand ("P" being primed strand, "nP"—non-primed strand) and the location of the PAM proximal protospacer base (in pT7 blue vector or T4 phage genome) is shown

into 990 μL of liquid LB and so on up to a $10^{-8}/10^{-10}$ dilution (5 μL of the last dilution should contain an estimated 10–100 plaque forming units).

5. Allow the drops containing phage stock dilutions to dry and incubate at 37 °C overnight.

6. Determine efficiency of plaquing (e.o.p.) as a ratio of most dilute phage titers where individual phage plaques are observed on lawns of cells expressing phage targeting crRNA and nontargeting control cell lawns (*see* **Note** 7). An e.o.p. of 1 indicates that there is no interference. An e.o.p. around $10^{-4}$–$10^{-5}$ can be attained with some crRNAs/phage pairs. Intermediate e.o.p. values (ca. $10^{-2}$ are also often observed). Results of a representative phage sensitivity spot test experiment are shown in Fig. 3.

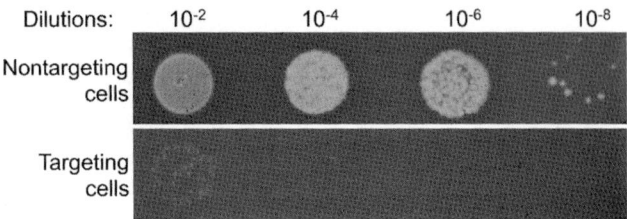

**Fig. 3** Phage susceptibility test. Results of phage-titering on two plates seeded with lawns of KD263 cells expressing a phage-targeting crRNA from a spacer acquired during the adaptation experiment (*bottom*) or a control, non-targeting crRNA (*top*) are shown. The dilutions of phage stock are idnicated above. The images have been contrasted to appear phage plaques look white on a dark background of bacterial lawns

## 4 Conclusions

We have successfully used the procedure described above to obtain numerous *E. coli* strains containing CRISPR spacers targeting multiple bacteriophages and plasmids. The availability of such strains opens up new interesting areas of research, for example, allowing one to determine the requirements for efficient CRISPR-Cas interference besides a spacer-protospacer match and the presence of a functional PAM sequence. The procedure can be made even more powerful by using libraries of spacer capture plasmids containing the entire genome of a phage or an episome being studied. Various more specialized applications, such as creating *E. coli* strains capable of targeting of specific regions of their own DNA and thus leading to defined genomic lesions are also possible.

## 5 Notes

1. The kinase must be inactivated before the ligation to avoid high vector background due to vector phosphorylation and subsequent self-ligation. The procedure described above is only needed if non-phosphorylated oligonucleotides were purchased from the vendor and is not necessary if 5′-end phosphorylated oligonucleotides were used to create the duplex.

2. It is not necessary to incubate competent cells for the outgrowth period after heat shock for ampicillin-resistant plasmids; however, this step may increase cloning efficiency.

3. Carbenicillin is preferred as a selection agent for clones containing ampicillin-resistant plasmids as it is more stable than ampicillin and so less secondary-growth "satellite" colonies that do not contain a plasmid are formed.

4. There is no need to add IPTG and X-gal when making a spacer capture plasmid or transforming KD263 competent cells with an already prepared spacer capture plasmid, below.

5. Some phage-encoded proteins are highly toxic. When creating a spacer capture plasmid try to avoid phage DNA fragments containing intact phage genes known to encode toxic proteins.

6. While the procedure described above is very efficient, not all phage-derived spacers may lead to efficient interference. To increase the proportion of cells harboring interference-capable spacers, one can infect cells after growth in the presence of inducers with a phage of interest at a high multiplicity of infection, collect surviving cells, restreak for single colonies on an LB agar plate and identify clones with expanded cassettes as described above.

7. Induced cell cultures can be prepared in advance stored at 4 °C for up for a week and used for phage susceptibility tests.

## Acknowledgments

This work was supported by National Institutes of Health grant R01 GM10407, Molecular and Cell Biology Program grant from the Russian Academy of Sciences Presidium, and Ministry of Education and Science of Russian Federation project 14.B25.31.0004 and Russian Science Foundation grant 14-14-00988 (to KS) and Russian Foundation for Basic Research grant 14-04-00916 (to Ekaterina Savitskaya).

## References

1. Pul U, Wurm R, Arslan Z, Geissen R, Hofmann N, Wagner R (2010) Identification and characterization of E. coli CRISPR-cas promoters and their silencing by H-NS. Mol Microbiol 75:1495–1512
2. Pougach K, Semenova E, Bogdanova E, Datsenko KA, Djordjevic M, Wanner BL, Severinov K (2010) Transcription, processing and function of CRISPR cassettes in Escherichia coli. Mol Microbiol 77:1367–1379
3. Westra ER, Pul U, Heidrich N, Jore MM, Lundgren M, Stratmann T, Wurm R, Raine A, Mescher M, Van Heereveld L, Mastop M, Wagner EG, Schnetz K, Van Der Oost J, Wagner R, Brouns SJ (2010) H-NS-mediated repression of CRISPR-based immunity in Escherichia coli K12 can be relieved by the transcription activator LeuO. Mol Microbiol 77:1380–1393
4. Brouns SJ, Jore MM, Lundgren M, Westra ER, Slijkhuis RJ, Snijders AP, Dickman MJ, Makarova KS, Koonin EV, van der Oost J (2008) Small CRISPR RNAs guide antiviral defense in prokaryotes. Science 321:960–964
5. Datsenko KA, Pougach K, Tikhonov A, Wanner BL, Severinov K, Semenova E (2012) Molecular memory of prior infections activates the CRISPR/Cas adaptive bacterial immunity system. Nat Commun 3:945
6. Semenova E, Jore MM, Datsenko KA, Semenova A, Westra ER, Wanner B, van der Oost J, Brouns SJ, Severinov K (2011) Interference by clustered regularly interspaced short palindromic repeat (CRISPR) RNA is governed by a seed sequence. Proc Natl Acad Sci U S A 108:10098–10103
7. Shmakov S, Savitskaya E, Semenova E, Datsenko KA, Severinov K (2014) Pervasive generation of oppositely-oriented spacers during CRISPR adaptation. Nucleic Acids Res 42:5907–5916
8. Lee SY, Rasheed S (1990) A simple procedure for maximum yield of high-quality plasmid DNA. Biotechniques 9:676–679

# Chapter 10

## Exploring CRISPR Interference by Transformation with Plasmid Mixtures: Identification of Target Interference Motifs in *Escherichia coli*

Cristóbal Almendros and Francisco J.M. Mojica

### Abstract

Plasmid transformation into a bacterial host harboring a functional CRISPR-Cas system targeting a sequence in the transforming molecule can be specifically hindered by CRISPR-mediated interference. In this case, measurements of transformation efficacy will provide an estimation of CRISPR activity. However, in order to standardize data of conventional assays (using a single plasmid in the input DNA sample), transformation efficiencies have to be compared to those obtained for a reference molecule in independent experiments. Here we describe the use of a transforming mixture of plasmids that includes the non-targeted vector as an internal reference to obtain normalized data which are unbiased by empirical variations.

**Key words** CRISPR-Cas systems, Electroporation, Plasmid transformation, Interference efficiency, *Escherichia coli*, Target interference motif, Protospacer adjacent motif

## 1 Introduction

CRISPR-Cas systems typically cleave DNAs complementary to repeat-intervening spacers that derive from sequences known as protospacers (for a comprehensive introduction on the CRISPR-Cas systems *see* ref. 1). Owing to this spacer-guided cleavage, CRISPR-Cas may lead to target degradation, interfering with the spread of transmissible genetic elements carrying the corresponding protospacer. Moreover, sequences partially matching the protospacer could also be targeted as a few mismatches with the spacer sequence are tolerated. However, the efficiency of interference greatly depends on the extent and location of the complementary bases and, moreover, most systems require the presence of specific motifs, generally named Protospacer Adjacent Motifs (PAMs) or more specifically TIMs for Target Interference Motifs, flanking the target [2]. Additional factors related to, for example, the genetic background and sequence, or perhaps the structure, of the DNA elements involved could also affect CRISPR activity [3]. The assessment of factors affecting interference is essential for the functional characterization of each particular CRISPR-Cas system. This sort of analysis can be performed by alternative approaches depending on the identity of the target molecule, as for example (1) cell lysis or lysogenization in the case of viruses, (2) host viability for chromosomal targets, and (3) conjugation or transformation efficiencies when dealing with plasmids. Among them, artificial transformation of plasmids stands as a feasible and versatile option to monitor CRISPR activity against diverse cloned targets under assorted circumstances.

Transformation efficiencies are commonly estimated from transformation reactions performed with input DNA samples containing a single plasmid. Then, the number of cells (inferred from the number of colony forming units) that have acquired that plasmid (transformants) per microgram of input DNA has to be normalized with respect to equivalent data obtained in an independent experiment performed with a reference plasmid. However, empirical conditions of separate experiments may greatly vary, notably when using different batches of competent cells [4–6]. Furthermore, offhand procedural differences may have significant effects on transformability, resulting in inconsistent transformation efficiencies. These variations can be particularly relevant when the interference activities to be compared are similar. In an alternative approach that circumvents the bias of transformation variables, cells are subjected to transformation with a mixture, rather than with a single plasmid, composed of one or more targeted plasmids along with the non-targeted vector [7]. For each experiment, the relative number of transformants carrying a given target plasmid will provide an estimation of interference against it (i.e., reduced numbers of transformants imply more effective interference). Moreover, these data can be normalized respect to the non-targeted vector serving as an internal reference to correlate separate assays performed with different accompanying plasmids or in different strains or conditions.

Here we describe an electrotransformation procedure devised to assess interference against input targeted artificial plasmids in a natural *Escherichia coli* isolate (strain LF82) [8] harboring a constitutively active CRISPR-Cas system [7]. Each transforming DNA sample consists of a mixture of two plasmids, one carrying a protospacer identical to a CRISPR spacer of the host (target plasmid) and the other being the non-targeted vector (internal control). Different mixtures vary in the sequence flanking one end of the protospacer (predicted TIM positions) in the target plasmid. Freshly prepared electrocompetent cells are electrotransformed with each mixture. The proportion of colonies transformed with the target plasmid with respect to those that have incorporated the reference vector allow for the identification of TIMs with diverse susceptibility to CRISPR-mediated interference.

Variations of this methodology could also be conceived to evaluate activity of a genomic CRISPR-Cas system in any competent bacterial or archaeal species, against distinct protospacers carried by diverse plasmids and under different genetic or environmental conditions. Furthermore, immunity against genomic protospacers driven by CRISPR systems carried in transforming plasmids could also be addressed.

## 2 Materials

Growth media and sterile solutions are sterilized by autoclaving for 20 min at 121 °C and 15 lb/sq.in. of pressure, unless otherwise specified.

1. *Escherichia coli* strain LF82 [8].
2. Control vector: pCR2.1® (Invitrogen).
3. Target plasmids: a set of 13 pCR2.1-derived recombinant plasmids. All of them carry a sequence identical to a particular CRISPR spacer of LF82 strain but differ in the PAM region [7].
4. Pure water: Elix water (Millipore), resistivity 15 MΩ-cm at 25 °C.
5. Ultrapure water: Milli-Q water (Millipore), resistivity 18.2 MΩ-cm at 25 °C.
6. Luria–Bertani (LB) broth: 10 g Bacto-Tryptone, 5 g yeast extract, 10 g NaCl/L, pH 7.0.
7. Kanamycin stock solution (50 mg/mL). Sterilize by filtration through a 0.22-μm filter. Store at −20 °C.
8. LB agar medium with kanamycin: LB broth containing 16 g/L agar and 50 μg/mL kanamycin. After autoclaving LB agar medium, allow it to cool to 60 °C and add kanamycin stock solution.
9. Sterile Petri dishes.
10. High Pure Plasmid Isolation Kit (Roche).
11. Refrigerated centrifuge: Labofuge 400R (Heraeus Instruments).
12. Incubator set at 37 °C.
13. NanoDrop ND-1000 (NanoDrop Technologies).
14. *Bam*HI restriction enzyme (Roche).
15. Supercoiled DNA Ladder (Invitrogen).
16. High DNA Mass Ladder (Invitrogen).

17. 6× DNA loading buffer: 30 % glycerol, 0.25 % xylene cyanol FF, 0.25 % bromophenol blue. Store at 4 °C.
18. Electrophoresis grade agarose: SeaKem LE Agarose (Lonza).
19. 10× TBE: 1 M Tris–HCl, 0.9 M boric acid, 10 mM EDTA pH 8.0.
20. Horizontal electrophoresis equipment and accessories.
21. Ethidium bromide (EtBr) stock solution: 10 mg/mL in ultrapure water.
22. Imaging system: Typhoon 9410 (GE Healthcare Life Sciences).
23. Gel analysis software: TotalLab 1D Software (TotalLab Ltd.).
24. Rotatory shaker: C25 Incubator Shaker (New Brunswick Scientific).
25. UV/Vis spectrophotometer: Ultraspec 2000 (Pharmacia Biotech).
26. 10- and 50-mL polypropylene tubes.
27. LB-TA: LB broth containing 0.01 % Tween 80 and 3 % potassium acetate.
28. Sterile 10 % glycerol solution in pure water.
29. Vortex.
30. Microfuge tubes (1.5-mL capacity).
31. Sterile Pasteur pipettes.
32. Sterile electroporation cuvettes with 2-mm gap, 400-μL capacity (Eppendorf).
33. Electroporator device: Electroporator 2510 (Eppendorf).
34. SOC medium: 20 g Bacto-Tryptone, 5 g yeast extract, 0.5 g NaCl, 2.5 mM KCl, 10 mM $MgCl_2$, 20 mM glucose/L, pH 7.0. Glucose (sterilized by filtration through a 0.22-μm filter) and $MgCl_2$ (autoclave-sterilized) are added to the room temperature medium after autoclaving.
35. Saline solution: sterile 0.9 % NaCl in pure water.
36. Sterile PCR plastic tubes (0.2-mL capacity).
37. PCR reagents: 5 mM dNTPS (1.25 mM each), 10× Taq Polymerase Buffer with 15 mM $MgCl_2$, 5 U/μL Taq Polymerase (Roche), 10 μM M13 forward primer, and 10 μM M13 reverse primer (these primers anneal to regions flanking the plasmid cloning site).
38. PCR thermal cycler: Mastercycler Gradient (Eppendorf).
39. 1 kb Plus DNA Ladder (Invitrogen).
40. UV transilluminator and camera: UVIdoc system (UVItec).

# 3 Methods

## 3.1 Plasmid Purification and Quantification

1. Transform control vector and each target plasmid separately into *Escherichia coli* strain LF82 (*see* **Note 1**). Use either the electrotransformation procedure described in Subheading 3.2 or any alternative method of transformation working with the strain under study. Transformation efficiency is not crucial at this stage.

2. Select transformed colonies by growth on LB agar medium containing kanamycin (LF82 strain is sensitive to kanamycin and the plasmids used encode a kanamycin resistance gene).

3. Purify plasmids from transformed colonies with High Pure Plasmid Isolation Kit (*see* **Note 2**).

4. Estimate the DNA concentration and purity of plasmid solutions using a NanoDrop ND-1000 (*see* **Note 3**).

5. Digest 350 ng of each plasmid DNA with *Bam*HI (or any other single cutter restriction enzyme) to linearize the molecules.

6. Transfer 50, 100 and 150 ng of the linearized plasmids, 400 ng of untreated samples and appropriate volumes of DNA markers (High DNA Mass™ and Supercoiled DNA Ladders) to separate microfuge tubes.

7. Add 1/6 of 6× loading buffer to each tube and mix. Load samples into a 1 % agarose gel in 1× TBE buffer and run an electrophoresis with an applied field of 10 V/cm at room temperature for 3–4 h or until suitable separation is achieved.

8. Stain gel with 0.5 µg/mL EtBr for 10 min and destain in pure water for 15–30 min.

9. Digitalize the gel using an imaging system (i.e., Typhoon 9410) and analyze bands with a gel analysis software (i.e., TotalLab 1D Software).

10. Compare intensities of bands among linearized samples and with DNA Mass Ladders (containing linear DNA species of known mass) to confirm mass estimations obtained in **step 4**.

11. Identify bands of supercoiled (monomer and multimer molecules), nicked (relaxed open circular) and full length linear forms of untreated (non-digested) samples using the loaded DNA markers as reference. Determine the proportion of forms within each sample and discard those plasmid preparations with apparent discrepancies with respect to the rest. This is particularly important in the case of plasmids used in the same transforming mixture (*see* **Note 4**).

## 3.2 Electrotransformation

1. Inoculate a single colony of *E. coli* strain LF82 into a sterile 10-mL tube containing 5 mL of LB broth and incubate at 37 °C with shaking over-night.

2. Inoculate 1 mL of the over-night culture into a 250-mL flask containing 100 mL of sterile LB-TA (*see* **Note 5**) and incubate at 37 °C with vigorous shaking (200 rpm in a rotary shaker) to an $ABS_{600}$ of approx 0.5.

3. Cool on ice for 10 min.

4. Split the culture into two sterile 50-mL polypropylene tubes.

5. Pellet the cells by centrifugation at $3,500 \times g$ for 10 min at 4 °C. Decant the supernatant and resuspend the pellet of each tube in 30 mL of sterile ice-cold pure water. Repeat this step twice.

6. Harvest cells by centrifugation at $3,500 \times g$ for 10 min at 4 °C. Decant the supernatant and resuspend cell pellet of each tube in 0.25 mL of sterile ice-cold 10 % glycerol.

7. Mix by vortexing and dispense 50 μL aliquots in ice-cold sterile microfuge tubes (*see* **Note 6**).

8. Add to one of the tubes containing ice-cold electrocompetent cells 10 ng (*see* **Note 7**) of a target plasmid and the same mass of control vector, from Subheading 3.1, **step 3** (*see* **Note 8**). Choose plasmid samples with equivalent proportions of topological forms (*see* **Note 4**).

9. Mix by vortexing for 10 s and incubate for 1 min on ice.

10. Transfer the DNA-cells mixture with a sterile Pasteur pipette to the bottom of an ice-cold electroporation cuvette.

11. Set the electroporator to 2.45 kV, 25 μF, 200 Ω.

12. Place the cuvette in the electroporator and deliver a pulse. The time constant should be of approximately 5.0.

13. Immediately after the pulse, remove the cuvette from the device and add 1 mL of SOC medium at room temperature.

14. Transfer the cell suspension to a sterile 10-mL tube and incubate at 37 °C with gently shaking (approx 100 rpm) for 1 h.

15. Repeat **steps 8–14** for each target plasmid to be tested.

16. After 1 h of growth, prepare $10^{-1}$, $10^{-2}$, and $10^{-3}$ dilutions from the cultures in saline solution and plate 100 μL of each dilution onto LB agar medium containing kanamycin. Incubate plates at 37 °C over-night.

## 3.3 Screening of Transformed Colonies

The proportion of transformed colonies carrying the target plasmid with respect to those that have incorporated the control vector is estimated for each experiment by the following PCR-screening procedure.

1. Select 20 transformed colonies.
2. Resuspend each colony in a microfuge tube containing 20 µL of sterile ultrapure water.
3. Mix by pipetting up and down and transfer 1 µL to a PCR plastic tube.
4. Prepare a PCR master mix containing all reagents required for the PCR: 62.5 µL of 10× PCR buffer, 25 µL of 5 mM dNTPs, 25 µL of 10 µM reverse primer, 25 µL of 10 µM forward primer, 460 µL of sterile ultrapure water, and 2.5 µL of Taq DNA polymerase.
5. Transfer 24 µL of the master mix to (1) each of the 20 PCR tubes containing the cell suspension, (2) a PCR tube containing 1 µL of sterile ultrapure water (negative control reaction without template) and (3) two additional tubes each containing 50 ng of either target plasmid or vector (positive control reactions) in a final volume of 1 µL.
6. Place the 23 tubes in a PCR thermal cycler and run the following program: (1) 95 °C for 5 min, (2) 35 cycles of denaturation at 95 °C for 30 s, annealing at 52 °C for 30 s, and extension at 72 °C for 1 min, and (3) a final extension at 72 °C for 2 min.
7. Transfer 5 µL of each PCR reaction and an appropriate volume of 1 kb Plus DNA Ladder to microfuge tubes. Add 1 µL of 6× loading buffer to each tube, mix and load samples into a 1.5 % agarose gel in 1× TBE buffer. Run an electrophoresis with an applied field of 10 V/cm at room temperature for 2–3 h or until suitable separation is achieved.
8. Stain and destain the gel as in Subheading 3.1, **step 8,** visualize DNA bands on an ultraviolet light box and photograph.
9. Compare the bands of the PCR products with the DNA marker to estimate their approximate size and infer from them the identity of the corresponding transformed plasmid.
10. Calculate the proportion of transformants carrying the target plasmid with respect to those that have incorporated the reference vector (*see* **Note 9**).
11. Proceed as in the previous steps of this subsection for each of the transformation experiments performed (i.e., at least three independent experiments with each plasmid pair).
12. Calculate the mean value of the plasmid proportion for the replicates of each plasmid pair (*see* **Note 10**). In the absence of interference, an equal proportion of colonies harboring either plasmid would be obtained. In the case of interference against the protospacer carrier, a lesser percentage of cells transformed with this plasmid would be observed, lower as interference activity increases (*see* Fig. 1).

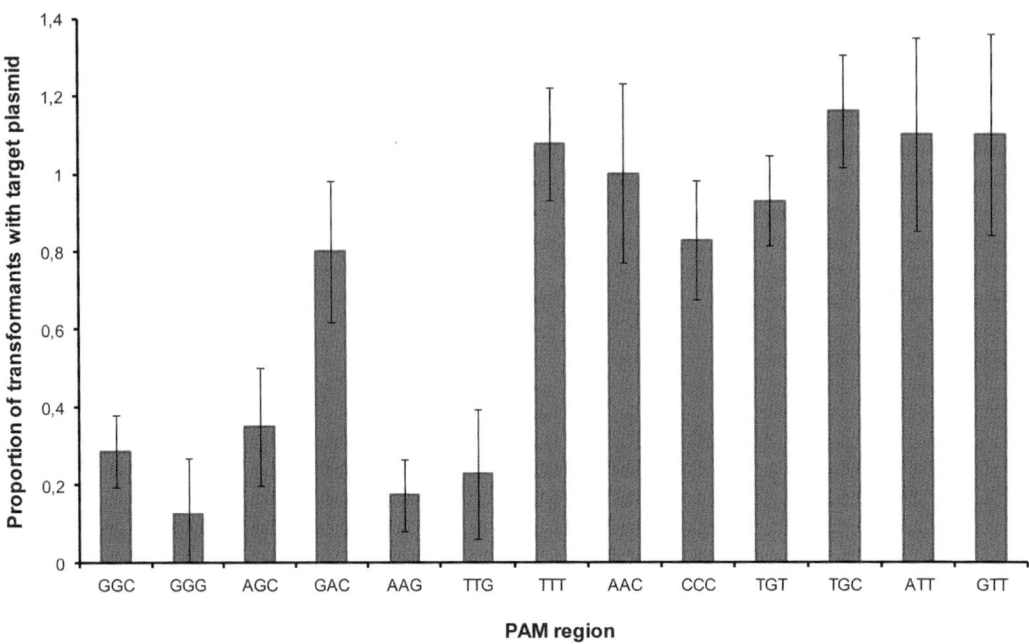

**Fig. 1** Histogram showing an example of results from transformation assays performed with plasmids mixtures as described in the text. Data correspond to the proportion of transformants carrying a target plasmid to transformant colonies carrying the control vector. Target plasmids differ in sequence at PAM positions (nucleotides indicated under the bar). The mean average from three independent transformation experiments for each target plasmid is shown with its standard deviation. Values significantly below 1 imply CRISPR interference. Five TIMs (GGC, GGG, AGC, AAG, and TTG; shown 5′–3′) were inferred from these results. Reproduced from ref. 7

## 4 Notes

1. In order to prevent restriction activity by restriction modification systems, which could potentially conceal CRISPR-mediated interference, plasmids are purified from the same strain to be used in the subsequent interference tests. We use high-copy-number plasmids to prevent complete loss of the incorporated target plasmid during growth of transformants due to CRISPR-mediated degradation. In this way, interference will be detected only when it happens soon after the plasmid gets into the cell (before it reaches a high copy number).

2. Use a method yielding high quality DNA and reproducible results. The presence of contaminating material or a low proportion of supercoiled plasmid molecules may diminish transformation efficiency [4, 5, 9]. We strongly recommend sequencing the protospacer and PAM region of the purified target plasmids to rule out the occurrence of mutations that could potentially affect CRISPR interference. Yet a low proportion of mutated sequences would not be detected by sequencing (*see* **Note 9**).

3. Calculate the DNA concentration for each plasmid preparation from the mean value of at least three measurements. Instead of with a NanoDrop, DNA quantification may be performed by using alternative devices such as a Qubit® fluorometer (Life Technologies) or a conventional UV spectrophotometer assuming 1 $ABS_{260}$ unit = 50 μg/mL. DNA samples with a ratio $ABS_{260/280}$ of about >1.8 are considered of good purity.

4. Interference activity against target plasmids could be influenced by their topology [10]. Moreover, the size (monomer versus multimers) and compactness (supercoiled versus open circular or linear) of the transforming molecule may affect transformation efficiency [9]. Thus, plasmids to be compared, notably those in the same transforming mixture, should have equivalent distribution of forms.

5. Electrocompetent *E. coli* LF82 cells obtained after growth in the commonly used LB broth yielded very low transformation efficiencies. The addition of Tween 80 and potassium acetate to LB (LB-TA) as indicated in ref. 6, greatly improves transformation efficiency.

6. Although electrocompetent cells can be stored at −70 °C for later use, immediate electroporation is preferred as we obtain reduced transformation efficiencies with long-term stored samples.

7. It is very important to use a low concentration of DNA in the cell-DNA mixture to hinder plasmid co-transformation.

8. At least three independent electrotransformation experiments should be run for each plasmid pair, using different plasmid preparations and stocks of freshly prepared electrocompetent cells.

9. Even in the case mutations had not been detected by sequencing the transforming samples (*see* **Note 2**), sequencing of the PCR products is encouraged to check for the presence of interference-resistant target mutants. Even though their occurrence is not desirable, acceptable comparisons could still be achieved taking into account the proportion of mutant colonies observed.

10. The significance of these data should be assessed by statistical analysis. If unacceptable, perform additional replicates.

# Acknowledgements

This work was supported by a grant from the Spanish Ministerio de Economía y Competitividad (BIO2011-24417). We thank Arlette Darfeuille-Michaud (Clermont Université, Université d'Auvergne, France) for strain LF82.

## References

1. Mojica FJM, Garrett RA (2013) Discovery and seminal developments in the CRISPR field. In: Barrangou R, van der Oost J (eds) CRISPR-Cas systems: RNA-mediated adaptive immunity in bacteria and archaea. Springer, Berlin
2. Shah SA, Erdmann S, Mojica FJM, Garrett RA (2013) Protospacer recognition motifs: mixed identities and functional diversity. RNA Biol 10:891–899
3. Brouns SJ, Jore MM, Lundgren M et al (2008) Small CRISPR RNAs guide antiviral defense in prokaryotes. Science 321:960–964
4. Zabarovsky ER, Winberg G (1990) High efficiency electroporation of ligated DNA into bacteria. Nucleic Acids Res 18:5912
5. Dower WJ, Miller JF, Ragsdale CW (1988) High efficiency transformation of *E. coli* by high voltage electroporation. Nucleic Acids Res 16:6127–6145
6. Shi X, Karkut T, Alting-Mees M et al (2003) Enhancing *Escherichia coli* electrotransformation competency by invoking physiological adaptations to stress and modifying membrane integrity. Anal Biochem 320:52–155
7. Almendros C, Guzmán NM, Díez-Villaseñor C et al (2012) Target motifs affecting natural immunity by a constitutive CRISPR-Cas system in *Escherichia coli*. PLoS One 7:e50797
8. Miquel S, Peyretaillade E, Claret L et al (2010) Complete genome sequence of Crohn's disease-associated adherent-invasive *E. coli* strain LF82. PLoS One 5:e12714
9. Hanahan D (1983) Studies on transformation of *Escherichia coli* with plasmids. J Mol Biol 166:557–580
10. Westra ER, van Erp PBG, Kunne T et al (2012) CRISPR immunity relies on the consecutive binding and degradation of negatively supercoiled invader DNA by Cascade and Cas3. Mol Cell 46:595–605

# Chapter 11

# Electrophoretic Mobility Shift Assay of DNA and CRISPR-Cas Ribonucleoprotein Complexes

**Tim Künne, Edze R. Westra, and Stan J.J. Brouns**

## Abstract

The Electrophoretic Mobility Shift Assay is a straightforward and inexpensive method for the determination and quantification of protein–nucleic acid interactions. It relies on the different mobility of free and protein-bound nucleic acid in a gel matrix during electrophoresis. Nucleic acid affinities of crRNA-Cas complexes can be quantified by calculating the dissociation constant ($K_d$). Here, we describe how two types of EMSA assays are performed using the Cascade ribonucleoprotein complex from *Escherichia coli* as an example.

**Key words** Electrophoretic mobility shift assay, EMSA, Gel shift, Binding assay, CRISPR, RNA guide, Affinity, Cascade, Protein–DNA interaction

## 1 Introduction

All CRISPR systems share the common feature of encoding a crRNA-guided ribonucleoprotein complex targeting complementary nucleic acids. Type I systems encode Cascade/crRNA complexes, Type II systems Cas9/crRNA complexes and Type III systems Cmr/crRNA or Csm/crRNA complexes (reviewed in ref. 1). Invader detection by these complexes is a key step of the CRISPR-dependent immune response. The binding behavior of these complexes is a key determinant of the activity and specificity of the respective systems and can reveal mechanistic features, such as the seed sequence [2–4]. Examining the binding behavior of proteins with nucleic acids can be done using various techniques, such as Surface Plasmon Resonance (Biacore), single molecule TIRF (total internal reflection microscopy), Microscale Thermophoresis, or Electrophoretic Mobility Shift Assays (EMSA) [5–7].

Usually EMSA is the method of choice, as it is a relatively straightforward and inexpensive method that generally provides robust data that is easy to interpret. EMSAs can be used for simple qualitative analysis, such as identifying target and non-target nucleic acids. Importantly, it can also be used for quantitative analysis, which can reveal binding stoichiometry and affinity [8]. To this end, protein and target are brought to binding equilibrium over a range of molar ratios and separated by gel electrophoresis. Free nucleic acid generally migrates faster through the gel than protein bound nucleic acid. This shift in migration is dependent on the bulkiness of the protein and the combination of protein pI and electrophoresis conditions.

EMSAs are easy to perform and do not require specialized equipment. A wide range of conditions can be used, as long as they are compatible with electrophoresis. Furthermore, any nucleic acid can be used as substrate as long as it can be visualized after electrophoresis; the nucleic acid size range spans from single stranded short oligonucleotides to plasmids of several thousand nucleotides. Furthermore, EMSA can be combined with footprinting analyses or competition binding experiments [9–12].

Despite these advantages of EMSA over more specialized techniques, EMSA also has some disadvantages. The main drawback of EMSAs is the fact that the chemical environment of electrophoresis differs from the environment of equilibration. Hence, the binding equilibrium can change at electrophoresis conditions. However, when the sample enters the gel matrix, interactions are usually stabilized by a caging effect, preventing or slowing down further changes [13, 14]. Still, low-affinity interactions can be lost during electrophoresis, while they are maintained in solution. This could lead to underestimation of binding affinity.

Here, we provide two different protocols for EMSA: Plasmid EMSA using agarose gel electrophoresis and short oligonucleotide EMSA using poly acrylamide gel electrophoresis (PAGE). The choice between these two protocols is determined mainly by the nature of the target nucleic acid. The use of plasmids better mimics biologically relevant conditions and allows one to address the influence of DNA topology on binding affinity. However, synthetic probes offer more experimental flexibility. Plasmids are best separated on agarose gels, while shorter nucleic acids are better separated on PAGE gels. When using intermediate sized nucleic acids, PAGE gels are preferred as they offer better resolution. Short probes usually need to be radiolabelled, since they cannot be sufficiently visualized by intercalating dyes. Here, $^{32}$P 5′ end labelling is the most widely used technique, but internal or 3′ labelling is also possible. When using larger target molecules isotope labelling is generally not required, instead standard intercalating dyes are used, which are compatible with both agarose and PAGE gels.

## 2 Materials

### 2.1 Agarose EMSA

1. 37 °C incubator or water bath and microcentrifuge.
2. 11–14 horizontal gel electrophoresis system (Biometra Horizon 11–14) or comparable.
3. Electrophoresis power supply.
4. UV imager (Syngene GBox or comparable).
5. Purified protein (1–10 mg/mL) (*see* **Notes 1** and **2**).
6. Plasmid DNA (60 ng/µL).
7. 5× Equilibration buffer: 100 mM HEPES pH 7.5, 375 mM NaCl, 5 mM DTT (*see* **Note 3**).
8. Optional: 100 µM stabilizing probe (*see* **Note 4**).
9. Optional: Competitor nucleic acid (*see* **Note 5**).
10. 1× sodium boric acid (SB) buffer: 8.6 mM sodium borate, 45 mM boric acid, pH 8.3 (*see* **Note 6**).
11. Agarose, molecular biology grade (Sigma).
12. 6× DNA loading dye (Thermo Scientific) and DNA size marker (Gene ruler 1 kb, Thermo Scientific).
13. SYBR Safe (Thermo Scientific) or ethidium bromide (Sigma).
14. Reagent grade water.

### 2.2 PAGE EMSA

1. Isotope facilities.
2. Programmable heat block or incubator and microcentrifuge.
3. Vertical PAGE apparatus and casting setup, including glass plates, spacers, combs, clamps, casting stand and running unit (Bio-Rad or comparable).
4. Electrophoresis power supply.
5. Phosphor screen (GE Healthcare) or autoradiography film (Kodak) or comparable.
6. Phosphor Imager (Bio-Rad PMI).
7. Purified protein (1–10 mg/mL) (*see* **Notes 1** and **2**).
8. DNA oligonucleotides (100 µM).
9. 5× Equilibration buffer: 100 mM HEPES pH 7.5, 375 mM NaCl, 5 mM DTT (*see* **Note 3**).
10. Optional: 100 µM stabilizing probe (*see* **Note 4**).
11. Optional: Competitor nucleic acid (*see* **Note 5**).
12. Polynucleotide kinase (PNK) (Thermo scientific). This comes with buffer A (forward reaction) and buffer B (exchange reaction).
13. Phenol, chloroform, isoamyl alcohol mix (25:24:1) (Roth).

14. Nucleotide removal kit (Qiagen) or Sephadex G50 columns (GE Healthcare).
15. ExoI enzyme (Thermo Scientific), supplied with 10× ExoI buffer.
16. 5× Tris–borate–EDTA (TBE) buffer: 445 mM Tris, 445 mM boric acid, 10 mM EDTA (see **Note 6**).
17. 30 % w/v acrylamide–bisacrylamide (29:1) stock solution (Sigma).
18. 10 % ammonium persulfate solution (APS), made by dissolving APS powder (Sigma) in reagent grade water.
19. TEMED ($N,N,N',N'$-Tetramethylethylenediamine) (Bio-Rad).
20. Optional: Gel dryer (Model 583 gel dryer, Bio-Rad or comparable).
21. Blotting paper (Whatman) or comparable.
22. Plastic food wrap (Saran Wrap®).
23. Reagent grade water.

# 3 Methods

## 3.1 Agarose EMSA

### 3.1.1 Protein–DNA Equilibration

1. Set up a pipetting scheme as in Table 1. The amounts shown are based on Cascade (Mw = 405 kDa) and pUC-1 (Mw = 1,739 kDa). The amount of DNA per reaction is fixed, while the protein is titrated (see **Note 2**). For good visualization of the DNA, use 360 ng plasmid per reaction. The amount of protein is calculated based on the desired molar ratio. As a negative control, include a sample lacking protein. The amounts can be modified according to the molecular weight of the ribonucleoprotein complex or the target plasmid, in order to keep the same molar ratio. Optional: Include a competitor nucleic acid by premixing this with your target plasmid (see **Note 5**). The amount of 5× equilibration buffer in the final reaction is calculated to yield a final 1× concentration (taking into account the salts present in the protein solution). Reactions are brought to a total volume of 30 µL with reagent grade water.

2. Make fresh working stock dilutions of your protein (see **Note 1**) in 1× equilibration buffer to fit your requirements.

3. Pipette everything on ice, add water and buffer first, then add protein solution, and last add plasmid solution. Vortex the reaction mixture for 10 s and briefly spin in a microcentrifuge.

4. Incubate at 37 °C for 30 min to allow the reaction to equilibrate (see **Note 2**). In the meanwhile, start preparing the gels.

## Table 1
### Pipetting scheme examples for plasmid EMSA for agarose gel

| Molar ratio Cascade–DNA | 0 | 1 | 2 | 4 | 8 | 16 | 32 | 48 | 64 | 80 | 96 | 150 | 200 | 250 | 300 | 350 | 400 |
|---|---|---|---|---|---|---|---|---|---|---|---|---|---|---|---|---|---|
| Plasmid (μL) | 6.0 | 6.0 | 6.0 | 6.0 | 6.0 | 6.0 | 6.0 | 6.0 | 6.0 | 6.0 | 6.0 | 6.0 | 6.0 | 6.0 | 6.0 | 6.0 | 6.0 |
| Cascade working stock dilution factor | | 100× | 100× | 100× | 20× | 20× | 4× | 4× | 4× | 2× | 2× | 1× | 1× | 1× | 1× | 1× | 1× |
| Cascade (μL) | 0.0 | 1.9 | 3.7 | 7.5 | 3.0 | 6.0 | 2.4 | 3.6 | 4.8 | 3.0 | 3.6 | 2.8 | 3.7 | 4.7 | 5.6 | 6.5 | 7.5 |
| 5× equilibration buffer (μL) | 6.0 | 5.6 | 5.3 | 4.5 | 5.4 | 4.8 | 5.5 | 5.3 | 5.0 | 5.4 | 5.3 | 5.4 | 5.3 | 5.1 | 4.9 | 4.7 | 4.5 |
| Water (μL) | 18.0 | 16.5 | 15.0 | 12.0 | 15.6 | 13.2 | 16.1 | 15.1 | 14.2 | 15.6 | 15.1 | 15.8 | 15.0 | 14.2 | 13.5 | 12.8 | 12.0 |
| Total volume (μL) | 30.0 | 30.0 | 30.0 | 30.0 | 30.0 | 30.0 | 30.0 | 30.0 | 30.0 | 30.0 | 30.0 | 30.0 | 30.0 | 30.0 | 30.0 | 30.0 | 30.0 |
| Cascade stock (M) = $1.1 \times 10^{-5}$ | | | | | | | | | | | | | | | | | |
| Plasmid stock (M) = $3.4 \times 10^{-8}$ | | | | | | | | | | | | | | | | | |

5. Optional: Add 1 µL of a 100 µM stabilizing probe to the reaction after equilibration (*see* **Note 4**). Incubate for another 20 min at 37 °C.

6. Optional: At this point, additional procedures, such as enzymatic footprinting can be carried out (*see* **Note 7**).

7. Add 6 µL 6× DNA loading dye to each sample and store on ice until loading on gel (see below).

*3.1.2 Preparing and Running the Gel*

1. Prepare a 0.8 % (*see* **Note 8**) SB buffer agarose gel by mixing 0.88 g agarose with 110 ml 1× SB buffer (*see* **Note 6**). Dissolve the agarose by heating the solution in a microwave. Make sure the agarose is completely dissolved. Do not include an intercalating dye when casting the gel as this can affect DNA binding. Cast the gel in an 11–14 gel tray or comparable system. Use a comb for 20 µL slots.

2. Assemble the electrophoresis unit and fill the container with 1× SB buffer (*see* **Note 6**) until it covers the gel.

3. Load half of each sample (18 µL) in the slots and add the DNA size marker in the first and the last lane of the gel (this allows to check if the gel ran uniformly). Store the other half of the samples in the freezer (−20 °C) as a backup.

4. Run the gel at 20 mA for 18 h (based on Cascade binding to a ~3 kb plasmid; less time is needed to separate smaller protein–DNA complexes).

5. Remove the gel from the electrophoresis chamber and put it in a plastic tray.

6. Stain the gel by covering it with 1× SB buffer containing 1:10,000 SYBR Safe or ethidium bromide for 30 min (important: mix the 1× SB buffer and SYBR Safe or ethidium bromide well before applying it on the gel. Poorly mixed solutions can yield stains on the gel).

7. Rinse the gel and destain it in $dH_2O$ for 15 min.

8. Visualize the DNA using a UV imager (*see* Table 3 for troubleshooting). Make sure not to saturate the signal anywhere in the gel. In case of quantification, save the file in an appropriate format for your image analysis software (.sgd file for GBox or .tif file for cross platform analysis). Figure 1 is a typical example of a plasmid EMSA on agarose.

9. Continue to image analysis (Subheading 3.3).

## 3.2 PAGE EMSA

*3.2.1 5' $^{32}P$ Labelling of Short Oligonucleotide Substrate*

1. To prepare dsDNA oligonucleotides mix the following in a microcentrifuge tube (annealing mix) (Note: You can perform this protocol using cold ATP in parallel to conveniently measure DNA concentrations afterwards): 1 µL 10 µM forward oligonucleotide, 1 µL 10 µM reverse oligonucleotide, 2 µL

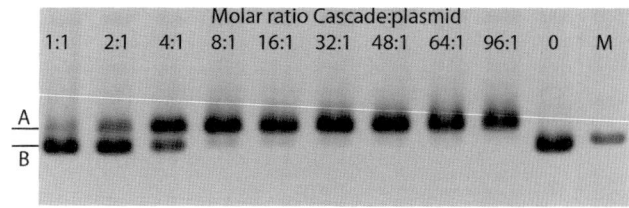

**Fig. 1** Agarose EMSA of Cascade with plasmid DNA. Purified Cascade and the plasmid DNA have been incubated to equilibrium at the molar ratios indicated and free plasmid (*upper band* (*A*)) has been separated from Cascade-bound plasmid (*lower band* (*B*)) on a 0.8 % sodium borate agarose gel. Each lane contains a total of 180 ng plasmid DNA and Cascade protein according to indicated molar ratios of Cascade–plasmid. The gel was run at 8 mA for 18 h and post-stained with SYBR Safe for 30 min

10× PNK buffer (use either Buffer A, for non-phosphorylated oligonucleotides or buffer B, for phosphorylated oligonucleotide), 16 µL $H_2O$.
2. If you are using single stranded DNA, replace the reverse oligonucleotide volume with reagent grade water and Skip to **step 5**.
3. Heat to 95 °C for 5 min.
4. Slowly cool down to 37 °C (>30 min).
5. Add 1 µL PNK and 2 µL γ-$^{32}$P ATP (Caution: Exposure to radiation is hazardous, follow safety procedures of your institution).
6. Incubate for 1 h at 37 °C.
7. Clean up using nucleotide removal kit or a sephadex G-50 column (preferred method).
8. Elute in 44 µL reagent grade water.
9. Add 5 µL 10× ExoI buffer and 1 µL ExoI (Important: Skip this step when using a single stranded DNA substrate).
10. Incubate for 30 min at 37 °C.
11. Add 50 µL phenol–chloroform–isoamyl alcohol mix.
12. Vortex thoroughly, spin for 3 min at maximum (>12,000 × $g$) speed in a microcentrifuge.
13. Recover the aqueous phase (upper phase).
14. Optional: Determine the DNA concentration of the cold sample (if available), or hot sample (Caution: Only use specified equipment for the use with radiolabelled samples).
15. Store at −20 °C.

### 3.2.2 Protein–DNA Equilibration

1. Set up a pipetting scheme as indicated in Table 2: The amounts shown are based on Cascade (Mw = 405 kDa) and a short dsDNA oligonucleotide. The amount of DNA per reaction is fixed, and protein is titrated. Use 1 μL of 4× diluted labelled oligonucleotide per reaction (the amount depends on the activity of the sample). The amount of protein is calculated based on the desired molar ratio (*see* **Note 2**). As a negative control include a sample with no protein. Amounts can be modified according to the molecular weight of the ribonucleoprotein complex or the target oligonucleotide, in order to keep the same molar ratio. Optional: Include an unlabelled competitor nucleic acid by premixing this with your oligonucleotide (*see* **Note 5**). The amount of 5× equilibration buffer in the final reaction is calculated to yield a final 1× concentration (taking into account the salts in the protein solution). Reactions are brought to a total of 30 μL with reagent grade water.

2. Make fresh stock dilutions of your protein (*see* **Note 1**) in 1× reaction buffer to fit your requirements.

3. Pipette everything on ice, add water and buffer first, then add protein and add the oligonucleotide last. Vortex for 10 s and spin down in a microcentrifuge.

4. Incubate at 37 °C for 30 min to allow the reaction to equilibrate (*see* **Note 2**). In the meanwhile start preparing the gels.

5. Optional: Add 1 μL of a 100 μM stabilizing probe to the reaction after equilibration (*see* **Note 4**). Incubate for another 20 min at 37 °C.

6. Optional: At this point, additional procedures, such as enzymatic footprinting can be carried out (*see* **Note 7**).

7. Add 6 μL 6× DNA loading dye to each sample and store on ice until loading on gel (see below).

### 3.2.3 Preparing and Running the Gel

1. Prepare a 5 % (*see* **Note 8**) TBE PAGE gel, preferably in a large format (e.g., 20 × 15): 8.3 mL 30 % w/v acrylamide–bisacrylamide (29:1) stock solution, 10 mL 5× TBE, 31 mL $H_2O$, 0.8 mL APS.

2. Assemble the glass plates in a casting setup. Fill with water, to check for leakage and then remove water.

3. Add 50 μL TEMED to the gel solution and pour the gel. Insert the appropriate comb for 20 μL samples.

4. Assemble the gel tray and wash the slots with running buffer (0.5× TBE).

5. Pre-run the gel for 20 min at 40 mA.

## Table 2
### Pipetting scheme examples for oligonucleotide EMSA for PAGE gel

| Molar ratio Cascade–DNA | 0 | 1 | 2 | 4 | 8 | 16 | 32 | 48 | 64 | 80 | 96 | 150 | 200 | 250 | 300 | 350 | 400 |
|---|---|---|---|---|---|---|---|---|---|---|---|---|---|---|---|---|---|
| Oligonucleotide (μL) | 1.0 | 1.0 | 1.0 | 1.0 | 1.0 | 1.0 | 1.0 | 1.0 | 1.0 | 1.0 | 1.0 | 1.0 | 1.0 | 1.0 | 1.0 | 1.0 | 1.0 |
| Cascade working stock dilution factor | | 1000× | 1000× | 1000× | 100× | 100× | 20× | 20× | 20× | 10× | 10× | 10× | 10× | 10× | 10× | 10× | 10× |
| Cascade (μL) | 0.0 | 3.2 | 6.3 | 12.6 | 2.5 | 5.0 | 2.0 | 3.0 | 4.0 | 2.5 | 3.0 | 4.7 | 6.3 | 7.9 | 9.5 | 11.0 | 12.6 |
| 5× equilibration buffer (μL) | 6.0 | 5.4 | 4.7 | 3.5 | 5.5 | 5.0 | 5.6 | 5.4 | 5.2 | 5.5 | 5.4 | 5.1 | 4.7 | 4.4 | 4.1 | 3.8 | 3.5 |
| Water (μL) | 23.0 | 20.4 | 18.0 | 12.9 | 21.0 | 19.0 | 21.4 | 20.6 | 19.8 | 21.0 | 20.6 | 19.2 | 18.0 | 16.7 | 15.4 | 14.2 | 12.9 |
| Total volume (μL) | 30.0 | 30.0 | 30.0 | 30.0 | 30.0 | 30.0 | 30.0 | 30.0 | 30.0 | 30.0 | 30.0 | 30.0 | 30.0 | 30.0 | 30.0 | 30.0 | 30.0 |
| Cascade stock (M) = $1.1 \times 10^{-5}$ | | | | | | | | | | | | | | | | | |
| Oligonucleotide stock (labelled) (M) = $3.5 \times 10^{-8}$ | | | | | | | | | | | | | | | | | |

6. Load the samples and run for 15 min at 30 mA until the tracking dye has migrated into the gel.

7. Run the gels for 3–4 h at 20 mA or until the cyan blue dye reaches ¼ of the gel. Carefully rinse and dry the assembly, separate the glass plates, and carefully transfer the gel to a blotting paper, wrap the gel and paper in plastic food wrap. Prevent air bubbles and wrinkles.

8. Optional: Dry the gel on a paper membrane using a gel-dryer (gives better resolution but there is a risk of breaking the gel).

9. Expose the gel to a phosphor screen or autoradiography film in an exposure cassette. Short time exposures (~2 h) can be performed at room temperature or at 4 °C. Longer exposures of undried gels can be performed at −20 °C to prevent diffusion. Make sure not to saturate the signal anywhere on the phosphor screen or autoradiography film. Scan the image using a phosphor imager (*see* Table 3 for troubleshooting). Proceed to image analysis (Subheading 3.3).

## 3.3 Image Analysis and Quantification

In many cases, EMSA results will need to be analyzed quantitatively to obtain the affinity values ($K_d$) associated with a protein–DNA interaction. Such quantitative analysis requires that the intensities of the bands as well as the background intensity are quantified.

1. To this end, import the picture in an image analysis software (e.g., Genetools for GBox .sgd files).

2. Use the program to automatically or manually quantify the intensity of unshifted and shifted bands in the gel. Apply appropriate background correction for each lane to correct for uneven exposure of the gel.

3. Once the intensities of shifted and unshifted bands have been obtained, the affinity of the interaction can be calculated as described below. Analysis of more complex binding behavior can be done as described in [9]. However, in the case of Cas ribonucleoprotein complexes binding to protospacer sequences, we can assume that each protein (complex) binds only one specific binding site per target molecule, which greatly simplifies the analysis. To obtain binding affinities using this assumption, calculate the fraction of bound substrate ($y$) and the free protein concentration ($x$) for each sample. '$y$' is calculated by dividing the shifted band intensity by the total intensity present in that lane. '$x$' is calculated by multiplying the total protein concentration (as added in the reaction) with the fraction of unbound DNA ($1-y$). Perform nonlinear regression using the formula $y = x/(K_d + x)$ to determine the dissociation constant $K_d$.

**Table 3**
**Troubleshooting**

| Problem | Possible cause | Potential solutions |
|---|---|---|
| No bands visible on gel | Too little or no nucleic acid in reaction | Check nucleic acid concentration, test sensitivity of applied visualization method |
| | Nucleic acid is degraded | Check substrate integrity on gel. Replace reagents when suspecting nuclease contamination. If possible exclude metal cations and include chelating agent (e.g., EDTA). If working with RNA, work RNase free or include commercial RNase inhibitors. |
| | Labelling failed or insufficient exposure time | Check functionality of labelling method. If necessary, adapt protocol. Increase exposure time. |
| No shifted band present | Protein concentration too low | Verify protein concentration, check protein for purity. Increase protein concentration in EMSA. |
| | Protein is inactive or cofactor missing | Check protein on SDS-PAGE for integrity. Repurify protein, possibly adapt protocol to get more active preparation. Test cofactors, e.g., divalent metal ions. |
| | Protein-bound nucleic acid migrates at same speed as free nucleic acid | Check migration of protein alone in gel. Use different pH in equilibration and electrophoresis buffer or use a different electrophoresis buffer. If protein is very small compared to nucleic acid, use smaller nucleic acid. |
| Bands are generally smeared | Gel overheating | Check gel concentration and running buffer conductivity. Lower the above or use lower voltage during electrophoresis. |
| | High sample conductivity | Reduce salt content in samples. |
| | Bad gel quality | Check even polymerization/solidification. Use fresh, clean components. Degas PAGE gels before polymerization. |
| Only protein bound band is smeared | Too high conductivity in protein solution | Reduce salt concentration in protein stock solution, or concentrate protein and add less volume to reactions. |
| | Complex dissociates during electrophoresis | Minimize time of sample in loading well. Start with higher voltage to run samples into the gel. Minimize overall electrophoresis time. |

# 4 Notes

1. Protein: Use purified Cas ribonucleoprotein complexes. Make sure to accurately determine the protein concentration. If additives are required for protein storage that are undesired in the EMSA equilibration, exchange the buffer beforehand by dialysis or a buffer exchange column. Alternatively, keep additive concentration low and protein concentration high to minimize the volume that is added in the reactions. Ideally, the

protein is dissolved in a buffer very similar to the equilibration buffer. Make working stock dilutions of the protein in 1× Equilibration buffer.

2. Equilibration: Test a range of molar ratios of protein–DNA (e.g., 0.5:1 up to 400:1). In later experiments, choose ratios that cover the whole dynamic range of binding (i.e., ranging from all DNA in the unbound state to all DNA protein-bound). Test the incubation time needed to reach equilibrium for each protein and each different type of substrate (plasmid, short oligonucleotides etc.). Do this by testing several time points (e.g., 5 min, 30 min, 60 min); when there is no change in the bound fraction between two time points, equilibrium is reached. Typically an incubation time of 30 min at 37 °C is used.

3. Equilibration buffer: Choose a buffer for protein–DNA complex equilibration that gives efficient complex formation, is relatively close to physiological conditions and is compatible with the buffer of your protein solution. Do not use salt concentrations that are considerably higher than in your electrophoresis buffer, as this leads to a higher conductivity of the sample compared to the gel and electrophoresis buffer, which will lead to distorted bands. If additives are required for the function of the protein (e.g., metal ions), these should be added to the equilibration buffer. Most common buffers can be used (e.g., Tris, MOPS, HEPES, Phosphate) and total salt concentration is typically around 100 mM.

4. Stabilizing probe: When working with an R-loop forming complex (i.e., the situation where the crRNA base pairs with the target DNA strand, while the non-target strand is displaced and remains single stranded), stabilizing DNA oligonucleotides can be added after equilibration to prevent changes during subsequent steps [15]. These probes are complementary to the crRNA as well as to the displaced strand. By binding to the displaced strand these probes prevent reannealing of the target and non-target strands and thereby inhibit complex dissociation. Furthermore free complex is inactivated by binding of the probe to the crRNA. This step is advisable when additional steps, such as enzymatic footprinting, are performed, or when the interaction of the complex and the target is unstable and tends to dissociate during electrophoresis (e.g., Cascade binding to relaxed plasmid DNA [12]). Choose to add an amount of probe resulting in a tenfold excess of probe, compared to target- or protein molecules.

5. Competitor nucleic acid: In case of a high nonspecific nucleic acid binding affinity of the protein (independent of crRNA sequence), which might obscure specific binding, it is advis-

able to add unlabelled competitor nucleic acid. It should have the same nonspecific binding affinity to the protein as the target, while the target should have a higher specific binding affinity to the protein. Concentrations of competitor have to be optimized empirically to find a ratio that allows clear discrimination of specific and nonspecific binding. If the gel is non-selectively stained by an intercalating dye, make sure the competitor has a different size than the target, to distinguish the target and the competitor.

6. Electrophoresis conditions: Do not leave samples in loading slots over a longer period of time, as complexes might dissociate. The choice of electrophoresis buffer can have an effect on the relative mobility of the protein–DNA complex and the free nucleic acid. Especially with large nucleic acid targets in agarose EMSAs it is possible that the mobility of free- and protein-bound nucleic acid is very similar, yielding poor resolution. In this case it is important to choose electrophoresis conditions such that the protein is not negatively charged. Negative protein charges lead to a co-migration of protein with nucleic acid to the anode, decreasing resolution. Hence, the resolution can be improved by changing to a running buffer with a lower pH. Commonly used buffers are TAE and TBE, while we successfully applied sodium borate buffer [16], improving the resolution of the shift. Make sure to use the same buffer in all involved steps. Although long electrophoresis times increase the risk of complex dissociation, we get the best results with overnight runs at low currents. We have not seen differences in bound fractions between short or long electrophoresis, while longer runs produced sharper bands.

7. Footprinting is generally an independent alternative to EMSA, but can be used in concert with it. Particularly useful, in the case of crRNA-Cas complexes, is the use of ssDNA specific Nuclease P1 [10] or an endonuclease cleaving in the known binding site (protospacer) on the target DNA [12]. Footprinting allows detection and quantification of protein–DNA interactions in solution, without relying on their stability in subsequent electrophoresis

8. An agarose gel percentage of 0.8 % is typically used and will give satisfactory results with most large nucleic acid targets. A PAGE gel percentage of 5 % is typically used; this can be adjusted to yield the best resolution with the chosen nucleic acid target (see Table 4). Be aware that higher gel percentages might not allow large proteins to enter the gel matrix, trapping protein–DNA complexes in the loading slots or leading to dissociation.

**Table 4**
**DNA size separation by gel percentage**

| DNA size range (base pairs) | Acrylamide (%) |
|---|---|
| 100–1,000 | 3.5 |
| 80–500 | 5.0 |
| 60–400 | 8.0 |
| 40–200 | 12.0 |
| 10–100 | 20.0 |

| DNA size range (base pairs) | Agarose (%) |
|---|---|
| 1,000–30,000 | 0.5 |
| 800–12,000 | 0.7 |
| 500–10,000 | 1.0 |
| 400–7,000 | 1.2 |

## Acknowledgements

This work was financially supported by a KNAW Beijerinck premium and NWO Vidi grant to S.J.J.B. (864.11.005). ERW received funding from the People Programme (Marie Curie Actions) of the European Union's Seventh Framework Programme (FP7/2007-2013) under REA grant agreement n$^0$ (327606).

## References

1. Westra ER et al (2012) The CRISPRs, they are a-Changin': how prokaryotes generate adaptive immunity. Annu Rev Genet 46:311–339
2. Hale CR et al (2009) RNA-guided RNA cleavage by a CRISPR RNA-Cas protein complex. Cell 139:945–956
3. Jinek M et al (2012) A programmable dual-RNA–guided DNA endonuclease in adaptive bacterial immunity. Science 337:816–821
4. Semenova E et al (2011) Interference by clustered regularly interspaced short palindromic repeat (CRISPR) RNA is governed by a seed sequence. Proc Natl Acad Sci 108:10098–10103
5. Helwa R, Hoheisel J (2010) Analysis of DNA–protein interactions: from nitrocellulose filter binding assays to microarray studies. Anal Bioanal Chem 398:2551–2561
6. Jerabek-Willemsen M et al (2011) Molecular interaction studies using microscale thermophoresis. Assay Drug Dev Technol 9:342–353
7. Monico C et al (2013) Optical methods to study protein-DNA interactions in vitro and in living cells at the single-molecule level. Int J Mol Sci 14:3961–3992
8. Fried MG (1989) Measurement of protein-DNA interaction parameters by electrophoresis mobility shift assay. Electrophoresis 10:366–376
9. Fried MG, Daugherty MA (1998) Electrophoretic analysis of multiple protein-DNA interactions. Electrophoresis 19:1247–1253
10. Jore MM et al (2011) Structural basis for CRISPR RNA-guided DNA recognition by Cascade. Nat Struct Mol Biol 18:529–536
11. Westra ER et al (2010) H-NS-mediated repression of CRISPR-based immunity in *Escherichia coli* K12 can be relieved by the transcription activator LeuO. Mol Microbiol 77:1380–1393
12. Westra ER et al (2012) CRISPR immunity relies on the consecutive binding and degradation of negatively supercoiled invader DNA by Cascade and Cas3. Mol Cell 46:595–605
13. Cann JR (1989) Phenomenological theory of gel electrophoresis of protein-nucleic acid complexes. J Biol Chem 264:17032–17040
14. Fried MG, Bromberg JL (1997) Factors that affect the stability of protein-DNA complexes during gel electrophoresis. Electrophoresis 18:6–11
15. Westra ER et al (2012) Cascade-mediated binding and bending of negatively supercoiled DNA. RNA Biol 9:1134–1138
16. Brody JR, Kern SE (2004) Sodium boric acid: a Tris-free, cooler conductive medium for DNA electrophoresis. Biotechniques 36:214–217

# Chapter 12

## Expression and Purification of the CMR (Type III-B) Complex in *Sulfolobus solfataricus*

**Jing Zhang and Malcolm F. White**

### Abstract

Protein purification is an important technique that allows us to characterize the structural and biochemical properties of either an individual protein or a multi-protein complex. However, expression and purification of one subunit of a complex in the absence of its binding partners has often proven difficult to achieve due to the issues such as instability and mis-folding. This is the case for the components of the CRISPR-Cas interference complexes, which degrade invading nucleic acids in a sequence homology-dependent manner in many prokaryotic species. Here, we describe the expression of a tandem-tagged subunit of the Type III-B (CMR) complex in *Sulfolobus solfataricus* and subsequent isolation and purification of the whole complex by affinity purification of the tagged subunit.

**Key words** Affinity purification, Hyperthermophile, Tandem tags, Protein complex, Viral shuttle vector

## 1 Introduction

Many proteins are produced in recombinant form, particularly in *Escherichia coli*, for biochemical characterization and crystallographic studies. For proteins of hyperthermophiles, this approach may not, however, be successful because a number of these proteins fold into their native state only under natural conditions of high temperature and/or in the presence of their native cofactors. Furthermore, proteins from hyperthermophilic crenarchaea are frequently modified by methylation of lysine residues [1]. Therefore, expression of tagged proteins in their native thermophilic host is an alternative approach that also allows the identification of associated cofactors and co-purification of large protein complexes. The crenarchaeote *S. solfataricus* is a model organism for hyperthermophilic archaea that grows optimally at 78 °C [2]. A virus-based shuttle vector system has been developed for high-level gene expression in *S. solfataricus* [3]. Although the viral vector is rather large and unstable in bacterial cells, the cloning procedure is greatly facilitated by using a small entry vector that has an arabinose-inducible promoter, paraS, upstream of the multiple cloning site and a tandem Streptavidine-his10 tag downstream. The gene of interest is cloned into the entry vector in frame with the tandem tags then transferred into the viral shuttle vector together with the paraS promoter and the tags. The activity of the promoter is tightly controlled by arabinose, therefore preventing adverse effect on growth caused by toxic proteins before sufficient biomass is obtained.

Using this method, we have expressed a tagged version of individual components of the Type III-B (CMR) CRISPR interference complex [4]. Subsequently, the CMR complex has been successfully purified from *S. solfataricus* mainly by affinity purification. Here we describe methods required to clone and express the CMR complex, tagged on one subunit, in *S. solfataricus* and to purify the intact tagged complex to homogeneity. Using this method, we routinely obtained approximately 5 mg of protein from 100 g of biomass. This method can also be applied to other protein complexes such as the Type III-A (CSM) complex [5].

## 2 Materials

Prepare all solutions using sterilized water and analytical grade reagents. Prepare and store all reagents at room temperature (unless indicated otherwise).

### 2.1 Brock Medium Components

An *S. solfataricus pyrEF lacS* mutant (M16) [6] and its derivatives were grown at 75 °C and pH 3 in Brock medium [7] with supplements as indicated.

1. Fe solution (1,000×): 20 g/l $FeCl_3$. Weigh 2 g $FeCl_3$ and transfer to a 100-ml graduated cylinder. Add water to a total volume of 100 ml.

2. Trace element solution (2,000×): 3.6 g/l $MnCl_2 \cdot 4H_2O$, 9 g/l $Na_2B_4O_7 \cdot 10H_2O$, 0.44 g/l $ZnSO_4 \cdot 7H_2O$, 0.1 g/l $CuCl_2 \cdot 2H_2O$, 0.06 g/l $NaMoO_4 \cdot 2H_2O$, 0.06 g/l $VOSO_4 \cdot 2H_2O$, 0.02 g/l $CoSO_4 \cdot 7H_2O$. Add about 100 ml water to a 1-L graduated cylinder. Weigh the above components accordingly and transfer to the cylinder. Make up to 1 L with water.

3. Brock I (1,000×): 70 g/l $CaCl_2 \cdot 2H_2O$. Weigh 70 g $CaCl_2 \cdot 2H_2O$ and transfer to a 1-L graduated cylinder. Add water to a total volume of 1 L. Autoclave the solution.

4. Brock II (100×): 130 g/l $(NH_4)_2SO_4$, 25 g/l $MgSO_4 \cdot 7H_2O$, 1.5 ml 50 % $H_2SO_4$. Add about 500 ml water to a 1-L graduated cylinder. Then add 1.5 ml 50 % $H_2SO_4$ to adjust the pH of the water in the cylinder. Weigh 130 g $(NH_4)_2SO_4$ and 25 g

$MgSO_4 \cdot 7H_2O$ and transfer to the cylinder. Add water to a total volume of 1 L. Autoclave the solution.

5. Brock III (200×): 56 g/l $KH_2PO_4$, 100 ml trace element solution, 1.5 ml 50 % $H_2SO_4$. Add about 500 ml water to a 1-L graduated cylinder. Add 1.5 ml 50 % $H_2SO_4$ to adjust the pH of the water. Then add 100 ml trace element solution. Weigh 56 g $KH_2PO_4$ and transfer to the cylinder. Add water to a total volume of 1 L. Autoclave the solution.

6. Tryptone solution (20 %): Weigh 10 g of Tryptone in a 50 ml tube. Add water to a total volume of 50 ml.

7. Uracil solution (5 mg/ml): Weigh 50 mg of Uracil in a 50 ml tube. Add water to a total volume of 50 ml. Aliquot in 1 ml aliquots and store at –20 °C (*see* **Note 1**).

8. N-Z-Amine® solution (20 %): Weigh 100 g of protein hydrolysate N-Z-Amine® and transfer to 500-ml beaker. Add water to a total volume of 500 ml. Store at 4 °C in a glass bottle.

9. Arabinose solution (20 %): Weigh 200 g of d-arabinose and transfer to a 1-L beaker. Add water to a total volume of 1 L. Store at 4 °C in a glass bottle.

10. Glucose solution (20 %): Weigh 10 g of d (+) -Glucose in a 50 ml tube. Add water to a total volume of 50 ml.

11. Brock medium: Add about 500 ml water to a 1-L graduated cylinder. Add 125 µl 50 % $H_2SO_4$ to the cylinder. Then add 10 ml Brock II, 5 ml Brock III, 1 ml Brock I and 1 ml Fe solution to the cylinder. Make up with water to 1 L (*see* **Note 2**).

12. M16 growth medium: Brock medium supplemented with 0.1 % tryptone and 10 µg/ml uracil. Add 500 µl tryptone solution and 200 µl uracil solution to 100 ml Brock medium.

13. Unselective medium: Brock medium supplemented with 0.1 % tryptone, 10 µg/ml uracil and 0.2 % glucose. Add 500 µl tryptone solution, 200 µl uracil solution and 1 ml glucose solution to 100 ml Brock medium.

14. Selective medium: Brock medium supplemented with 0.2 % N-Z-Amine® and 0.2 % glucose. Add 500 µl N-Z-Amine® solution and 1 ml glucose solution to 100 ml Brock medium.

15. Expression medium: Brock medium supplemented with 0.2 % N-Z-Amine® and 0.2 % arabinose. Add 500 µl N-Z-Amine® solution and 1 ml arabinose solution to 100 ml Brock medium.

## 2.2 Protein Purification Buffers

1. Buffer A: 20 mM HEPES pH 7.5, 250 mM NaCl, 30 mM Imidazole.

2. Buffer B: 20 mM HEPES pH 7.5, 250 mM NaCl, 1 M Imidazole.

3. Buffer C: 20 mM Tris–HCl pH 8.0, 250 mM NaCl.

## 3 Methods

### 3.1 Cloning Gene of Interest into the Entry Vector pMZ1

1. Amplify the gene of interest from *S. solfataricus* P2 genome with forward and reverse oligonucleotides containing NcoI and BamHI restriction sites respectively. The ATG sequence of the NcoI site should be in frame with the coding sequence. The gene should not contain internal sites for NcoI, BamHI, AvrII, or EagI, as these are used in the sub-cloning process. If this is a problem, these sites can first be removed by site directed mutagenesis or a synthetic gene can be designed.

2. Digest the PCR product and the pMZ1 vector (Fig. 1a) with NcoI and BamHI restriction enzymes. Electrophorese the restricted PCR product and vector in a 0.8 % agarose gel. Excise the corresponding gel bands and purify the DNA using QIAquick gel purification kit (QIAGEN). Elute the DNA in water.

3. Ligate the NcoI/BamHI restricted PCR product and pMZ1 vector from the previous step using T4 DNA ligase. Incubate the ligation reaction at room temperature overnight.

4. Transform the ligation reaction mixture into 200 μl DH5α (or similar) chemical competent cells and plate on LB plates supplemented with 100 μg/ml ampicillin. Grow several transformants in liquid LB medium supplemented with 100 μg/ml ampicillin at 37 °C overnight. Extract plasmids from these overnight cultures and digest with NcoI and BamHI to confirm the presence of the chosen gene (Fig. 2a).

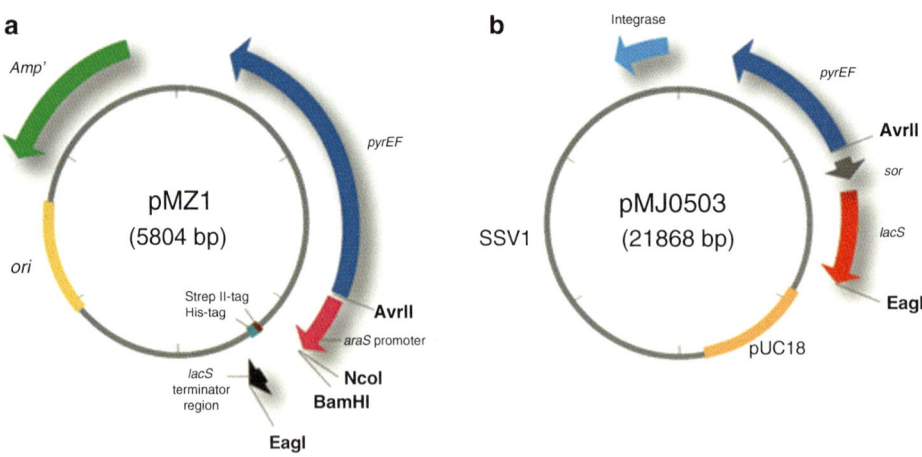

**Fig. 1** Maps of the entry vector pMZ1 and the viral vector pMJ0503. (**a**) The entry vector pMZ1. The gene of interest is cloned into pMZ1 at NcoI and BamHI sites. (**b**) The viral vector pMJ0503. The gene of interest together the *araS* promoter and the tandem tags are cloned into pMJ0503 at AvrII and EagI sites

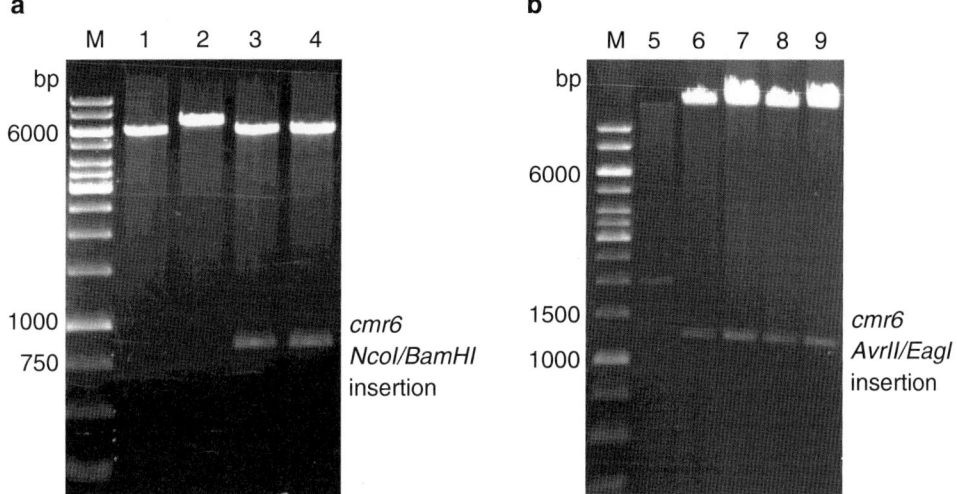

**Fig. 2** Cloning *cmr6* into pMZ1 and subsequently pMJ0503. (**a**) *Lane M*, markers; *Lane 1–4*, pMZ1-*cmr6* clones A to D digested by NcoI and BamHI. Clones (A) and (B) (*lane 1* and *2*) are not correct due to the lack of the insert bands. (**b**) *Lane M*, markers; *Lane 5*, pMJ0503 digested by AvrII and EagI; *Lane 6–9*, pMJ0503-*cmr6* clones A to D digested by AvrII and EagI. All clones show the insert bands of correct size

## 3.2 Sub-cloning Gene of Interest into the Viral Vector pMJ0503

1. Digest the entry vector pMZ1 derivative that contains the gene of interest with AvrII and EagI restriction enzymes to release the DNA fragment for sub-cloning into the viral vector. Digest the viral vector pMJ0503 (Fig. 1b) with AvrII and EagI enzymes (*see* **Note 3**). Electrophorese the restricted DNA in a 0.6 % agarose gel. Excise the corresponding gel bands and purify the DNA as previously described.

2. Ligate the AvrII/EagI restricted DNA fragment and the pMJ0503 vector as previously described.

3. Transform the ligation reaction mixture into 200 μl DH5α chemical competent cells and plate on LB plates supplemented with 50 μg/ml ampicillin (*see* **Note 4**). Grow several transformants in liquid LB medium supplemented with 50 μg/ml ampicillin at 37 °C overnight (*see* **Note 5**). Extract plasmids from these overnight cultures and digest with AvrII and EagI to confirm the presence of the chosen gene (Fig. 2b).

## 3.3 Preparation of S. solfataricus M16 Competent Cells

1. Grow M16 cells in 50 ml of M16 growth medium in a 100-ml flask at 75 °C to $A_{600} = 0.3$. Cool down the cells on ice. From this step on, cells are kept on ice all the time.

2. Centrifuge 50 ml of cells in a 50 ml tube at $2,500 \times g$ for 20 min. Discard the supernatant.

3. Resuspend the cell pellet in 50 ml of 20 mM sucrose. Repeat **step 2**.

4. Resuspend the cell pellet in 25 ml of 20 mM sucrose. Repeat **step 2**.

5. Resuspend the cell pellet in 600 µl of 20 mM sucrose. At this step, cells are ready to be used in electroporation.

## 3.4 Electroporation

1. Mix 50 µl competent cells with 3 µl (100–300 ng) plasmid DNA. Transfer the mixture into a cuvette with 2 mm gap distance.

2. Electroporate the cell and DNA mixture at the following setting: Voltage 1,500 V, Capacity 25 µF, Resistance 400 Ω (*see* **Note 6**). The time constant should be around 10 ms.

3. Immediately after electroporation, add 1 ml of Brock medium into the cuvette. Transfer the mixture into a 1.5 ml tube and incubate at 75 °C for 1 h for recovery. Open the lid of the tube for aeration every 15 min during incubation.

## 3.5 Selection of Transformants

Selection of transformants is based on marker genes *pyrEF* for the complementation of uracil auxotrophic mutants (M16).

1. Pre-warm 50 ml non-selective medium in a 100-ml flask in an incubator at 75 °C.

2. After the 1 h recovery period, transfer the cells from the 1.5 ml tube to the flask with 50 ml pre-warmed non-selective medium.

3. Grow the cells in the flask at 75 °C for 2–3 days until the $A_{600}$ reaches 0.6.

4. Transfer 1 ml of cells from non-selective medium into 50 ml pre-warmed selective medium in a fresh 100 ml flask.

5. Grow the cells in the flask at 75 °C for 4–5 days until $A_{600}$ reaches 0.6.

## 3.6 Expression of the Tagged Subunit of the Complex

1. Transfer cells from selective medium into expression medium in 2-L flasks. Use 10 ml of cells per 1.5 L expression medium (*see* **Note 7**). Grow cells for 3–5 days until $A_{600}$ reaches 0.8.

2. Harvest the cells by centrifuging at $6,600 \times g$ for 10 min. Discard the supernatant.

3. Collect the cell pellet in a 50 ml tube. Pellets can be stored at −20 °C for future purification.

## 3.7 Purification of the Tagged Protein Complex

1. Resuspend 40 g of cell pellet in 40 ml Buffer A. Disrupt the cells by sonicating at around 10 µm amplitude three times for 5 min with a rest of 10 min on ice between each sonication cycle.

2. Centrifuge the sonicated sample at $160,000 \times g$ for 1 h. Collect the supernatant and filter it through 0.45 µm syringe-driven filters. At this step, the filtered cell lysate is ready to be loaded onto a Ni-NTA column (Histrap FF 5 ml, GE Healthcare).

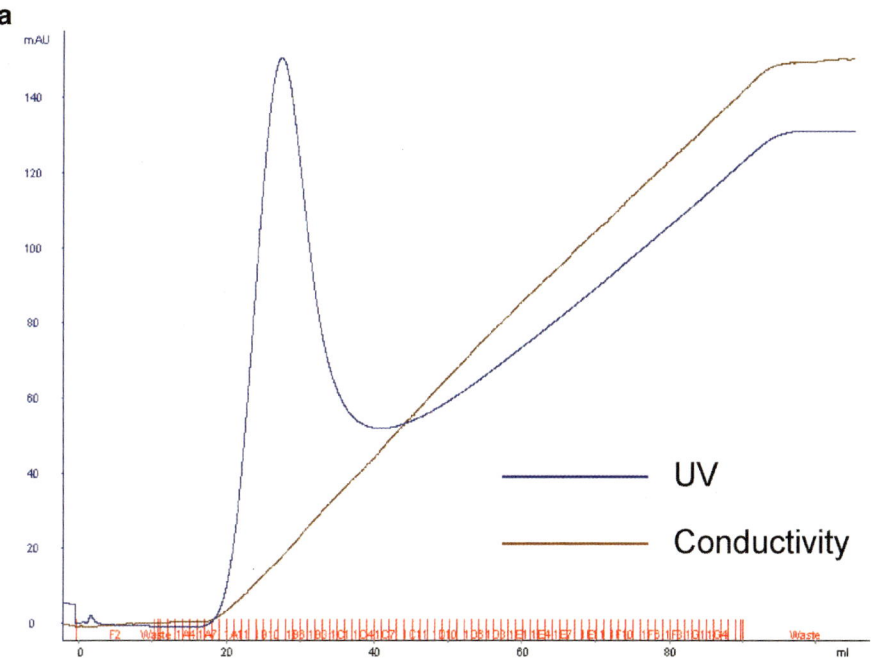

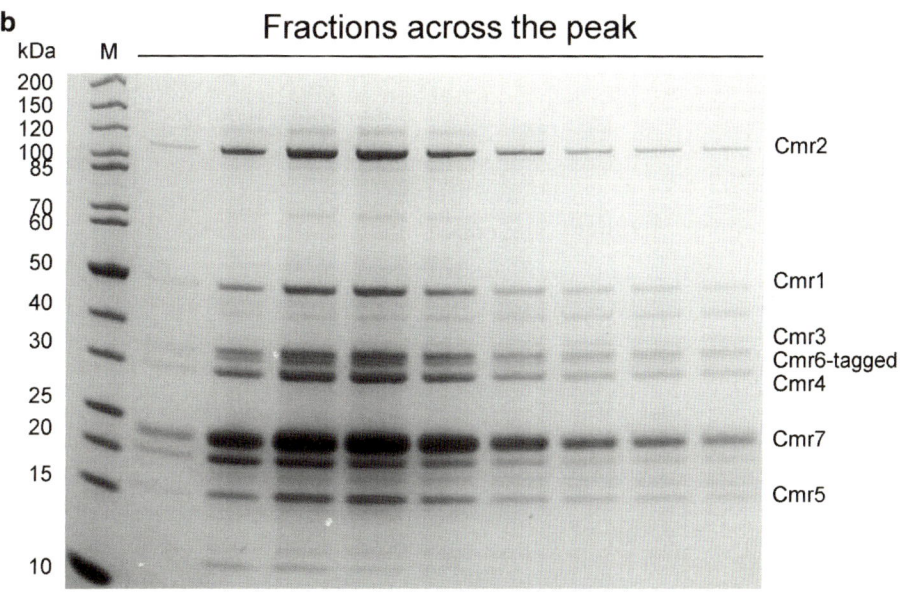

**Fig. 3** Affinity purification of tagged Cmr6. (**a**) The chromatogram of the Histrap column; (**b**) SDS-PAGE gel showing CMR complex-containing fractions across the peak

3. Equilibrate the Ni-NTA column in buffer A. Load the sample onto the column at a speed of 5 ml/min. Wash the column with 50 CV of buffer A. Elute the bound protein with a linear gradient (0–1 M) of buffer B and collect fractions.

4. Examine samples from each fraction across the peak in SDS-PAGE gels (*see* **Notes 8** and **9**) (Fig. 3). Pool the fractions

containing the CMR complex and concentrate the pooled sample using a 30 kDa cut-off concentrator (Amicon) to about 1 ml for size exclusion chromatography in the next step.

5. Equilibrate the size exclusion column (HiPrep™ 16/60 Sephacryl™ S-500 HR, GE Healthcare) in buffer C. Load the concentrated sample onto the S500 column at 0.2 ml/min. Then wash the column with 2 CV buffer C at 1 ml/min and collect fractions.

6. Examine samples from each fraction across the peak in SDS-PAGE gels (Fig. 4). Pool the fractions containing the CMR complex and concentrate as previously described. Store the protein at 4 °C.

## 4  Notes

1. Uracil precipitates out of the solution at room temperature but it is soluble in water at high temperature. We thaw aliquots of uracil solution at 75 °C to avoid precipitation of uracil.

2. Adding Brock components in such an order reduces $FeCl_3$ precipitating with Brock I.

3. Do not leave AvrII and EagI double digestion over 2 h. Prolonged incubation may increase enzyme star activity and cause unspecific cleavage of the plasmid DNA. Alternatively, one can digest the plasmid with the two enzymes sequentially and inactivate the first enzyme by heating before adding the other.

4. The viral vector pMJ0503 is unstable in DH5α cells. Rearrangements of the vector are occasionally observed when cells are grown at 37 °C and with 100 μg/ml ampicillin [3, 8, 9]. To overcome the problem, one can use the ElectroMAX™ Stbl4™ competent cells (Life Technologies) for transformation and grow the cells at 30 °C.

5. We invariably observed that transformants give rise to either large or small colonies. The small colonies often contain plasmids with the expected insert, whereas the large colonies usually lack the insert. We suspect that insertion of the gene of interest may affect the expression of the ampicillin resistant gene downstream in the same vector and therefore lead to smaller colonies.

6. We also used an electroporator, on which the only parameter that could be changed was the voltage. We found that when the voltage was set at 1,500 V, using a 1 mm cuvette instead of 2 mm gave us higher success rates.

7. We usually grow 100 L of cells in expression medium for approximately 5 mg protein yield.

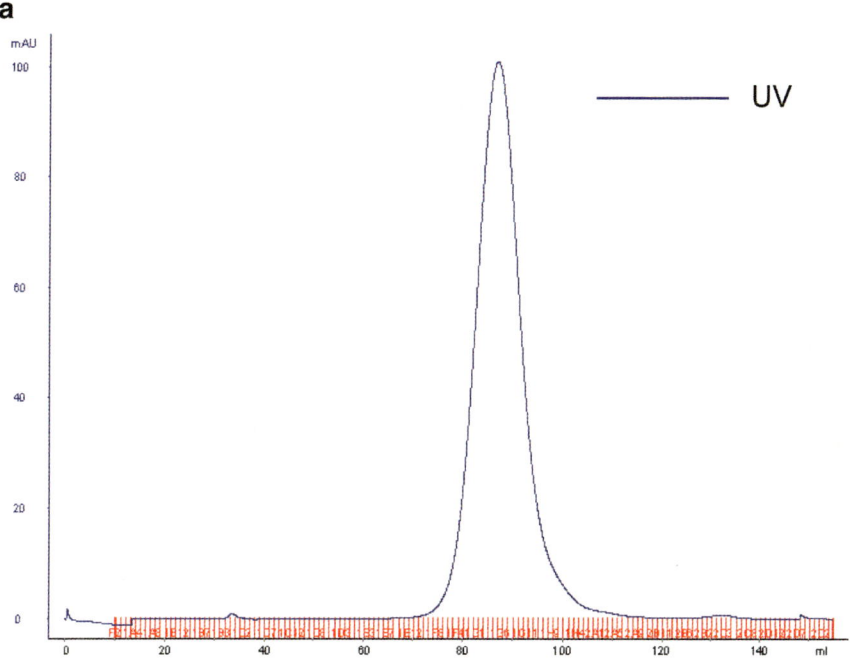

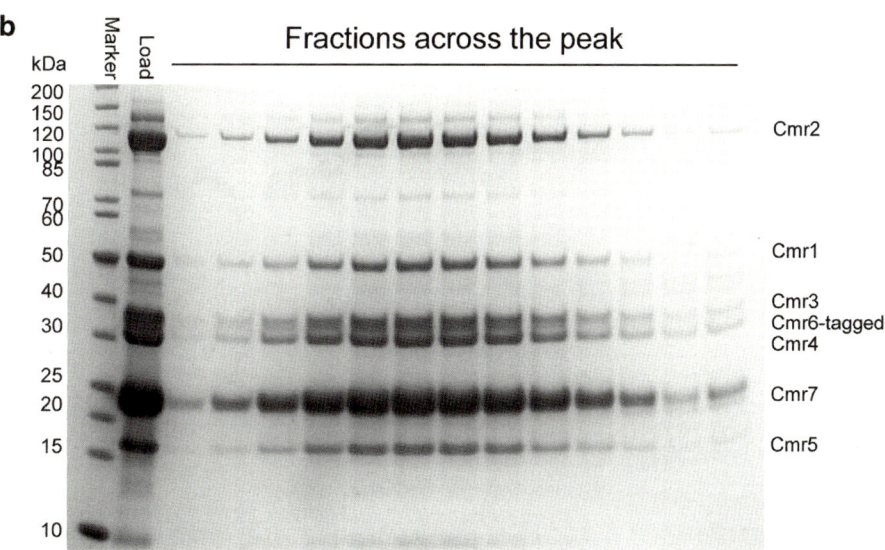

**Fig. 4** Gel filtration of the Cmr6-tagged CMR complex. (**a**) The chromatogram of the gel filtration column; (**b**) SDS-PAGE gel showing CMR complex-containing fractions across the peak

8. When there are multiple copies of a particular subunit present in the protein complex, tagging this subunit will likely give rise to a mixture of tagged and untagged forms in the complex. If this is a problem, for example if a particular mutation of that subunit is required, then it may be worth considering deletion of the endogenous gene.

9. In some cases, a subunit with the tandem tag on its C-terminus can disrupt the complex. We modified the entry vector by deleting the C-terminal tag sequences and adding a 10× His-tag to the N-terminus. With the N-terminal tag, we obtained full complex again.

## Acknowledgement

We would like to thank Dr Sonja Albers for kindly providing us with the vectors and the M16 strain and to Dr Shirley Graham for providing traces and gel photos for the purification step. This work was supported by BBSRC grants BB/J005665/1 and BB/K000314/1 to MFW.

## References

1. Botting CH, Talbot P, Paytubi S, White MF (2010) Extensive lysine methylation in Hyperthermophilic Crenarchaea: potential implications for protein stability and recombinant enzymes. Archaea, 2010, Article ID 106341
2. Jonuscheit M, Martusewitsch E, Stedman KM, Schleper C (2003) A reporter gene system for the hyperthermophilic archaeon *Sulfolobus solfataricus* based on a selectable and integrative shuttle vector. Mol Microbiol 48:1241–1252
3. Albers SV, Jonuscheit M, Dinkelaker S, Urich T, Kletzin A, Tampe R, Driessen AJM, Schleper C (2006) Production of recombinant and tagged proteins in the hyperthermophilic archaeon *Sulfolobus solfataricus*. Appl Environ Microbiol 72:102–111
4. Zhang J, Rouillon C, Kerou M, Reeks J, Brügger K, Graham S, Reimann J, Cannone G, Liu H, Albers S-V et al (2012) Structure and mechanism of the CMR complex for CRISPR-mediated antiviral immunity. Mol Cell 45:303–313
5. Rouillon C, Zhou M, Zhang J, Politis A, Beilsten-Edmands V, Cannone G, Graham S, Robinson CV, Spagnolo L, White MF (2013) Structure of the CRISPR interference complex CSM reveals key similarities with Cascade. Mol Cell 52:124–134
6. Martusewitsch E, Sensen CW, Schleper C (2000) High spontaneous mutation rate in the hyperthermophilic archaeon *Sulfolobus solfataricus* is mediated by transposable elements. J Bacteriol 182:2574–2581
7. Brock T, Brock K, Belly R, Weiss R (1972) Sulfolobus: a new genus of sulfur-oxidizing bacteria living at low pH and high temperature. Archiv Mikrobiol 84:54–68
8. Aravalli RN, Garrett RA (1997) Shuttle vectors for hyperthermophilic archaea. Extremophiles 1:183–191
9. Berkner S, Grogan D, Albers SV, Lipps G (2007) Small multicopy, non-integrative shuttle vectors based on the plasmid pRN1 for *Sulfolobus acidocaldarius* and *Sulfolobus solfataricus*, model organisms of the cren-archaea. Nucleic Acids Res 35:e88

# Chapter 13

## Procedures for Generating CRISPR Mutants with Novel Spacers Acquired from Viruses or Plasmids

Marie-Ève Dupuis, Rodolphe Barrangou, and Sylvain Moineau

### Abstract

CRISPR-Cas systems provide immunity in bacteria and archaea against nucleic acids in the form of viral genomes and plasmids, and influence their coevolution. The first main step of CRISPR-Cas activity is the immune adaptation through spacer(s) acquisition into an active CRISPR locus. This step is also mandatory for the final stage of CRISPR-Cas activity, namely interference. This chapter describes general procedures for studying the CRISPR adaptation step, accomplished by producing bacteriophage-insensitive mutants (BIMs) or plasmid-interfering mutants (PIMs) using various spacer acquisition analyses and experiments. Since each bacterial or archaeal species (and even strain) needs specific conditions to optimize the acquisition process, the protocols described below should be thought of as general guidelines and may not be applicable universally, without modification.

Because *Streptococcus thermophilus* was used as the model system in the first published study on novel spacer acquisition and in many studies ever since, the protocols in this chapter describe specific conditions, media, and buffers that have been used with this microorganism. Details for other species will be given when possible, but readers should first evaluate the best growth and storage conditions for each bacterium—foreign element pair (named the procedure settings) and bear in mind the specificity and variability of CRISPR-Cas types and subtypes. Also, we suggest to be mindful of the fact that some CRISPR-Cas systems are not "naturally" active in terms of the ability to acquire novel CRISPR spacers, and that some systems may require specific conditions to induce the CRISPR-Cas activity for spacer acquisition.

**Key words** Acquisition, Adaptation, Adaptive immunity, Bacteriophage insensitive mutants (BIMs), CRISPR, Plasmid interfering mutants (PIMs), Protospacer, Protospacer adjacent motif (PAM), Spacer, Cas

## 1 Introduction

### 1.1 CRISPR-Cas Systems

CRISPR-Cas systems (Clustered Regularly Interspaced Short Palindromic Repeats loci, together with CRISPR-associated proteins) are the most recently identified defense systems against invasive nucleic acids such as virus and plasmids, and have been identified in many bacteria (45 %) and most archaea (84 %) [1]. Although they were first observed in an *Escherichia coli* intergenic sequence in 1987 [2], their function remained elusive for two decades. This genomic structure was implicated in phage resistance in *S. thermophilus* in 2007 [3] and with plasmid uptake control in 2010 [4]. Subsequently, additional biological functions have been documented for CRISPR-Cas systems, although adaptive immunity remains their primary role. We recommend several reviews that discuss various aspects of CRISPR-Cas systems, their functions and applications [5–8].

CRISPR-Cas systems are composed of a CRISPR locus coupled with specific genes (*cas* genes) that function together to provide adaptive immunity against foreign genetic elements (*see* Subheading 1.2). A broad diversity of CRISPR-Cas architectures exist, that associate various combinations of CRISPR repeats together with a wide array of *cas* genes [9, 10]. Generally, a CRISPR locus consists of multiple short direct repeats separated by non-repeated sequences, called spacers. In many systems these spacers derive from mobile genomic elements such as phages and plasmids (*see* Table 1). From these elements, an identical sequence, called the protospacer, is selected and inserted into a CRISPR locus during the CRISPR-Cas acquisition step at the same time as a new repeat (*see* Subheading 1.3). This novel spacer acquisition is typically polarized at the leader 5′-end of the locus, and the novel repeat is derived from the leader-end template repeat. Spacers may be "native" (i.e., they were originally present in the parental wild type strain, previously identified within the CRISPR locus of a sequenced genome during a comparative study [11–14] or metagenomic survey [15–17]), or "induced" (i.e., the acquisition occurred following in vivo or in vitro experiments [4, 18, 19]).

Although several classifications of CRISPR-Cas systems have been proposed over time, the milestone study by Makarova et al. in 2011 [9] established three main types (I, II, III) and ten subtypes (IA, IB, IC, ID, IE, IF, IIA, IIB, IIIA, and IIIB) that have withstood the test of time. Nevertheless, some peculiar systems occasionally do not fit within these subtypes, and this classification is accordingly updated as new information is gathered. Indeed, some modifications have already been proposed, such as the new type II-C subtype [20].

Some of the authors of this chapter are a member (MED) or Curator (SM) of the Félix d'Hérelle Reference Center for Bacterial Viruses (www.phage.ulaval.ca). We have used microbiological approaches since 2005 (i.e., experiments involving bacterial and phage amplifications, in vitro infections, natural selection of spacers and protospacers, etc.) to better understand and characterize CRISPR-Cas systems. Thus, we present here general guidelines for phage and plasmid manipulations for experiments with CRISPR-Cas systems, as well as more "specific" guidelines for troubleshooting and ensuring reproducible assays.

## Table 1
## List of media used for BIM and PIM production and related references

| Subject | Details | Reference |
|---|---|---|
| BIMs of *Streptococcus thermophilus* | Strain: DGCC7710<br>Bacterial growth: 42 °C<br>Medium: LM17 (M17 + 0.5 % Lactose)<br>Phages: 2972 and 858<br>Infection at $OD_{600}$ 0.3 at 42 °C<br>Phages added at $10^7$ PFU/mL with 50 mM $CaCl_2$ | Barrangou et al. Science (2007) [3] |
| BIMs of *Streptococcus thermophilus* | Strains: DGCC7710 and SMQ-301<br>Bacterial growth: 42 °C<br>Medium: LM17 (M17 + 0.5 % Lactose)<br>Phages: 2972 and 858<br>Infection at $OD_{600}$ 0.3 at 42 °C<br>Phages added at MOI of 5 with 10 mM $CaCl_2$ | Deveau et al. J Bacteriol. (2008) [18] |
| BIMs of *Streptococcus thermophilus* | Strains: DGCC7710 and LMD-9<br>Bacterial growth: 42 °C<br>Medium: LM17 (M17 + 0.5 % Lactose + 0.5 % Sucrose)<br>Phages: 858, 3821, and 4241<br>Infection at $OD_{600}$ 0.3 at 42 °C<br>Phages added at $10^7$ PFU/mL with 50 mM $CaCl_2$ | Horvath et al. J Bacteriol. (2008) [37] |
| BIMs of *Streptococcus thermophilus* | Strain: DGCC7710<br>Bacterial growth: 42 °C (or 37 °C for overnight)<br>Medium: LM17 (M17 + 0.5 % Lactose)<br>Phages: 2972 and 858<br>Infection at $OD_{600}$ 0.6 at 42 °C<br>Phages added at MOI of 5 with 10 mM $CaCl_2$ | Garneau (2009) [49] |

(continued)

**Table 1**
(continued)

| Subject | Details | Reference |
|---|---|---|
| BIMs and PIMs of *Streptococcus thermophilus* | Strain: DGCC7710<br>Bacterial growth: 42 °C (or 37 °C for overnight)<br>Medium: LM17 (M17 + 0.5 % Lactose)<br>Phage: 2972<br>Infection at $OD_{600}$ 0.5 at 42 °C<br>Phages added at MOI of 5 with 10 mM $CaCl_2$<br>Electroporation of pNT1, harboring chloramphenicol marker.<br>Strains cultured over 5-days (total of nine passages).<br>Plating on LM17-agar with 5 µg/mL chloramphenicol. | Garneau et al. Nature (2010) [4] |
| BIMs of *Streptococcus thermophilus* | Strains: CSK938, CSK939, and CSM944<br>Bacterial growth: 37 °C<br>Medium: Elliker (Elliker + 1 % beef extract)<br>Phages: 5000, 5002, 5077, 5027, 5093, 5196<br>Infection with overnight culture at 37 °C<br>Phages added at MOI ≥1 with 10 mM $CaCl_2$ | Mills et al. J Appl. Microbiol. (2010) [42] |
| BIMs of *Streptococcus thermophilus* | Strains: DGCC7710 and SMQ-301<br>Bacterial growth: 42 °C (or 37 °C for overnight)<br>Medium: LM17 (M17 + 0.5 % Lactose)<br>Phages: 2972, 858, and DT1<br>Infection at $OD_{600}$ 0.6 at 42 °C<br>Phages added at MOI of 0.01, 0.1, and 1 with 10 mM $CaCl_2$ | Dupuis (2011) [50] |
| BIMs of *Streptococcus thermophilus* | Strain: DGCC7710<br>Bacterial growth: 42 °C (or 37 °C for overnight)<br>Medium: LM17 (M17 + 0.5 % Lactose)<br>Phage: 2972<br>Infection at $OD_{600}$ 0.6 at 42 °C<br>Phages added at MOI of 5 with 10 mM $CaCl_2$ | Magadan et al. PLoS One (2012) [21] |

| | | |
|---|---|---|
| BIMs of *Streptococcus thermophilus* | Strain: DGCC7710<br>Bacterial growth: 42 °C<br>Medium: milk solution (1 % nonfat dry milk + dH$_2$O)<br>Phage: 2972<br>Infection at OD$_{600}$ 0.2–0.5 at 42 °C<br>Phages added at MOI of 2 and 10 | Paez-Espino et al. Nat Commun. (2013) [41] |
| BIMs of *Streptococcus thermophilus* | Strain: DGCC7710<br>Bacterial growth: 42 °C<br>Medium: LM17 (M17 + 0.5 % Lactose)<br>Phage: 2972<br>Infection at OD$_{600}$ 0.6 at 42 °C<br>Phages added at MOI of 5 with 10 mM CaCl$_2$ | Dupuis et al. Nat Commun. (2013) [48] |
| BIMs of *Streptococcus thermophilus* | Strain: DGCC7710<br>Bacterial growth: 37 °C<br>Medium: sterile milk (10 % milk powder + water)<br>Phage: 2972<br>Infection at 37 °C for 24 h<br>Phages added at MOI of ≈27 | Sun et al. Environ Microbiol. (2013) [26] |
| BIMs and PIMs of *Escherichia coli* | Strain: K12 with modifications for acquisition activity (BW40114, BW40119, KD29, KD27, BW40302, BW40305, KD60)<br>Bacterial growth: 37 °C<br>Medium: LB (LB + 1 mM IPTG + 1 mM arabinose)<br>Phage: M13<br>Infection at OD$_{600}$ 0.4–0.5 at 37 °C with agitation<br>Phages added at MOI of 10<br>Transformation of pG8_dir or pG8_rev, with ampicillin marker.<br>Plating on LB-agar with 25 μg/mL streptomycin, 25 μg/mL kanamycin, 34 μg/mL chloramphenicol, and 100 μg/mL ampicillin. | Datsenko et al. Nat Commun. (2012) [43] |

(continued)

Table 1
(continued)

| Subject | Details | Reference |
|---|---|---|
| PIMs of *Escherichia coli* | Strain: K12 with modifications for acquisition activity (W3110 derivate Δ*hns*, JW1225)<br>Bacterial growth: 37 °C<br>Medium: LB and 2YT<br>Transformation (chemically competent) of pRSF-1b, with kanamycin marker<br>Plating on LB-agar with 100 μg/mL kanamycin. | Swarts et al. PLoS One (2012) [28] |
| PIMs of *Escherichia coli* | Strain: BL21-AI or IYB5101<br>Bacterial growth: 37 °C with aeration<br>Medium: LB (LB + 0.1 mM IPTG + 0.2 % arabinose)<br>Culture (10–16 h) of strains harboring pCas1 + 2, with streptomycin marker.<br>Plating on LB-agar with 50 μg/mL streptomycin. | Yosef et al. NAR (2012) [44] |
| PIMs of *Escherichia coli* | Strain: BL21AI harboring 2 plasmids (pWUR397, pWUR399)<br>Bacterial growth: N/A<br>Medium: LB (LB + 0.1 mM IPTG + 0.2 % arabinose)<br>Transformation of p1, p2, or pT7Blue plasmid harboring ampicillin marker.<br>Plating on LB-agar with 25 μg/mL streptomycin, 25 μg/mL kanamycin, 34 μg/mL chloramphenicol, and 150 μg/mL ampicillin. | Savitskaya et al. RNA Biol. (2013) [19] |
| BIMs of *Streptococcus mutans* | Strain: OMZ381<br>Bacterial growth: 37 °C, anaerobic<br>Medium: THY (Todd-Hewitt + 0.3 % yeast extract)<br>Phage: M102<br>Infection of exponentially growing culture (serial dilutions) at 37 °C during 30 min<br>Phages added: $10^7$ PFU | van der Ploeg Microbiology (2009) [51] |

| | | |
|---|---|---|
| BIMs of *Pseudomonas aeruginosa* | Strain: PA14<br>Bacterial growth: 37 °C<br>Medium: LB<br>Phage: DMS3<br>Infection of overnight culture at 37 °C for overnight infection<br>Phages added at MOI ≈ 10–100 | Cady et al. J Bacteriol. (2012) [52] |
| BIMs of *Sulfolobus solfataricus* | Strain: P2<br>Bacterial growth: 78 °C, anaerobically<br>Medium: TSY (0.2 % tryptone, 0.1 % yeast extract, 0.2 % sucrose)<br>Phages: virus mixture<br>Infection of 5-day culture at 78 °C<br>Samples taken every 24 h for DNA isolation and growth on TSY medium Gel-rite plates. | Erdman et al. Mol Microbiol. (2012) [53] |
| PIMs of *Streptococcus agalactiae* | Strains: A909RF, 2602 V/R, and BM110<br>Bacterial growth: N/A<br>Medium: N/A<br>Conjugation assays with the plasmid pG+host5 (of strain GMP201), harboring erythromycin marker.<br>Plating on agar medium with 10 mg/mL erythromycin, 100 mg/mL rifampicin, and 10 mg/mL fusidic acid. | Lopez-Sanchez et al. Mol Microbiol. (2013) [54] |

## 1.2 CRISPR-Cas Activity

It is known that CRISPR-Cas systems act in three main stages [9]: (1) acquisition of new repeat-spacer unit(s) in the CRISPR locus (the adaptation step), (2) small RNA biogenesis from the CRISPR-Cas system (the expression step), and (3) resistance to the foreign element for defense (the interference step). These three stages have been described in various systems and organisms. Although the ability to generate small interfering RNAs and to interfere with invasive elements are both critical for CRISPR immunity, we define CRISPR-Cas activity here as the ability to acquire novel spacers naturally following exposure to invasive elements, given that adaptive immunity is the primary biological function of CRISPR-Cas systems.

## 1.3 Spacer Acquisition

In the first step, adaptation (immunization) occurs when spacer(s) are acquired in an active CRISPR locus, which drives the subsequent interference stage. The spacer sequence is identical to the protospacer found in the genome of the foreign element and a new repeat is also added in the CRISPR locus. Protospacer-associated motifs (PAM), flanking the sequence sampled in the invasive element, have been found in many type I and type II systems, and is believed to be involved in the spacer acquisition process [4, 18, 21–28].

CRISPR-Cas type II systems have been well studied because they contain naturally active systems for acquisition and interference (*see* Table 1). They also consist of a small(er) complement of *cas* genes, notably the universal *cas1* and *cas2* genes, as well as the type II signature gene *cas9*. Related studies have shown that the systems can be readily exploited to generate BIMs and PIMs and are very convenient and useful for genome editing applications, notably with the ability to re-program DNA targeting and direct sequence-specific cleavage [29–36]. For the Gram-positive bacterium *S. thermophilus*, two different type II systems are active in several strains [37] and the three main steps have been confirmed experimentally [4, 18, 21, 38].

## 1.4 Spacer Acquisition Procedures

For spacer acquisition analyses, different procedures are available depending on the purpose of the studies. First, it is possible to generate BIMs or PIMs with virulent phages and transformable plasmids, respectively (*see* Table 1). We recommend proceeding to preliminary experimentation to determine the specific procedures needed, even for strains of the same species. Second, spacers currently contained in the CRISPR loci may already match with the chosen foreign element and lead to interference [18]. A basic alignment search (such as BLASTn or CRISPRTarget) should be done with these native spacers before experimentation to avoid unwanted interference. Third, although obvious, it is critical to ensure that the materials are pure (phages, plasmids, and bacterial or archaeal strains) to reduce the difficulty of spacer content analysis. We also recommend using products from a single supplier (e.g., powdered media) and to use the same lot number for more reproducible assays. Indeed, we have seen that results can quantitatively differ when materials from different manufacturers and/or different lots are used. We speculate that medium conditions have an important impact on the molecular interactions and dynamics between bacterial hosts and either phages and/or plasmids.

Because the optimal growth media may differ between microbial species and because phage infection and plasmid replication may require specific conditions, the protocols described below should be considered only as general guidelines. Furthermore, *S. thermophilus* will be used as a model for determining specific procedural details for acquisition, but details for other species will be provided where known. Also, some studies may require a phage lysate with a very high titer (>$10^{11}$ phages per mL) so phage purification and concentration through a cesium chloride gradient may be needed. Finally, the plasmid transformation procedure (electroporation, conjugation, natural competence, etc.) is left to the discretion of the experimenters but should be compatible with the bacterial or archaeal strains employed. In addition, for antibiotic selection markers, resistance tests should be carried out to determine the optimal concentration (*see* Table 1).

## 1.5 BIM and PIM Production Procedures

There are two ways to produce BIMs (for phage infection) and PIMs (for transformed-cell growth): (1) growth in small volumes of liquid medium in tubes or large volumes in bottles, and (2) growth on solid medium in petri dishes. The choice will depend on the needs of subsequent analysis. For example, solid media procedures allow the number of BIMs or PIMs to be quantified and compared using different growth or experimental conditions while liquid-media procedures require a plating step before the isolation of colonies. However, some characteristics, such as the diameter of phage plaques, may be observed only with the use of solid-media. Also, CRISPR-escape mutant phages (CEMs) may be isolated when the phage/bacterial cell ratio is sufficiently high and the first purification step is already completed when the experiment uses solid media [18, 39]. Of note, phage purification should be performed at least three times before storage. Finally, it is our experience that for some phage–host systems it is more efficient to perform infection assays on plates.

Production of BIMs in liquid-media is more suitable for the analyses of the fitness cost or population and evolutionary dynamics, which may require long-term infection (over days or weeks). Liquid cultures also facilitate the process if further DNA or RNA extraction is planned. Also, infection in liquid medium may be done in microplates and followed with a spectrophotometer to evaluate specific conditions rapidly or to follow, over time, the infection of phage and/or CEMs, or to compare both. In that case, growth of the bacterial or archaeal cells may indicate resistance or sensitivity to phages and/or CEMs. Liquid media used for standard culture and infection may not be suitable if it produces fluorescence which interferes with the spectrophotometer readings.

## 2 Materials

### 2.1 Growth of S. thermophilus Bacterial Strains

1. LM17 media: *S. thermophilus* strains typically grow well in M17 medium supplemented with 0.5 % (w/v) lactose (LM17). Use ready-to-use powdered M17 to prepare LM17 in a liquid solution, sterilize in an autoclave and store at 4 °C or preferably at room temperature until use (*see* **Notes 1** and **2**). Aliquot 10 mL of liquid media into 18 mm glass tubes and sterilize by autoclaving (*see* **Note 3**).

2. Use recommended media for other species (*see* **Note 4** and Table 1). When media components are heat-labile, sterilize liquid media through a 0.2 μm filter or add filtered-sterilized component to autoclaved media.

3. LM17 agar plates (including bottom plates for phage assays): add agar at 1–1.5 % (w/v) to a liquid solution of LM17, sterilize in an autoclave and pour media into sterile petri dishes (e.g., 9 cm diameter) (*see* **Notes 5–7**).

4. LM17 top-agar (for the double agar overlay method): add agar at 0.75 % (w/v) to a liquid solution of LM17, sterilize in an autoclave and hold at 55 °C for immediate use or allow it to solidify and store at room temperature for later use (*see* **Notes 4** and **8**). Sterilize glass tubes (13 mm) by autoclaving. These tubes will serve for holding aliquots of top agar. Use a block heater (or water bath) to keep the top media at 55 °C before plating (*see* **Note 9**).

5. Long-term bacterial storage: Sterilize glycerol in 2 mL stock tubes containing 150–250 μL of glycerol by autoclaving with the cap fitted loosely and close them tightly after autoclaving. Store at room temperature. Wash cells twice with the appropriate growth medium (LM17), resuspend the resulting bacterial pellet in the same medium and add to the sterile glycerol to a final concentration of 15–25 % (v/v) (*see* **Notes 10–13**). Use ethanol-, isopropanol- or ethylene glycol-dry ice slush to flash-freeze the bacterial strain stocks and store at −80 °C.

### 2.2 Growth of S. thermophilus Phages

1. Phage buffer: 0.05 M Tris–HCl, pH 7.4, 0.1 M NaCl, 0.008 M $MgSO_4$. Many phages, such as those infecting *S. thermophilus* strains, may be diluted and stored in phage buffer (*see* **Note 14**). For high-titer phage lysate procedures, use the phage buffer for dialysis and for phage storage.

2. LM17 media containing CaCl$_2$: Prepare a 2 M CaCl$_2$ stock solution and sterilize smaller volumes by autoclaving when needed (*see* **Note 15**). Many *S. thermophilus* phages amplify better in LM17 media containing CaCl$_2$ (0.01 M). Some phages strictly require CaCl$_2$. A range of CaCl$_2$ (or other cofactors) concentrations should be tested to determine the optimum concentration for complete phage infection (*see* **Note 16**). Add the sterile stock to sterile liquid or solid media. When working with phages, these bottles of cofactors are a common source of contamination and they should be replaced very regularly (daily or weekly).

3. Glycine: Prepare a glycine stock solution (2 M) and add at 0.25 % (v/v) to sterile LM17 liquid medium (with or without CaCl$_2$ for *S. thermophilus* phages) before plating and/or use. The addition of glycine is sometimes required to improve phage plaque visualization on a plate (*see* **Note 17**).

4. When needed, add at 0.25–1.25 % to sterile liquid media or heated agar-media before plating. The glycine destabilizes the peptidoglycan cell wall of bacteria and promotes cell lysis [40].

5. Short-term phage storage: phage lysates may be preserved at 4 °C in phage buffer or the infection-medium (LM17-CaCl$_2$) (*see* **Note 18**).

6. Long-term phage storage: add sterile glycerol to phage lysates to a final concentration of 15–25 % and store at −80 °C (*see* Subheading 2.1 **step 5** and **Notes 19–21**). Similar to bacteria, key phage isolates should also be lyophilized. For phage sensitive isolates, we suggest to freeze or to lyophilize phage-infected cells (*see* **Note 22**). The latter is particularly useful for shipping to collaborators.

*2.3 Assays in Microplates*

1. Media: tests should be done to make sure the media is compatible with the microplate reader.

2. Cofactors: when a cofactor is necessary, tests should be done to make sure the addition of the cofactors will not change the media turbidity.

3. Plastic cover: to stick to or to seal the microplate border to avoid contamination of the phage or bacteria with aerosols.

4. Ready-to-use microplates: fill with medium in advance and stored at −20 °C. The microplates should be thawed without agitation (e.g., overnight at room temperature) before use. We recommend using the same medium lot number for linked assays.

*2.4 PCR Amplification and Sequence Analysis*

1. Design a set of primers for each CRISPR locus. Since the length of the PCR product is limited, the position of the primers depends on many parameters, such as the spacer and repeat length, the number of native repeat-spacer-units, and the

number of inserted/deleted units (*see* **Notes 23** and **24**). The conservation of sequences flanking CRISPR loci also varies widely across systems. *See* Table 2 for a list of primers for particular strains.

## 3 Methods

### 3.1 Bacteriophage-Insensitive Mutant Production

#### 3.1.1 BIM Production Using Solid Medium

1. Inoculate 10 mL of fresh LM17 medium at 1 % (v/v) with an overnight culture and grow to an optical density at 600 nm ($OD_{600}$) of 0.2–0.8 (*see* **Notes 25–28**). Use the volume of culture required and re-incubate the strain for additional sub-culturing, if needed.

2. Inoculate top-agar medium (0.75 % agar held at 55 °C) with 500 μL of the subcultured strain. Use cultures at an $OD_{600}$ of 0.4, 0.8 and 1.2. Add 100 μL of an appropriate phage dilution to achieve multiplicities of infection (MOI, which is the ratio of viral particles to the target host numbers) of 1, 0.1 and 0.01 (*see* **Notes 29** and **30**). Set up in triplicate (i.e., a total of nine conditions in triplicate).

3. Pour inoculated top-agar media on a solid agar media (1–1.5 % (w/v) agar), containing the appropriate cation, if necessary (for *S. thermophilus*: $CaCl_2$ at 10 mM).

4. Incubate agar plates for 24–48 h at the appropriate temperature until single isolated colonies appear (*see* **Note 31**).

5. Test the isolated colonies by PCR amplification (*see* Subheading 3.2.1) and phage sensitivity (*see* **Note 21**) to determine, respectively, the CRISPR-dependent resistance and the level of phage resistance (*see* **Note 32**).

6. For BIMs, the phage sensitivity should be tested by spot test or complete test (*see* **Note 21**). The efficiency of plaquing (EOP) is calculated from the ratio of the phage titer obtained on the BIM divided by the titer obtained on the wild-type strain. For *S. thermophilus*, the EOP of a BIM that has acquired a new repeat-spacer unit in a CRISPR locus is approximately between $10^{-5}$ and $10^{-6}$ for 2972-like phages.

#### 3.1.2 BIM Production Using Liquid Medium

1. Amplify the phages in liquid media using a standard procedure (*see* **Notes 33** and **34**). Add the phage at an $OD_{600}$ of 0.3–0.4 for *S. thermophilus*.

2. Incubate the tubes overnight at 37 °C (*see* **Notes 35** and **36**). BIMs should emerge after the massive lysis of phage sensitive cells.

3. Centrifuge the culture and resuspend the cell pellet in 10 mL of fresh pre-warmed media for faster BIM growth.

# Table 2
## PCR primers to amplify each entire locus of several microbila species

| Locus | Primer name | Primer sequence (5'–3') |
|---|---|---|
| *Streptococcus thermophilus* [3, 4, 18, 21, 26, 37, 41, 48] | | |
| CRISPR1 | CR1_UP | TGCTGAGACAACCTAGTCTCTC |
| | CR1_DOWN | TAAACAGAGCCTCCCTATCC |
| CRISPR2 | CR2_UP | GCCCCTACCATAGTGCTGAAAAATTAG |
| | CR2_DOWN | CCAAATCTTGTGCAGGATGGTCG |
| CRISPR3 | CR3_LeadF1_UP | CTGAGATTAATAGTGCGATTACG |
| | CR3_TrailR2_DOWN | GCTGGATATTCGTATAACATGTC |
| CRISPR4 | CR4_UP | CCTCATAGAGCTTTGAAAGATGCTAGAC |
| | CR4_DOWN | CTATCTTTAAGATATGCTGCTTACAACGGC |
| *Sulfolobus solfatarius* [53] | | |
| CRISPR Locus A | Forward | AGCTTCTGACCCGCTCCTGA |
| | Reverse | GCACATCATCAAACAATGGTAAGCC |
| CRISPR Locus B | Forward | AGGGGTTTGTGGGATGGGTTGTG |
| | Reverse | ACAACTACCACCACTACCACGG |
| CRISPR Locus C | Forward | TCGCTTATCTCTCTCATGCGCCATT |
| | Reverse | TGTCCCGTTTTTGTAAGTGGGGG |
| CRISPR Locus D | Forward | AGTTCCACCCCCGAAGCTCCT |
| | Reverse | AGCCGGGACAAGTTTCACAAATTGA |
| CRISPR Locus E | Forward | ATAGGGAAAGAGTTCCCCCG |
| | Reverse | TGACTCTAGTGCAATCTTCGA |
| CRISPR Locus F | Forward | CGGCGTTATAATGGGTATCGGAATCGG |
| | Reverse | GCTCACTATCTCACCCCTATCAATACCC |
| *Escherichia coli* [28, 43] | | |
| CRISPR I | Ec_LDR_F | AAGGTTGGTGGGTTGTTTTTATGG |
| | Ec_I_sp2 | CGGCATCACCTTTGGCTTCGGCTG |
| CRISPR II | Ec_II F | AACATAATGGATGTGTTGTTTGTG |
| | Ec_II R | GAAATGCTGGTGAGCGTTAATG |
| CRISPR 2.1 | BG3474 | AAATGTTACATTAAGGTTGGTG |
| | BG3475 | GAAATTCCAGACCCGATCC |
| CRISPR 2.3 | BG3414 | GGTAGATTTTAGTTTGTATAGAG |
| | BG3415 | CAACAGCAGCACCCATGAC |
| *Streptococcus mutans* [51] | | |
| CRISPR1 | Cru | CATGATTCAAGCTGATCCTA |
| | Crd | CGACTCATCTCATTAAATCC |
| | Cru2 | GAGATGAATGGCGCGATTAC |
| | Crd2 | CGGAACGGTCATTACCTA |

(continued)

**Table 2**
**(continued)**

| Locus | Primer name | Primer sequence (5′–3′) |
|---|---|---|
| CRISPR2 | Cau | AATCTGGGTTGCACATCAAA |
|  | Cad | TATAAGCTGATGGCGTTAAG |
|  | Cau2 | TCACAAACGGTGCAACTAC |
|  | Cas2 | GTTGCTTTCATGGACGAACTC |
| *Streptococcus agalactiae* [54] | | |
| CRISPR1 | PCRF | GAGACACAGAGCGACACTATC |
|  | PCRR | CATTTCTTTCTCCACTATTATAAC |
|  | SEQF | GAAGACTCTATGATTTACCGC |
|  | SEQR | CAGCAATCACTAAAAGAACCAAC |
| CRISPR2 | 1F | GATGTCAATACGAAGACTGTAGC |
|  | 1R | ACAAATCAATCGCATCTC |
|  | 2F | CGTGTGGAAACTATTGG |
|  | 2R | GAAACTACACACCCGCAAAG |

4. Incubate overnight at 37 °C and repeat **steps 3** and **4** twice more, as needed (*see* **Notes 37–39**).

5. Streak the "final" culture on solid media (1.5 % agar) and incubate overnight at 42 °C, until single isolated colonies appear.

6. Sample single isolated colonies (with a sterilized truncated tip or toothpick) and inoculate a 10 mL tube of fresh media. Incubate overnight at 42 °C.

7. Confirm the CRISPR-linked resistance by verifying the repeat-spacer content at each locus of the strains (*see* **Note 40**). Purify each BIM independently (*see* **Note 41**) before the sequence analysis (*see* Subheading 3.2) and prepare stocks for storage (*see* Subheading 2.1).

8. For pure BIMs, test the phage sensitivity using a spot test or complete test to calculate the EOP (*see* **Note 21**).

*3.1.3 BIM Production Using Microplates*

1. Prepare microplates by filling each well with the appropriate medium or thaw a previously prepared microplate stored at −20 °C (*see* **Note 42**). For an example plate layout *see* Fig. 1.

2. Prepare a pre-culture by incubating the bacteria overnight at 37 °C without shaking for S. *thermophilus*.

3. Prepare the microplate by adding the appropriate concentration of any required cofactors to each well. For S. *thermophilus* phage, the $CaCl_2$ cofactor should be added to a final concentration of 10 mM (*see* **Note 43**).

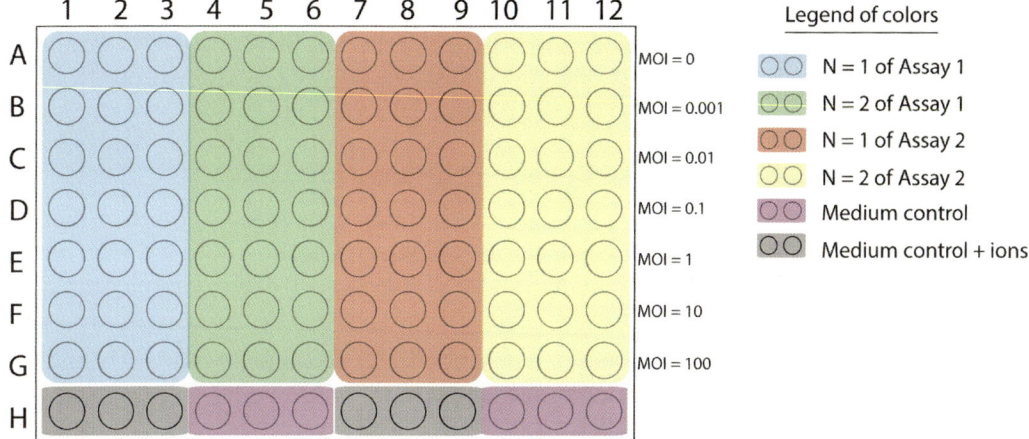

**Fig. 1** Example of the design of a microplate to isolate CRISPR BIMs. Assay 1 is BIM production with the addition of phage at an $OD_{600}$ of 0.2. Assay 2 is the same as assay 1 but with the addition of phage at the start of the assay ($OD_{600}$ of ≈0). Rows A to G represent increases of the MOI from 0 to 100. Row A functions as the wells for bacterial growth control since no phages are added. In row H, *purple* wells are blank controls and *grey* wells are blank controls containing the appropriate ions (example, $Ca^{2+}$). The resulting curve should show the decrease in bacterial population during phage infection then another increase due to the growth of BIMs

4. Prepare at least duplicate samples for each of the different conditions to validate the readings (*see* **Note 44**). Include appropriate control wells for each set of conditions (*see* **Note 45**). Controls include wells containing medium and cofactor only as well as the positive control for bacterial growth. Carry out each assay in triplicate (*see* **Note 46**).

5. Add the phages and then the bacterial culture to all wells except for the control wells (*see* **Notes 47–49**).

6. Incubate the microplate in the reader when possible or in a standard incubator until the end of the assay (*see* **Note 50**).

7. Single BIM colonies must be isolated for further characterization. *See* **Notes 41** and **21**, respectively, for BIM purification and phage sensitivity evaluation steps.

*3.1.4 PIM Production*

1. Transform a plasmid into the strain being studied and add the resulting mixture to 10 mL of fresh media. Incubate cultures and prepare stocks from the resulting cultures (*see* **Note 51**). For *S. thermophilus* DGCC7710 [4], the plasmid pNT1 can be transformed by electroporation, then the derivatives grown on LM17 agar containing 5 μg/mL of chloramphenicol.

2. Inoculate 10 mL of fresh pre-warmed LM17 medium at 1 % (v/v) for 12 h per passage to achieve the number of passages required for the study. For *S. thermophilus* DGCC7710-pNT1 [4], a total of nine transfers can be done over the course of 5 days (*see* **Note 52**).

3. Dilute a sample of the culture and spread it on solid agar medium. Screen 100 colonies for the appropriate antibiotic sensitivity. Plasmid stability is analyzed by comparing the number of colonies growing on LM17 agar that have or do not have the marker. For *S. thermophilus* DGCC7710-pNT1 [4], the marker is chloramphenicol sensitivity.

4. Analyze the CRISPR content of each locus for each marker-sensitive colony using PCR amplification coupled with amplicon sequencing (*see* Subheading 3.2). These colonies are good candidates for PIMs.

5. Make stocks of confirmed PIMs with one or more new inserted spacers targeting the plasmid.

6. To verify the transformation interference, transform a selected PIM (containing a new spacer) by electroporating a second time with the same plasmid or for first time with another plasmid carrying the same target sequence. Calculate the efficiency from the number of colony-forming units (CFU) per microgram of plasmid DNA.

7. Verify the plasmid content of the PIM before and after the plasmid transformation using a DNA extraction protocol and PCR amplification.

## 3.2 Spacer Analysis

1. With confirmed phage-resistant BIM pure cultures, verify the content of each CRISPR locus using PCR amplification (*see* Subheading 3.2.1) followed by PCR product sequencing (*see* Subheading 3.2.2) and manual sequence analysis. Observing amplicons of a larger size typically reflects novel spacer acquisition. With known CRISPR repeat and CRISPR spacer sizes, it is possible to calculate (or estimate) the number of acquired spacers within a CRISPR locus.

2. Each CRISPR locus should be amplified in an independent PCR reaction (*see* **Note 53**).

3. The PCR products may or may not need to be purified before sequencing, depending on the sequencing procedure to be used. Carry out trials when determining the steps of the procedure to determine whether this step may be removed to save time and materials.

4. The mode of sequencing is at the discretion of the user and subject to the availability of sequencing platforms, but we recommended sequencing at least once with each forward and reverse primer to get two reads per CRISPR locus per BIM.

5. The repeat-spacer unit is then analyzed manually to find insertions and/or deletions. The BIM sequences are compared with the sequences of the wild-type strain, which is also treated as a sample during BIM spacer analysis.

### 3.2.1 PCR-Amplification of CRISPR Loci

1. Design a set of primers for each CRISPR locus. As the length of the PCR product is limited, the position of the primers depends on many parameters, such as the spacer and repeat length, the number of native repeat-spacer-units, and the number of inserted/deleted units. (*see* **Notes 23** and **24**). *See* Table 2 for a list of some primer examples.

2. Carry out a standard PCR amplification. We recommend trying more than one set of primers and testing with a temperature gradient (if possible with the available PCR apparatus) to find the best conditions for each locus.

3. The insertion/deletion of units is confirmed visually by the size of the migrated PCR product on the appropriate agarose gel and comparison with the wild-type strain (*see* **Note 54**).

4. The PCR products may or may not be purified before sequencing to obtain the sequence of the complete locus.

### 3.2.2 CRISPR Sequence Analysis

1. Use both sequences of each locus (with reverse and forward primer) to determine the consensus sequence for each BIM and control (wild-type strain). For *S. thermophilus*, primers are listed in Table 2 and an example of the sequence treatment is presented in Fig. 2.

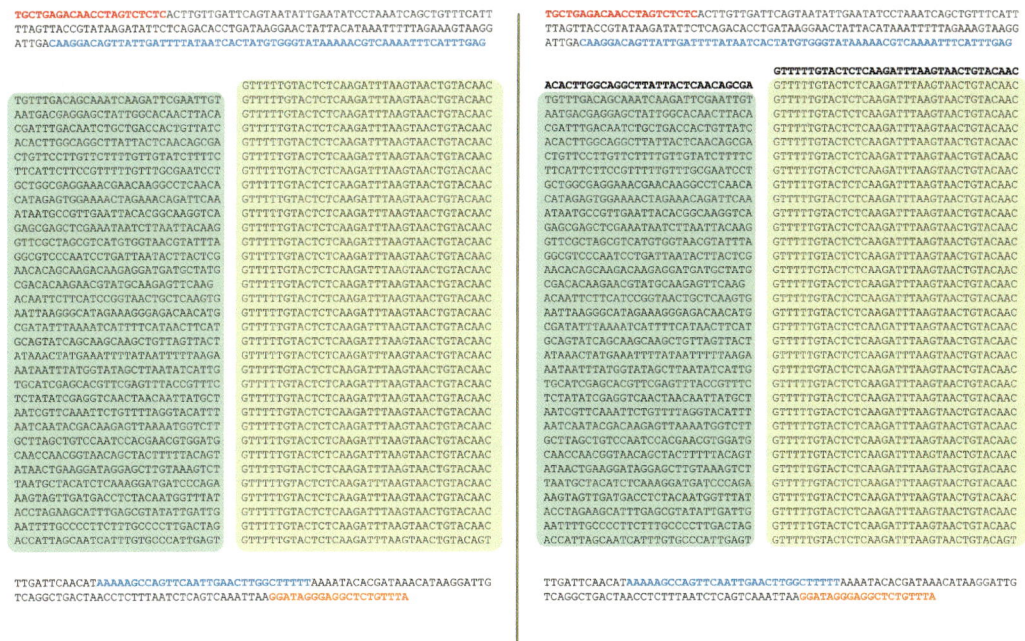

**Fig. 2** Example of sequence analysis with CRISPR1 of *Streptococcus thermophilus* DGCC7710. BIM sequence analysis is easier using a text program. *Left side*: the CRISPR1 locus of *Streptococcus thermophilus* DGCC7710. *Right side*: a BIM with a new spacer-unit inserted into its CRISPR1 locus. Sequences are presented from the forward primer CR1_UP (in *red*) to the reverse primer CR1_DOWN (in *orange*). The leader and the terminator are *blue*. The *green* and *yellow boxes* are the spacers and repeats, respectively. On *the right side*, the new spacer and repeat unit is *bolded*

2. Enter the sequence into a text program such as Microsoft Word®, highlight all repeat sequences (*see* **Notes 55** and **56**) and insert a line break after all of them. For *S. thermophilus* CRISPR 1 and 3 loci, the repeat sequences are, respectively: 5′-GTTTTTGTACTCTCAAGATTTAAGTAACTGTAC AAC-3′ and 5′-GTTTTAGAGCTGTGTTGTTTCGAATGT TCC AAAAC-3′.

3. Find the 5′- and 3′-external sequences of the loci (i.e., the leader and terminator in many cases of CRISPR loci) and mark them with bold text. For *S. thermophilus* CRISPR 1 locus, the leader and terminator sequences are, respectively: 5′-CAAGGA CAGTTATTGATTTTATAATCACTATGTGGGTA TAAAAACGTCAAAATTTCATTTGAG-3′ and 5′-AAAAAG CCAGTTCAATTGAACTTGGCTTTTT-3′.

4. For all spacers (at the left of each repeat), identify the ones that already exist in the native locus (from the wild type parental strain) and locate all insertion and/or deletion events. It is suggested to name the new spacer following the nomenclature for BIM acquisition (S+#, e.g., S1, S2, S3, and S4, [18]) (*see* **Note 57**).

5. List all spacers found for each BIM-phage or PIM-plasmid pair. The list may be shared, or not, depending on the interest of the community (*see* **Note 58**).

6. If the sequence of the phage or plasmid is known, locate the associated protospacer sequence. Identify the protospacer-associated motif (for type I and II), which is flanking the protospacer. RNA biogenesis or interference experiments may be done with the same or another invader (phage or plasmid) harboring a targeted protospacer with a good protospacer-associated motif (*see* **Note 59**).

In summary, the aforementioned procedures, which are primarily derived from the *S. thermophilus* model organism, where type II CRISPR-Cas systems are active, will allow users to readily and rapidly carry out experiments that generate CRISPR mutants following exposure to invasive genetic elements. Specifically, BIMs can be readily generated using phages to challenge the host, while PIMs can readily be generated using plasmids. Novel spacer acquisition reflecting CRISPR immunization events can be monitored using PCR-based screening, to monitor CRISPR locus size and specifically select mutants with an increase in the number of CRISPR spacers within a locus focusing on the leader end of the array, where polarized novel spacer acquisition is typically observed. We urge the users to ensure that the aforementioned procedures are adapted to their model organism, and that the protocols be optimized as needed.

## 4 Notes

1. The results may differ according to the media suppliers. We recommend that you do not mix different suppliers and do not change the supplier for linked experiments. Some pre-mixed dehydrated formulas are better than others—use the highest quality available to your laboratory. It is also better to use the same lot number for linked assays for better reproducibility.

2. In some papers, other growth media were used for assays with *S. thermophilus* strains, e.g., milk [26, 41] and Elliker broth [42].

3. Sterilized liquid or solid media are typically stable for several weeks at room temperature, but "old" media may significantly influence the results of some procedures, particularly for duplicate and triplicate experiments. We recommend preparing solutions close to the time of use in an assay and to verify the expiration date of the product to avoid variation in experimental results.

4. The most common media is LB medium for *E. coli* and *Pseudomonas aeruginosa*, THY medium for *Streptococcus mutans* and TSY medium for *Sulfolobus solfatarius*. In theory, any medium suitable for growth of your microorganism of interest may be used in a BIM or PIM production assay. For species and strains with several media choices, the different media should be tested to find the best one for bacterial mutant growth.

5. The choice between LM17 containing 1 % (w/v) or 1.5 % agar (w/v) depends on the use. A higher percentage of agar is needed for streaking (to avoid scratching the agar) and for recuperating lysis plaques (with truncated sterile tips, for example). For estimating phage titer, 1 % agar is sufficient.

6. Depending on the purpose, square petri dishes may be used. They make it easier to evaluate a phage titer using spot-tests but are usually more expensive.

7. Agar plates may be kept for a few days at room temperature but stored at 4 °C for longer periods. If needed, leave dishes open for 10 min under a laminar flow hood to remove excess humidity before use.

8. Top media may be cooled and re-warmed a couple of times, but we suggest avoiding too many cycles to reduce phage contamination and preserve media quality. When preparing a large amount of top media it is better to prepare many small bottles rather than a few big bottles.

9. When the temperature of the top medium decreases (but is still liquid), the top agar may solidify before plating is complete.

The bacterial lawn and/or the dispersion of the plaques or bacterial colonies will be significantly affected. Reheating above 100 °C to dissolve agar (and subsequent cooling at 55 °C) may be warranted.

10. Be mindful of bacterial contamination (from related or unrelated strains) from aerosols and the lab environment. For CRISPR-experimentation, sequencing of CRISPR loci is the best confirmation for strain identity (see Subheading 3.2 for details). It is not uncommon to observe spacers derived from aerosolized phages within a laboratory environment, especially in deep sequencing analyses.

11. Preserve stocks at −80 °C and avoid too many cycles of defrosting-freezing during the inoculation from storage stock to maximize long-term bacterial viability.

12. Even though overnight cultures are adequate for preparing bacterial stocks, we recommend using a mid-exponential phase culture for long term storage.

13. We recommend storing cultures (minimum of two tubes) in liquid nitrogen in addition to a series at −80 °C. Finally, if possible, key bacterial strains should be lyophilized.

14. For stock solutions of phage buffer, prepare a 10× solution, sterilize and store it at room temperature. Prepare a 1× solution by diluting the 10× stock with sterile water in a pre-sterilized bottle or with non-sterile water or bottle and sterilize the buffer again.

15. Some ion solutions may form precipitates during sterilization. These solutions may be sterilized by filtration. To avoid the problem of ion precipitates, we suggest adding sterile ion solutions to sterile media rather than adding ions to media before sterilization.

16. A way to test the adequate concentration of a cofactor, it is to select 4–6 different concentrations of it (for example with $CaCl_2$, 0, 5, 10, 15, and 20 mM) to add to the liquid medium during phage amplification. The "standard" concentrations are 5–10 mM. Different cofactors ($Ca^{2+}$ from $CaCl_2$, $Mg^{2+}$ from $MgCl_2$, etc.) may also be tested. The optimum conditions are visualized by plotting a graph of the cofactor concentration (in mM) vs. the phage titer (in phages/mL) of the independent phage infections carried out using the same inoculated culture.

17. The addition of glycine may significantly increase the plaque sizes. Glycine may be added in plate, top-agar or both, depending on the phage.

18. Remove bacterial cells from the infection-medium using a 0.45 μm filter.

19. Experimenters should be attentive to phage contamination sources, such as aerosols produced during pipetting and the lab environment (clothes, benches, etc.). The use of an adequate (tested) disinfectant is also strongly encouraged to limit phage contamination in the laboratory settings.

20. We recommend preparing a series of at least two tubes of stocks for storage of each pure phage. Preserve at –80 °C and avoid too many cycles of defrosting-freezing to maintain phage infectivity over time.

21. To verify phage infectivity and/or titer (number of infectious viruses per volume), add approximately 300–500 µL of an exponentially growing or an overnight bacterial culture to 3 mL of $CaCl_2$-containing top agar and plate it on an agar-plate. After a short period to dry, carry out a spot test to evaluate the phage titer by pipetting 10 µL of different dilutions of the phage stocks or amplification on the top agar. Incubate under the appropriate bacterial growth and phage amplification conditions (for *S. thermophilus*, 42 °C without other specific conditions) until phage plaques or lysis zones appear. For precise phage titer evaluation, we recommend carrying out a complete test. In that case, 50–100 µL of a non-diluted phage lysate is added to approximately 300–500 µL of a growing or overnight bacterial culture and 3 mL of $CaCl_2$-containing top agar. Pour the top agar on an agar-plate and repeat with serially diluted phage lysate (e.g., $10^{-1}$ to $10^{-7}$). Each dilution should be plated in triplicate. Incubate until phage lysis plaques appear, count the number of plaques and evaluate the titer using the dilution producing a count between 30 and 300 plaque-forming units (PFU). The titer of the non-diluted phage lysate is calculated in PFU per mL.

22. It might be important in some assays to know the latent period of the phage (at least approximately) to determine when the infection should be stopped by flash-freezing. An alternate method may be required for phages that do not withstand standard phage storage procedures, but the technique should be validated and the resulting stock tested for phage infectivity.

23. For a small CRISPR locus, primers may be designed to anneal beyond the ends of the locus and with variations in the primer parameters, such as melting temperature, length, and nucleotide content. For a large CRISPR locus, the primer design possibilities are more limited since covering the entire sequence may require more than one PCR reaction. In many CRISPR-Cas systems, the insertion of a new repeat-spacer unit occurs in a specific side of the locus (mostly 5′-end), so we suggest selecting one primer set to cover that area (e.g., at the leader end of the CRISPR1 and CRISPR3 loci of *S. thermophilus* strain) and another set to completely cover the locus. For example, for the locus CRISPR1 in strain *S. thermophilus*

DGCC7710, we first amplify the leader end extremity with an outside primer annealing within the leader and an internal primer annealing within the seventh spacer sequence.

24. *S. thermophilus* DGCC7710 is presented as an example in Table 1. Four sets of primers were designed to amplify all four CRISPR loci in this strain. Usually, we use a second set of primers for the CRISPR1 locus (composed of 32 spacers) to help visualize the overall insertion/deletion of repeat-spacer units. These sets of primers may be used with many other *S. thermophilus* strains but should be tested first to verify the length of each PCR product.

25. We recommend preparing more than one tube (at different concentrations, such as 0.5, 1, and 2 %) to avoid exceeding the desired optical density ($OD_{600}$).

26. The maximum number of BIMs may depend on the growth state of the bacterial cells. Some species or strains may need a different optical density than *S. thermophilus*. It is best to test different bacterial concentrations (e.g., $OD_{600}$ = 0.2, 0.4, 0.6, 0.8, 1.0, and 1.2) to find the one with the greatest number of phage-resistant colonies. Then, test to verify the CRISPR active loci to determine the best $OD_{600}$ for CRISPR BIM production (*see* Subheading 3.2 for spacer analysis details).

27. An overnight culture may be used for CRISPR BIM production assays if the goal is to rapidly test CRISPR BIM production. Further screening of purified BIMs by CRISPR loci verification is still needed to confirm CRISPR BIM isolation. Still, it is best to determine the optimal $OD_{600}$ and MOI (*see* **Note 29**) for CRISPR BIM isolation.

28. For some CRISPR-Cas systems, it is necessary to activate the system. This is the case for *E. coli* and procedures are now well-established to obtain BIMs and PIMs with strains of that species [19, 28, 43, 44]. For *E. coli*, the absence of acquisition activity for CRISPR-Cas systems is due, for example, to the repression by the protein H-NS [45–47].

29. The maximal quantity of BIM may depend on the ratio of phage particles/bacterial cells (MOI). Some species or strains may need a different MOI than *S. thermophilus*. We recommend testing different MOIs (e.g., 0.01, 0.1, 1, 10, and 100) to find the one leading to the greatest number of phage-resistant colonies. Verify the CRISPR active loci to confirm the CRISPR-linked phage resistance and to find the optimal MOI for CRISPR BIM production (*see* Subheading 3.2 for spacer analysis details).

30. We recommend testing the $OD_{600}$ and MOI simultaneously even if these assays require a lot of materials (tubes, media, solutions, and phage and bacterial stocks and cultures). To reduce the amount of materials needed, you may use only

3 ODs (0.4, 0.8, and 1.2) and 3 MOIs (0.01, 0.1, and 1), which represents only 9 different sets of assay conditions. More precise assays may be carried out to determine the optimal conditions for CRISPR BIM production. For *S. thermophilus*, these conditions had to be determined for each phage–host system.

31. For *S. thermophilus*, it helps to reduce the incubation temperature (37 °C) compared to the optimal growth temperature (42 °C) to obtain more BIM colonies.

32. We found that the best conditions for *S. thermophilus* DGCC7710 infected by the virulent model phage 2972 were: $OD_{600}$ of 0.6–0.8, MOI of 0.1 and incubation overnight at 37 °C.

33. A standard phage amplification in liquid media consists of adding a volume of phage lysate or storage stock to a growing bacterial culture and incubating until the complete lysis of the host. For BIM production, the incubation of the infected-culture is prolonged to allow the growth of phage-resistant bacterial cells. For *S. thermophilus*, inoculate 10 mL of LM17 liquid medium with 100 μL of an overnight bacterial culture and incubate the culture at 42 °C until it reaches an $OD_{600}$ of 0.3. At that time, add 50 μL of 2 M $CaCl_2$ as well as 50–100 μL of a phage lysate (>$10^8$ PFU/mL. PFU, plaque forming units). Complete lysis usually occurs in 3–5 h.

34. We recommend using three controls: (1) a sterile 10 mL tube of media without any bacterial cells or phage particles (contamination control), (2) a 10 mL tube of media inoculated with the host strain (bacterial-growth control), and (3) a 10 mL tube of media inoculated with the host strain and infected with the phage (phage infection control). The phages infection control tube should be removed from the incubator after complete lysis and stored at 4 °C for further phage titer evaluation. If the phage infection did not proceed well, it will be difficult to calculate the phage-resistance of the BIMs and it is better to do a new assay for BIM production. Usually, a phage-sensitive *S. thermophilus* strain allows the phage infection to obtain a phage titer $\geq 10^9$ PFU/mL. See **Note 21** for details on phage titer estimation.

35. The optimal bacterial growing temperature for *S. thermophilus* is 42 °C.

36. *S. thermophilus* is a lactic acid bacterium and by lowering the temperature, we reduce lactic acid production. Furthermore, *S. thermophilus* does not tolerate long incubation at 42 °C in liquid medium.

37. The second culture of the BIM may affect CRISPR diversity in the bacterial population. Some BIMs may be under- or

over-represented. Depending on the goal of the study it may be better to avoid repeating these steps.

38. **Steps 1** and **2** may also be repeated with a CRISPR-resistant phage (a CEM, CRISPR-escaping mutant previously isolated from a CRISPR BIM using the same host-phage couple) or another phage able to infect the same host. This may increase the number of new spacers inserted into one or many active CRISPR loci. For *S. thermophilus*, BIMs of second generation were produced using phages 2972 and 858, which are approximately 91 % identical at the genome level [18].

39. **Steps 1–4** may be repeated any number of times, as for long-term infections [41]. But readers should be aware that for some species or strains, too many growth assays may alter other bacterial characteristics. For example, an industrial *S. thermophilus* strain, which is used to produce yoghurt, may lose desirable fermentation capacity. If the impact of CRISPR insertion is being evaluated for industrial strains, it may be best to limit rounds of infection/culture.

40. Even if CRISPR-resistance is confirmed, we recommend evaluating phage adsorption to the bacterial host to be sure that phage adsorption is not impaired. With *S. thermophilus*, we seldom isolated a CRISPR BIM that was also mutated in a phage receptor [50]. The presence of multiple phage resistance mechanisms may complicate the analysis of the CRISPR-Cas system.

41. Purification of each desirable BIM can be accomplished by streaking on a LM17 plate. Carry out at least three successive rounds of streaking and colony isolation before making bacterial stocks for long-term storage and use the same culture stocks for sequence analysis. Be alert for BIM cross-contamination of stock since these mutants usually differ, genetically, by only one or few spacers. We recommend verifying the CRISPR loci of a BIM before using them in any given assay.

42. We tested LM17 medium for BIM production but it was naturally fluorescent and significantly altered the resulting data from the microplate reader. We changed the growth media to bromocresol purple (BCP) medium (2 % tryptone, 0.5 % yeast extract, 0.4 % NaCl, 0.15 % Na-acetate, 40 mg/L of bromocresol purple).

43. The quantity of added cofactor may increase the readings from the microplate reader by too large a factor. In this case, use less than the optimal cofactor concentration determined for phage infection.

44. A way to assay duplicate samples is to separate the plate into two parts and to fill the wells the same way for both parts. *See* Fig. 1 for an example of the microplate design.

45. The controls should include: (1) a well containing only sterile medium and the cofactor (contamination control) and (2) a well containing sterile medium, the cofactor and the bacterial culture (bacterial growth control). Note that the assay does not need a phage infectivity control (*see* **Note 21**) because the reads will provide this information.

46. If the microplate has a sufficient number of wells, duplicates may be included in the same plate, but totally independent assays are recommended.

47. The volume of phage used will depend on the titer and the desired MOI. The volume of bacterial culture used will depend on the desired quantity of bacterial cells per well at the beginning of the assay. Also, the bacterial titer depends on the $OD_{600}$ (evaluated by plating on agar media) which should be determined before beginning the assay. The desired values for these parameters will depend on the purpose of the assay.

48. Because of the proximity of the wells to each other, use care when adding phages and bacterial culture to avoid contamination.

49. Phages may also be added before the beginning of the microplate incubation if the assay needs a specific $OD_{600}$. In that case, add phages quickly if time is an important parameter of the assay.

50. The procedure is best carried out using an incubating microplate reader.

51. Confirm the nucleotide sequence of the plasmid and the CRISPR content of the original strain.

52. For *S. thermophilus*, cultures were grown at 42 °C in the morning (during the day) and at 37 °C in the evening (for overnight incubations). As indicated previously, *S. thermophilus* does not tolerate well long incubation at 42 °C in liquid medium.

53. We attempted a multiplex PCR protocol using the four CRISPR loci of *S. thermophilus* DGCC7710 and a protocol using only the two active CRISPR loci from that strain. The results were not very convincing at least in our hands, therefore, we chose a procedure using independent reactions. But multiplex protocols may be designed for other species and strains if the selected sets of primers are compatible in a single reaction.

54. A migration shift is visible between the PCR product amplified from the locus of the BIM and the one amplified from the wild-type strain. For the CRISPR1-Cas system of *S. thermophilus* DGCC7710, an insertion of a single repeat-spacer unit results in a 66 nt increase in length.

55. We suggest highlighting the sequence as follows: repeats in green, native spacers in yellow, leader in blue, terminator in red, and the new repeats in grey. *See* Fig. 2 for details.

56. With alignment programs such as BioEdit, the repetitive nature of the locus and the random sequence of the new spacers complicates the analysis of many sequences. The alignment is not always correct and spaces may be inserted in the wrong locations. Highlighting the repeat in a text program will allow you to see the repeat with a non-consensus sequence (vs. natural variation of the sequence, a mutation during spacer acquisition, or a sequencing mistake) and to easily find spacers with one more or one less nucleotide.

57. We suggest calling the spacer S# and the linked protospacer PS# to avoid confusion.

58. A new phage challenge or plasmid transformation may be used to analyze other CRISPR-Cas steps such as the biogenesis of and interference by small RNAs.

59. Additional experiments on RNA biogenesis or interference may be done with the same or another invader (phage or plasmid) harboring a target. See other chapters for more details on these procedures.

## References

1. Grissa I, Vergnaud G, Pourcel C (2007) CRISPRFinder: a web tool to identify clustered regularly interspaced short palindromic repeats. Nucleic Acids Res 35:W52–W57

2. Ishino Y, Shinagawa H, Makino K, Amemura M, Nakata A (1987) Nucleotide sequence of the *iap* gene, responsible for alkaline phosphatase isozyme conversion in *Escherichia coli*, and identification of the gene product. J Bacteriol 169:5429–5433

3. Barrangou R, Fremaux C, Deveau H, Richards M, Boyaval P, Moineau S, Romero DA, Horvath P (2007) CRISPR provides acquired resistance against viruses in prokaryotes. Science 315:1709–1712

4. Garneau JE, Dupuis MÈ, Villion M, Romero DA, Barrangou R, Boyaval P, Fremaux C, Horvath P, Magadán AH, Moineau S (2010) The CRISPR/Cas bacterial immune system cleaves bacteriophage and plasmid DNA. Nature 468:67–71

5. Makarova KS, Wolf YI, Koonin EV (2013) Comparative genomics of defense systems in archaea and bacteria. Nucleic Acids Res 41: 4360–4377

6. Marchfelder A (2013) Special focus CRISPR-Cas. RNA Biol 10:655–658

7. Koonin EV, Makarova KS (2013) CRISPR-Cas: evolution of an RNA-based adaptive immunity system in prokaryotes. RNA Biol 10:679–686

8. Sorek R, Lawrence CM, Wiedenheft B (2013) CRISPR-mediated adaptive immune systems in bacteria and archaea. Annu Rev Biochem 82:237–266

9. Makarova KS, Haft DH, Barrangou R, Brouns SJ, Charpentier E, Horvath P, Moineau S, Mojica FJ, Wolf YI, Yakunin AF, van der Oost J, Koonin EV (2011) Evolution and classification of the CRISPR-Cas systems. Nat Rev Microbiol 9:467–477

10. Lange SJ, Alkhnbashi OS, Rose D, Will S, Backofen R (2013) CRISPRmap: an automated classification of repeat conservation in prokaryotic adaptive immune systems. Nucleic Acids Res 41:8034–8044

11. Yin S, Jensen MA, Bai J, Debroy C, Barrangou R, Dudley EG (2013) Evolutionary divergence of Shiga toxin-producing *Escherichia coli* is reflected in CRISPR spacer composition. Appl Environ Microbiol 79:5710–5720

12. Held NL, Herrera A, Whitaker RJ (2013) Reassortment of CRISPR repeat-spacer loci in *Sulfolobus islandicus*. Environ Microbiol 15: 3065–3076

13. Shariat N, Kirchner MK, Sandt CH, Trees E, Barrangou R, Dudley EG (2013) Subtyping of *Salmonella enterica* serovar Newport outbreak isolates by CRISPR-MVLST and determination of the relationship between CRISPR-MVLST and PFGE results. J Clin Microbiol 51:2328–2336

14. Cai F, Axen SD, Kerfeld CA (2013) Evidence for the widespread distribution of CRISPR-Cas system in the Phylum *Cyanobacteria*. RNA Biol 10:687–693
15. Emerson JB, Andrade K, Thomas BC, Norman A, Allen EE, Heidelberg KB, Banfield JF (2013) Virus-host and CRISPR dynamics in archaea-dominated hypersaline Lake Tyrrell, Victoria, Australia. Archaea 2013:370871
16. Minot S, Bryson A, Chehoud C, Wu GD, Lewis JD, Bushman FD (2013) Rapid evolution of the human gut virome. Proc Natl Acad Sci U S A 110:12450–12455
17. Zhang Q, Rho M, Tang H, Doak TG, Ye Y (2013) CRISPR-Cas systems target a diverse collection of invasive mobile genetic elements in human microbiomes. Genome Biol 14:R40
18. Deveau H, Barrangou R, Garneau JE, Labonté J, Fremaux C, Boyaval P, Romero DA, Horvath P, Moineau S (2008) Phage response to CRISPR-encoded resistance in *Streptococcus thermophilus*. J Bacteriol 190:1390–1400
19. Savitskaya E, Semenova E, Dedkov V, Metlitskaya A, Severinov K (2013) High-throughput analysis of type I-E CRISPR/Cas spacer acquisition in *E. coli*. RNA Biol 10:716–725
20. Chylinskim K, Le Rhunm A, Charpentier E (2013) The tracrRNA and Cas9 families of type II CRISPR-Cas immunity systems. RNA Biol 10:726–737
21. Magadán AH, Dupuis MÈ, Villion M, Moineau S (2012) Cleavage of phage DNA by the *Streptococcus thermophilus* CRISPR3-Cas system. PLoS One 7:e40913
22. Biswas A, Gagnon JN, Brouns SJ, Fineran PC, Brown CM (2013) CRISPRTarget: bioinformatic prediction and analysis of crRNA targets. RNA Biol 10:817–827
23. Shah SA, Erdmann S, Mojica FJ, Garrett RA (2013) Protospacer recognition motifs: mixed identities and functional diversity. RNA Biol 10:891–899
24. Peng W, Li H, Hallstrøm S, Peng N, Liang YX, She Q (2013) Genetic determinants of PAM-dependent DNA targeting and pre-crRNA processing in *Sulfolobus islandicus*. RNA Biol 10:738–748
25. Almendros C, Guzmán NM, Díez-Villaseñor C, García-Martínez J, Mojica FJ (2012) Target motifs affecting natural immunity by a constitutive CRISPR-Cas system in *Escherichia coli*. PLoS One 8:e50797
26. Sun CL, Barrangou R, Thomas BC, Horvath P, Fremaux C, Banfield JF (2013) Phage mutations in response to CRISPR diversification in a bacterial population. Environ Microbiol 15: 463–470
27. Fischer S, Maier LK, Stoll B, Brendel J, Fischer E, Pfeiffer F, Dyall-Smith M, Marchfelder A (2012) An archaeal immune system can detect multiple protospacer adjacent motifs (PAMs) to target invader DNA. J Biol Chem 287: 33351–33363
28. Swarts DC, Mosterd C, van Passel MW, Brouns SJ (2012) CRISPR interference directs strand specific spacer acquisition. PLoS One 7:e35888
29. Jinek M, Chylinski K, Fonfara I, Hauer M, Doudna JA, Charpentier E (2012) A programmable dual-RNA-guided DNA endonuclease in adaptive bacterial immunity. Science 337:816–821
30. Cong L, Ran FA, Cox D, Lin S, Barretto R, Habib N, Hsu PD, Wu X, Jiang W, Marraffini LA, Zhang F (2013) Multiplex genome engineering using CRISPR/Cas systems. Science 339:819–823
31. Mali P, Yang L, Esvelt KM, Aach J, Guell M, DiCarlo JE, Norville JE, Church GM (2013) RNA-guided human genome engineering via Cas9. Science 339(6121):823–826
32. Jiang W, Bikard D, Cox D, Zhang F, Marraffini LA (2013) RNA-guided editing of bacterial genomes using CRISPR-Cas systems. Nat Biotechnol 31:233–239
33. Cho SW, Kim S, Kim JM, Kim JS (2013) Targeted genome engineering in human cells with the Cas9 RNA-guided endonuclease. Nat Biotechnol 31:230–232
34. Horvath P, Barrangou R (2013) RNA-guided genome editing à la carte. Cell Res 23: 733–734
35. Wang H, Yang H, Shivalila CS, Dawlaty MM, Cheng AW, Zhang F, Jaenisch R (2013) One-step generation of mice carrying mutations in multiple genes by CRISPR/Cas-mediated genome engineering. Cell 153:910–918
36. Friedland AE, Tzur YB, Esvelt KM, Colaiácovo MP, Church GM, Calarco JA (2013) Heritable genome editing in *C. elegans* via a CRISPR-Cas9 system. Nat Methods 10:741–743
37. Horvath P, Romero DA, Coûté-Monvoisin AC, Richards M, Deveau H, Moineau S, Boyaval P, Fremaux C, Barrangou R (2008) Diversity, activity, and evolution of CRISPR loci in *Streptococcus thermophilus*. J Bacteriol 190:1401–1412
38. Deltcheva E, Chylinski K, Sharma CM, Gonzales K, Chao Y, Pirzada ZA, Eckert MR, Vogel J, Charpentier E (2011) CRISPR RNA maturation by trans-encoded small RNA and host factor RNase III. Nature 471:602–607

39. Levin BR, Moineau S, Bushman M, Barrangou R (2013) The population and evolutionary dynamics of phage and bacteria with CRISPR-mediated immunity. PLoS One Genet 9: e1003312
40. Lillehaug D (1997) An improved plaque assay for poor plaque-producing temperate lactococcal bacteriophages. J Appl Microbiol 83: 85–90
41. Paez-Espino D, Morovic W, Sun CL, Thomas BC, Ueda K, Stahl B, Barrangou R, Banfield JF (2013) Strong bias in the bacterial CRISPR elements that confer immunity to phage. Nat Commun 4:1430
42. Mills S, Griffin C, Coffey A, Meijer WC, Hafkamp B, Ross RP (2010) CRISPR analysis of bacteriophage-insensitive mutants (BIMs) of industrial Streptococcus thermophilus—implications for starter design. J Appl Microbiol 108:945–955
43. Datsenko KA, Pougach K, Tikhonov A, Wanner BL, Severinov K, Semenova E (2012) Molecular memory of prior infections activates the CRISPR/Cas adaptive bacterial immunity system. Nat Commun 3:945
44. Yosef I, Goren MG, Qimron U (2012) Proteins and DNA elements essential for the CRISPR adaptation process in *Escherichia coli*. Nat Acids Res 40:5569–5576
45. Pougach K, Semenova E, Bogdanova E, Datsenko KA, Djordjevic M, Wanner BL, Severinov K (2010) Transcription, processing and function of CRISPR cassettes in *Escherichia coli*. Mol Microbiol 77:1367–1379
46. Pul U, Wurm R, Arslan Z, Geissen R, Hofmann N, Wagner R (2010) Identification and characterization of *E. coli* CRISPR-cas promoters and their silencing by H-NS. Mol Microbiol 75: 1495–1512
47. Medina-Aparicio L, Rebollar-Flores JE, Gallego-Hernández AL, Vázquez A, Olvera L, Gutiérrez-Ríos RM, Calva E, Hernández-Lucas I (2011) The CRISPR/Cas immune system is an operon regulated by LeuO, H-NS, and leucine-responsive regulatory protein in *Salmonella enterica* serovar Typhi. J Bacteriol 193: 2396–2407
48. Dupuis MÈ, Villion M, Magadán AH, Moineau S (2013) CRISPR-Cas and restriction-modification systems are compatible and increase phage resistance. Nat Commun 4:2087
49. Garneau J (2009) Caractérisation du système CRISPR-cas chez *Streptococcus thermophilus*. Master thesis, University Laval, Quebec, 109 p. http://ariane2.bibl.ulaval.ca/ariane/?wicket:interface=:8::::
50. Dupuis MÈ (2011) Caractérisation du mode d'action du système CRISPR1/Cas de *Streptococcus thermophilus*. Master thesis, University Laval, Quebec, 113 p. http://ariane2.bibl.ulaval.ca/ariane/?wicket:interface=:5::::
51. van der Ploeg JR (2009) Analysis of CRISPR in *Streptococcus mutans* suggests frequent occurrence of acquired immunity against infection by M102-like bacteriophages. Microbiology 155:1966–1976
52. Cady KC, Bondy-Denomy J, Heussler GE, Davidson AR, O'Toole GA (2012) The CRISPR/Cas adaptive immune system of *Pseudomonas aeruginosa* mediates resistance to naturally occurring and engineered phages. J Bacteriol 194(5728):5738
53. Erdmann S, Garrett RA (2012) Selective and hyperactive uptake of foreign DNA by adaptive immune systems of an archaeon via two distinct mechanisms. Mol Microbiol 85:1044–1056
54. Lopez-Sanchez MJ, Sauvage E, Da Cunha V, Clermont D, Ratsima Hariniaina E, Gonzalez-Zorn B, Poyart C, Rosinski-Chupin I, Glaser P (2013) The highly dynamic CRISPR1 system of *Streptococcus agalactiae* controls the diversity of its mobilome. Mol Microbiol 85: 1057–1071

# Chapter 14

# Archaeal Viruses of the Sulfolobales: Isolation, Infection, and CRISPR Spacer Acquisition

## Susanne Erdmann and Roger A. Garrett

### Abstract

Infection of archaea with phylogenetically diverse single viruses, performed in different laboratories, has failed to activate spacer acquisition into host CRISPR loci. The first successful uptake of archaeal de novo spacers was observed on infection of *Sulfolobus solfataricus* P2 with an environmental virus mixture isolated from Yellowstone National Park (Erdmann and Garrett, Mol Microbiol 85:1044–1056, 2012). Experimental studies of isolated genetic elements from this mixture revealed that SMV1 (*Sulfolobus* Monocauda Virus 1), a tailed spindle-shaped virus, can induce spacer acquisition in CRISPR loci of *Sulfolobus* species from a second coinfecting conjugative plasmid or virus (Erdmann and Garrett, Mol Microbiol 85:1044–1056, 2012; Erdmann et al. Mol Microbiol 91:900–917, 2014). Here we describe, firstly, the isolation of archaeal virus mixtures from terrestrial hot springs and the techniques used both to infect laboratory strains with these virus mixtures and to obtain purified virus particles. Secondly, we present the experimental conditions required for activating SMV1-induced spacer acquisition in two different *Sulfolobus* species.

**Key words** CRISPR, Spacer acquisition, Archaeal Virus, Conjugative plasmid, Sulfolobales

## 1 Introduction

A wide range of phylogenetically diverse viruses have been characterized that infect archaea, the majority of which differ from both bacteriophages and eukaryal viruses in their morphological and genomic properties. These differences led to the hypothesis that archaeal viruses constitute a distinct phylogenetic lineage [1]. Many archaea carry CRISPR immune systems constituting complex mixtures of type I and type III systems, and these are likely to generate different modes of nucleic acid interference [2–5]. Moreover, their CRISPR arrays are generally large, carrying more than 200 spacers per cell. Most laboratory studies have focused on members of the crenarchaeal order Sulfolobales. These organisms frequent terrestrial acidic hot springs and have provided the main model archaeal organisms for laboratory investigations of both the diversity of archaeal viruses and their interplay with host CRISPR immune systems [6]. Here we describe methods used in the first successful induction of spacer acquisition in the laboratory, for the model archaea *Sulfolobus solfataricus* P2 and *Sulfolobus islandicus* REY14A [7]. This work was preceded by many unsuccessful attempts at inducing spacer acquisition in different laboratories, including our own. The successful spacer acquisition appears to be activated by a specific tailed-fusiform virus SMV1. Exceptionally, the virus can avoid CRISPR-directed spacer acquisition and DNA interference, but it is able to induce spacer acquisition from a coinfecting conjugative plasmid or virus [7, 8].

## 2 Materials

### 2.1 Media and Buffers

1. 10× *Sulfolobus* medium: 250 mM $(NH_4)_2SO_4$, 30 mM $K_2SO_4$, 15 mM KCl, 200 mM Glycine, 0.04 mM $MnCl_2 \cdot 4H_2O$, 0.1 mM $Na_2B_4O_7 \cdot 2H_2O$, 0.04 mM $ZnSO_4 \cdot 7H_2O$, 0.00125 mM $CuSO_4 \cdot 2H_2O$, 0.0006 mM $Na_2MoO_4 \cdot 2H_2O$, 0.0006 mM $VoSO_4 \cdot 5H_2O$, 0.000175 mM $CoSO_4 \cdot 7H_2O$, 0.0002 mM $NiSO_4 \cdot 6H_2O$, 10 mM $MgCl_2$, 3 mM $Ca(NO_3)_2$, 1 mM HCl, 0.07 mM $FeSO_4 \cdot 7H_2O$ in $ddH_2O$, pH adjusted to 3.5 with $H_2SO_4$ (autoclave at 121 °C).

2. TYS-medium: 4 mL 50 % sucrose, 10 mL 5 % tryptone, 5 mL 10 % yeast extract, 100 mL 10× *Sulfolobus* medium. Make up to 1 L with $dH_2O$ and adjust pH with $H_2SO_4$.

3. High-salt buffer: 250 mM $(NH_4)_2SO_4$, 30 mM $K_2SO_4$, 15 mM KCl, 200 mM glycine in $ddH_2O$, pH 5.0 (autoclave at 121 °C).

4. 2 % uranyl acetate: dissolve 0.2 g of uranyl acetate dihydrate (PLANO GmbH, Wetzlar, Germany) in 10 mL of $ddH_2O$ and centrifuge at $12,000 \times g$ directly before use to remove precipitates.

### 2.2 Filters

1. 0.2 μm filter: Nalgene™ Rapid-Flow™, 500 mL, membrane: 75 mm (Thermo Scientific).

2. 1 MDa filter: Vivaspin 20, 1,000,000 MWCO (Sartorius Stedim, Aubagne, France).

### 2.3 Sulfolobales Strains

1. *Sulfolobus solfataricus* P2 (DSM 1617; Accession number: AE0006641.1).

2. *Sulfolobus islandicus* REY15A (Accession number: CP002425.1).

### 2.4 Others

1. Parafilm: Pechiney plastic packaging company (Chicago, Illinois).

2. TEM-grids: carbon-coated copper grids, 200 mesh, (Electron Microscopy Sciences, Hatfield, USA).

3. Centrifuge tubes: Ultra-Clear Centrifuge Tubes, 14×89 mm (Beckman Instruments, Inc.; Spinco Division; Palo Alto, CA, USA).

4. Dialysis tubes: Spectra/Por® Dialysis Membrane, MWCO: 6–8,000 (Spectrum Laboratories, Inc.; Rancho Dominguez, CA, USA).

## 3 Methods

### 3.1 Sampling

Viruses infecting members of the order Sulfolobales can be isolated from terrestrial hot springs with a pH range between 1.5 and 4 and a temperature range of 65–90 °C.

1. Measure the pH and the temperature of the hot spring.

2. Take a mud or sediment sample with a spoon or another suitable tool, use hot spring water to keep the sample humid and store the sample in a water-tight container (e.g., 50 mL Falcon tube) (*see* **Note 1**).

3. Store samples at room temperature (*see* **Note 2**). Samples can be stored for at least 3–6 months.

### 3.2 Enrichment Culture

1. Take a small spoonful of sediment or about 1 mL of mud sample and inoculate into 50 mL of TYS-medium. The pH of the medium can be adjusted between 2 and 4 according to the pH measured in the hot spring, a small variation in pH can strongly influence the composition of the enrichment culture (Fig. 1).

2. Incubate the sample in appropriate containers and shake for at least 3 days at temperatures between 65 and 85 °C, dependent on the temperature measured in the hot spring. Growth can be

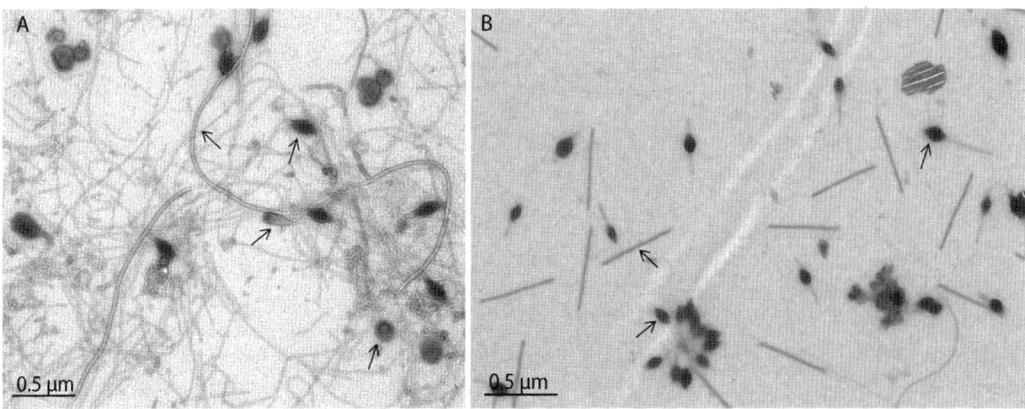

**Fig. 1** Electron micrographs of virus-like particles isolated from enrichment cultures of the same sample grown at (**a**) pH 2.5 and (**b**) pH 3.5. Samples were negatively stained with 2 % uranyl acetate. Different virus-like particles are labeled with *arrows* and size bars are included

monitored by $A_{600}$ measurements or by light microscopy if the culture is too opaque due to suspended sediment particles.

3. As soon as growth is detectable, take a sample for storage at −80 °C by harvesting 10 mL of culture, centrifuging at 6,000×*g*, and dissolving the cell pellet in High-salt buffer containing 15 % glycerol (*see* **Note 3**).

4. In order to obtain a sufficient quantity of virus particles for electron microscopy studies and for subsequent infection experiments, enrichment cultures should be scaled up to at least 250 mL by adding fresh TYS-medium to the 50 mL culture about 3–4 days after the first inoculation (*see* **Note 4**), with incubation for a further 3 days at the same temperature. Additional scaling up is recommended. Cultures should be diluted with fresh TYS-medium as soon as growth has reached stationary phase.

## 3.3 Isolation of Virus-Like Particles (VLP) from Enrichment Cultures

1. Spin down cells at 6,000×*g* for 10 min and remove residual cell debris and sediment from the supernatant by a second centrifugation at 10,000×*g* for 10 min.

2. Transfer the supernatant to a 0.2 μm filter and collect the flow-through in a sterile bottle. Usually about 50 % of the VLPs remain on top of the filter. Wash the membrane of the filter with about 500 μL fresh TYS-medium and transfer the sample to a 1.5 mL tube. Centrifuge this sample for 5 min at 10,000×*g*, to remove cells and other contaminants, and transfer the supernatant to a fresh 1.5 mL tube (Sample 1) (*see* **Note 5**). Add $NaN_3$ to a final concentration of 0.02 % and store the sample in the fridge at 4 °C (*see* **Note 6**).

3. Concentrate VLPs in the flow-through by using 1 MDa filters (Vivaspin 20 or similar columns). Transfer the flow-through to the top of the column and centrifuge at a maximum speed as defined by the manufacturer (3,000×*g* for swinging buckets, 6,000×*g* for fixed-angle rotors), until the liquid has passed through the membrane. Discard the flow-through and repeat this step with the remaining flow-through from the 0.2 μm filter until about 100–500 μL remain. Wash the membrane of the column with the remaining liquid and transfer the sample into a fresh 1.5 mL tube (Sample 2). Add $NaN_3$ to a final concentration of 0.02 % and store the sample at 4 °C (*see* **Note 7**).

## 3.4 Analysis of VLPs by Electron Microscopy

1. Fix a piece of Parafilm on the lab bench. Apply a 10 μL drop from each sample (Sample 1 and 2) and place a TEM-grid on top of it. Allow the VLPs to adsorb to the TEM-grid for about 10 min.

2. Wash the TEM-grids by transferring the grids on 20 μL drops of $ddH_2O$ and maintaining at room temperature for 10 s.

3. Stain the sample by transferring the grids on 10 μL drops of 2 % uranyl acetate. After 2 min, remove residual uranyl acetate by absorption with filter paper.

4. Analyze the TEM-grids for VLPs using a transmission electron microscope (Fig. 1).

### 3.5 Infection of Laboratory Strains

1. Harvest 2× 10 mL cells of a freshly grown culture of *Sulfolobus* (*see* **Note 8**) with an $A_{600}$ of 0.8 by centrifuging at 6,000 ×$g$ for 10 min.

2. Resuspend the cell pellets in 1 mL of fresh TYS-medium or High-salt buffer (*see* **Note 9**) and transfer the samples in two fresh 1.5 mL tubes.

3. Add 10 μL of the isolate Sample 2 to one of the samples and leave the second uninfected. Sample 1 has to be purified further to be usable for infection experiments (*see* Subheading 3.6). Mix the samples by inverting the tubes several times and incubate the tubes on a heating block at temperatures between 75 and 85 °C (*see* **Note 10**) for at least 2 h and a maximum of 6 h with careful mixing every 30 min.

4. Wash away unbound VLPs by centrifuging for 5 min at 6,000 ×$g$ and resuspend the cells in fresh TYS-medium.

5. Repeat the washing step and transfer the infected, and uninfected, cells into 50 mL preheated TYS-medium. Measure the $A_{600}$ to determine the starting cell concentrations and incubate the cultures, with shaking, at temperatures between 75 and 85 °C, dependent on the optimal growing conditions of the host strain.

6. Monitor cell growth by $A_{600}$ measurements every 6–12 h. Growth retardation of the infected culture relative to the uninfected culture can imply successful infection (*see* **Note 11**). Dilute the culture after growth has reached stationary phase (usually every 3 days) to an $A_{600}$ of 0.05 by transferring the appropriate volume of culture to fresh preheated TYS-medium. Growth retardation may be delayed by up to 12 days post infection, as was observed for SMV1-infected *S. solfataricus* P2 cultures [7]. Subsequent isolation of the VLPs can be performed as described above for the enrichment cultures (*see* Subheading 3.3).

### 3.6 Purification of Virus-Like Particles

VLPs isolated directly from cell cultures by filtration may be cross-contaminated with host cells, when using 0.2 μm filters, or by cell debris and other components including DNA or proteins, when using 1 MDa filters. To purify VLPs sufficiently for DNA-sequencing or protein analysis they must be subjected to additional purification steps.

1. Prepare a 0.45 mg/mL CsCl solution by dissolving 45 mg CsCl in 100 mL dd$H_2O$.

2. Add 0.45 mg/mL CsCl solution to the centrifuge tube to at least three quarters full and layer the virus solution onto the CsCl solution (*see* **Note 12**). Use different tubes for Samples 1 and 2.

3. Centrifuge at 200,000 × $g$ for 48 h in a SW41 rotor (Beckman, Fullerton, USA) and carefully transfer the tube into an appropriate rack.

4. Extract the white bands containing the virus particles, through the side of the tube using a syringe with an appropriate needle.

5. Transfer the sample to a dialysis tube (prepared according to the producer's instructions) and perform the dialysis in 2 L of 1 mM Tris–HCl, pH 6 (*see* **Note 13**) for at least 12 h to remove the CsCl.

6. Collect the sample, add $NaN_3$ to a final concentration of 0.02 % and store the sample at 4 °C. At this stage, an electron microscopy analysis of the purified particles is recommended to confirm their purity and integrity.

### 3.7 Screening for Newly Acquired Spacers

1. Infect the host strain of choice with the isolated virus mixture or single viruses as described in Subheading 3.5. Prepare an uninfected culture under the same conditions as a control and monitor growth by $A_{600}$ measurement. The experiments are performed in 50 mL cultures.

2. Dilute the culture after growth has reached stationary phase (usually every 3 days) to an $A_{600}$ of 0.05 by transferring the appropriate volume of culture to fresh preheated TYS-medium.

3. Take a 2 mL sample every 1–3 days in a 2 mL tube and harvest the cells by centrifuging at 6,000 × $g$ for 10 min. Discard the supernatant and isolate total DNA from the cell pellet (*see* **Note 14**).

4. Design primers binding within the leader region and the third to fifth spacer within each CRISPR array of the host (*see* **Note 15**) (Fig. 2a).

5. Perform PCR amplification of the CRISPR regions of your host strain. Use total DNA isolated from the cultures as template together with the designed primers. Repeat at least every 3 days.

6. Electrophorese the PCR products in 1 % agarose gels in pockets at least 1 cm in width for about 1 h at 100 V and stain the gel with ethidium bromide or another DNA staining agent (Fig. 2b).

7. Excise PCR products that are larger than the wild-type products from the gel (*see* **Note 16**) and extract the DNA using QIAquick Gel Extraction Kit (Qiagen) or another appropriate kit.

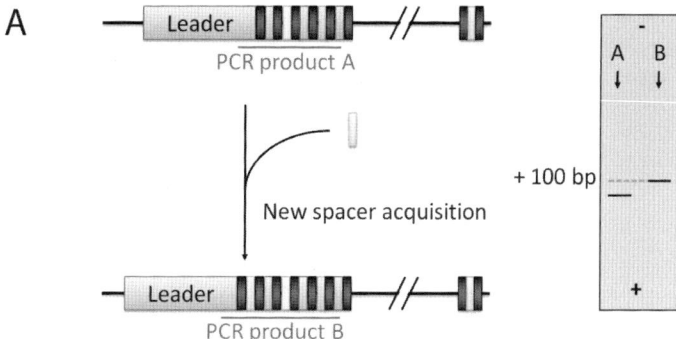
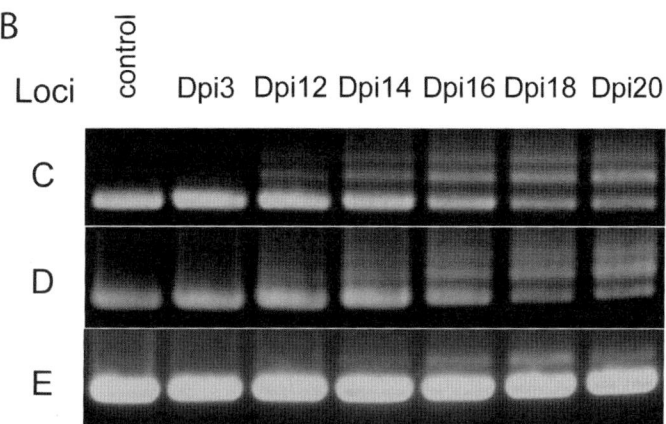

**Fig. 2** (**a**) Schematic presentation of the PCR amplification strategy used to screen infected cultures for newly acquired spacers. (**b**) PCR products amplified from leader proximal regions of CRISPR loci C, D, and E of *S. solfataricus* are shown after coinfection with SMV1 and pMGB1. *Dpi* days post infection. Control, uninfected *S. solfataricus*

8. Clone the purified PCR products using InsTAclone™PCR Cloning Kit (Thermo Fisher Scientific) following the manufacturer's protocol. Transform the ligated products into *E. coli* strain DH5α or other appropriate strains.

9. Screen the resulting *E. coli* clones by colony PCR for inserts of increased size using primers that bind adjacent to the cloning site of the vector. Then isolate plasmids of positive colonies using the QIAGEN Plasmid Mini Kit (Qiagen).

10. Sequence the plasmids using primers binding adjacent to the cloning site of the vector and analyze the sequence for newly acquired spacer(s).

## 3.8 Induction of Spacer Acquisition with SMV1

Induction of spacer acquisition in *S. solfataricus* P2 and *S. islandicus* REY15A was only observed upon coinfection with the spindle-shaped virus SMV1 and a second genetic element, either a conjugative plasmid [7] or a virus [8]. Infection and screening for newly acquired spacers was performed as described in Subheadings 3.5 and 3.7. De novo spacers in loci C, D and E of *S. solfataricus* P2 were detected about 12 days post infection after coinfection with SMV1 and the conjugative plasmid pMGB1 [7]. pMGB1 conjugation was only observed in the host *S. solfataricus* P2. De novo spacers in both locus 1 and locus 2 of *S. islandicus* REY15A were detected about 16 days post infection upon coinfection with SMV1 and the spindle-shaped virus STSV2 [8]. SMV1 can be obtained upon request from S.E. at the Danish Archaea Centre, University of Copenhagen, Denmark.

## 4 Notes

1. Mud or sediment samples contain a higher concentration of host cells than water samples, which can be crucial for obtaining enrichment cultures. High concentrations of viruses or host cells can also be obtained by filtering large volumes of water samples using filters with an exclusion of 0.2 μm or smaller.

2. Freezing is not recommended; temperature stress can change the host cell composition or destroy virus particles.

3. The composition of host cells in enrichment cultures changes very rapidly, and this influences the composition of the propagating viruses. Therefore, it is recommended to conserve samples at different stages of the enrichment culture (every 3 days) in order to be able to reproduce the obtained virus mixtures.

4. The culture should not be diluted to an $A_{600}$ of less than 0.05 because this can result in interruption of cell growth. If growth is very slow, scaling up can be achieved by harvesting the cells from the 50 mL cultures, centrifuging at $6{,}000 \times g$, and resuspending the cell pellet in 100–250 mL (dependent on the cell density) of fresh TYS-medium. Store the supernatant for isolation of virus-like particles.

5. Samples obtained from the top of the 0.2 μm filter can be used directly for electron microscopy analyses. Since contamination with cells from the enrichment culture is likely, it is not recommended to use these samples for further infection experiments.

6. Addition of $NaN_3$ prevents growth of contaminating bacteria. Samples can also be stored at room temperature or frozen at −80 °C containing 15 % glycerol, but efficient infectivity of virus particles stored at temperatures between 4 and 8 °C has been observed for 1 year, or more, after isolation.

7. Samples obtained from the 1 MDa filter should not contain cellular contaminants from the enrichment culture and they can be used directly for infection experiments.

8. *S. solfataricus* P2 and *S. islandicus* REY15A were used for infection experiments in our laboratory. Other members of the Sulfolobales can be effective virus hosts. Moreover, employing different strains for infection can facilitate isolation of single viruses from virus mixtures.

9. The buffer or medium used for infection can influence the infectivity of certain VLPs. Other media or buffers than those recommended can be tested. Using different buffers for infection may also help to isolate single viruses from virus mixtures.

10. The temperature used for infection can influence the infectivity of certain VLPs. Therefore, it is recommended to test different temperatures and the use of different infection temperatures may also help to isolate single viruses from virus mixtures.

11. Some viruses infecting the Sulfolobales do not induce strong growth retardation upon infection. Therefore, it is recommended to verify infection by isolating VLPs from the supernatant and to analyze them by electron microscopy.

12. Purification on CsCl gradients can result in disruption of virus particles and may lead to loss of viral proteins and reduced infectivity. Purification of spindle-shaped and linear viruses via CsCl gradient resulted in a reduction to about 50 % infectivity. To reduce virus particle disruption other purification methods including sucrose density gradients, preformed CsCl gradients [8] or CIM® Disc Virus Purification Pack (BIA Separations, GesmbH; Villach, Austria) can be tested.

13. The buffer used for dialysis and storage of purified virus particles can influence infectivity. 1 mM Tris–HCl, pH 6 has been proven to be appropriate for spindle-shaped and linear viruses.

14. Dry cell pellets can be stored at –20 °C before isolation of total DNA. Total DNA can be isolated using DNeasy® Blood &Tissue Kit (Qiagen, Westberg, Germany) or other appropriate kits and protocols.

15. The PCR products should not be larger than about 800 bp in order to achieve sufficient electrophoretic resolution of products containing newly acquired spacer from wild-type products.

16. Larger PCR products corresponding to newly acquired spacer were first observed 12–16 days post infection in SMV1-infected *S. solfataricus* or *S. islandicus* cultures and produced strong growth retardation of the cultures [7, 8].

## Acknowledgements

The work was supported by grants from the Danish Natural Science Research Council and Copenhagen University. We are grateful to Dr. Xu Peng and Soley Ruth Gudbergsdottir for helpful advice and discussions.

## References

1. Prangishvili D, Forterre P, Garrett RA (2006) Viruses of the Archaea: a unifying view. Nat Rev Microbiol 4:837–848
2. Gudbergsdottir S, Deng L, Chen Z, Jensen JVK, Jensen LR, She Q et al (2011) Dynamic properties of the *Sulfolobus* CRISPR/Cas and CRISPR/Cmr systems when challenged with vector-borne viral and plasmid genes and protospacers. Mol Microbiol 79:35–49
3. Manica A, Zebec Z, Teichmann D, Schleper C (2011) *In vivo* activity of CRISPR-mediated virus defence in a hyperthermophilic archaeon. Mol Microbiol 80:481–491
4. Zhang J, Rouillon C, Kerou M, Reeks J, Brügger K, Graham SJ et al (2012) Structure and mechanism of the CMR complex for CRISPR-mediated antiviral immunity. Mol Cell 45:303–313
5. Deng L, Garrett RA, Shah SA, Peng X, She Q (2013) A novel interference mechanism by a type IIIB CRISPR-Cmr module in *Sulfolobus*. Mol Microbiol 87:1088–1099
6. Garrett RA, Vestergaard G, Shah SA (2011) Archaeal CRISPR-based immune systems: exchangeable functional modules. Trends Microbiol 19:549–556
7. Erdmann S, Garrett RA (2012) Selective and hyperactive uptake of foreign DNA by adaptive immune systems of an archaeon via two distinct mechanisms. Mol Microbiol 86:757, Mol Microbiol 85: 1044–1056. Corrigendum
8. Erdmann S, Le Moine Bauer S, Garrett RA (2014) Inter-viral conflicts that exploit host CRISPR immune systems of *Sulfolobus*. Mol Microbiol 91:900–917

# Chapter 15

## Using the CRISPR-Cas System to Positively Select Mutants in Genes Essential for Its Function

Ido Yosef, Moran G. Goren, Rotem Edgar, and Udi Qimron

### Abstract

The clustered regularly interspaced short palindromic repeats (CRISPR) and CRISPR associated proteins (Cas) comprise a prokaryotic adaptive defense system against foreign nucleic acids. This defense is mediated by Cas proteins, which are guided by sequences flanked by the repeats, called spacers, to target nucleic acids. Spacers designed against the prokaryotic self chromosome are lethal to the prokaryotic cell. This self-killing of the bacterium by its own CRISPR-Cas system can be used to positively select genes that participate in this killing, as their absence will result in viable cells. Here we describe a positive selection assay that uses this feature to identify *E. coli* mutants encoding an inactive CRISPR-Cas system. The procedure includes establishment of an assay that detects this self-killing, generation of transposon insertion mutants in random genes, and selection of viable mutants, suspected as required for this lethal activity. This procedure enabled us to identify a novel gene, *htpG*, that is required for the activity of the CRISPR-Cas system. The procedures described here can be adjusted to various organisms to identify genes required for their CRISPR-Cas activity.

**Key words** Defense mechanism, Phage-host interaction, Non-cas genes, Autoimmunity, Positive selection

## 1 Introduction

Clustered regularly interspaced short palindromic repeats (CRISPR) and CRISPR associated (Cas) proteins comprise an important prokaryotic defense system against horizontally transferred DNA [1–3] and RNA [4]. This system, found across ~90 % of archaeal genomes and ~50 % of bacterial genomes, shows remarkable analogies to the mammalian immune system [5, 6], as well as to eukaryotic RNA-interference mechanisms [7, 8]. Three major types, and ten subtypes of CRISPR-Cas systems have been classified [9]. All types consist of a CRISPR array—short repeated sequences called "repeats" flanking short sequences called "spacers." The array is usually preceded by a leader, AT-rich DNA sequence that drives CRISPR array expression and is important for acquiring new spacers into the array [10, 11]. A cluster of *cas* genes encoding proteins that process the transcript, interfere with foreign nucleic acids, and acquire new spacers usually lies adjacent to the CRISPR array(s) [12–14]. RNA transcribed from the CRISPR array (pre-crRNA) is processed by Cas proteins into RNA-based spacers flanked by partial repeats (crRNAs). These crRNAs specifically direct Cas interfering proteins to target nucleic acids matching the spacers. The spacers are acquired from invading sequences, termed "protospacers," which are subsequently targeted. Spacer acquisition into the CRISPR array consequently results in guiding the system to cleave DNA molecules harboring the corresponding protospacers. This feature renders the system competent in adaptively and specifically targeting invaders.

Since nucleic acid cleavage mediated by the CRISPR-Cas system is dictated by spacer homology, the system can be manipulated to attack any nucleic acid sequence by genetic engineering of a rationally designed spacer into the CRISPR array. Such manipulation can be used to target the self chromosome, and in fact, was shown to target the self chromosome in various organisms, facilitating selection of deletion mutants and genetic engineering of complex genomes [15–22]. It was shown that in *E. coli*, and in other bacteria, targeting of DNA integrated into the genome results in massive killing of the bacterial culture [21, 23, 24, 25]. This was demonstrated, for example, in an *E. coli* strain lacking *hns*, a natural repressor of the system, in which the CRISPR-Cas system was constitutively active. This strain, encoding a λ prophage and transformed with a plasmid encoding an array against the prophage was killed in a CRISPR-dependent manner. This "self killing" activity of the CRISPR-Cas system served as a selection force in a genetic assay that identified genes that are essential for the system's function [24].

After establishing conditions that kill most bacteria having a self-spacer in a CRISPR-Cas-dependent manner, one has to introduce genetic diversity into this experimental system. This genetic diversity allows selection of mutants that are deficient in the self-killing activity, suggesting that these mutants are defective in genes essential for the CRISPR-Cas function. Transposon mutagenesis allows this diversity by simple generation of random insertion-mutants in various genes. Importantly, the transposon insertion site is easy to detect and therefore this mutant-generation system is preferred for many such assays.

Lastly, after isolating mutants resistant to this self-killing, it is essential to validate that the transposon insertion mutation caused the observed phenotype. For this, "clean" deletions having no remnants of the gene are made or obtained from known collections. These mutants are validated for resistance to the self-killing. To validate that the CRISPR-Cas system in these mutants is functional, and that no secondary mutation is responsible for the phenotype, the suspected genes are individually cloned on plasmids which are then transformed into the deletion mutant. Restoration of the phenotype is then validated. Using two types of gene disruption along with a complementation assay unequivocally indicates the essentiality of the suspected gene to the activity of the CRISPR-Cas system in self-killing.

The procedures described here in detail refer to *E. coli* harboring the type I-E CRISPR-Cas system. Nevertheless, they can be adjusted to other systems and other subtypes to identify other novel elements of this fascinating defense system.

## 2 Materials

Prepare all solutions using double-distilled water (DDW). Prepare and store all reagents at room temperature (RT) unless indicated otherwise.

### 2.1 Determining the CRISPR-Cas Self Killing Activity by Transformation Efficiency Assays

1. Glycerol stocks of *E. coli* IYB5237 and IYB5163 (*see* **Note 1**).
2. 50 mg/mL kanamycin (Kan): Dissolve 0.5 g Kan in 10 mL DDW. Filter with 0.22 μm syringe filter and store in aliquots of 1 mL at −20 °C.
3. 34 mg/mL Chloramphenicol (Cm): Dissolve 0.34 g of Cm in 10 mL 100 % ethanol. Store at −20 °C.
4. 20 % w/v L-arabinose: Dissolve 20 g L-arabinose in 100 mL DDW. Sterilize by autoclaving and store in 20 mL aliquots at 4 °C (*see* **Note 3**).
5. 1 M Isopropyl-beta-D-thiogalactopyranoside (IPTG): Dissolve 2.38 g in 8 mL DDW. Filter with 0.22 μm syringe filter and store in aliquots of 1 mL at −20 °C.
6. 10 % v/v Glycerol: Add 1 volume of molecular biology grade glycerol to 9 volumes of DDW. Mix, sterilize by autoclaving and store in 200 mL aliquots at 4 °C (*see* **Note 3**).
7. LB medium: Add 10 g Bacto-tryptone, 5 g yeast extract, 5 g NaCl to 800 mL DDW, stir until clear, adjust volume to 1 L with DDW, sterilize by autoclaving and store at RT.
8. 2YT medium: Add 16 g Bacto-tryptone, 10 g yeast extract, 5 g NaCl to 800 mL DDW, stir until clear, adjust volume to 1 L with DDW, sterilize by autoclaving and store at RT.
9. LB agar plates: Add 15 g agar to LB (*see* Subheading 2.1 **item 7**) and sterilize by autoclaving. Let cool to about 45 °C, add antibiotics and/or inducers as required, and pour 20 mL into Petri plate. Let solidify and leave to dry overnight at RT. Store at 4 °C for up to 1 month.

### 2.2 Transposon Mutagenesis Procedure

1. Glycerol stock of *E. coli* RE1093 and WM2672 (*see* **Notes 1** and **5**).
2. 10 mg/mL Tetracycline (Tet): Dissolve 100 mg of Tet in 10 mL of 70 % ethanol. Mix until all Tet dissolves completely. Store at −20 °C.
3. 100 mg/mL ampicillin (Amp): Dissolve 1 g of Amp in 10 mL DDW. Filter with 0.22 μm syringe filter and store in aliquots of 1 mL at −20 °C.
4. 50 mg/mL Kan: *see* Subheading 2.1 **item 2**.
5. 10 % v/v glycerol: *see* Subheading 2.1 **item 6**.
6. LB medium and LB agar plates: *see* Subheading 2.1 **items 7** and **9**.

### 2.3 Selection of Mutants Escaping Self-killing by the CRISPR-Cas System

1. Glycerol stock of transposon library from Subheading 3.2.4 **step 6**.
2. 10 mg/mL Tet: *see* Subheading 2.2 **item 2**.
3. 100 mg/mL Amp: *see* Subheading 2.2 **item 3**.
4. 50 mg/mL Kan: *see* Subheading 2.1 **item 2**.
5. 34 mg/mL Cm: *see* Subheading 2.1 **item 3**.
6. 20 % w/v L-arabinose: *see* Subheading 2.1 **item 4**.
7. 1 M IPTG: *see* Subheading 2.1 **item 5**.
8. 10 % v/v glycerol: *see* Subheading 2.1 **item 6**.
9. LB medium and LB agar plates: *see* Subheading 2.1 **items 7** and **9**.

### 2.4 Excluding False Positives

1. 10 mg/mL Tet: *see* Subheading 2.2 **item 2**.
2. 50 mg/mL Kan: *see* Subheading 2.1 **item 2**.
3. 34 mg/mL Cm: *see* Subheading 2.1 **item 3**.
4. 20 % w/v L-arabinose: *see* Subheading 2.1 **item 4**.
5. 1 M IPTG: *see* Subheading 2.1 **item 5**.
6. 10 % v/v glycerol: *see* Subheading 2.1 **item 6**.
7. LB medium and LB agar plates: *see* Subheading 2.1 **items 7** and **9**.
8. LongAmp Taq 2X Master Mix.
9. Thermocycler.
10. Sterile toothpicks.

### 2.5 Identifying the Transposon Insertion Site by PCR

1. 10 mg/mL Tet: *see* Subheading 2.2 **item 2**.
2. 100 mg/mL Amp: *see* Subheading 2.2 **item 3**.
3. 50 mg/mL Kan: *see* Subheading 2.1 **item 2**.
4. 34 mg/mL Cm: *see* Subheading 2.1 **item 3**.

5. LB medium: *see* Subheading 2.1 **item 7**.
6. TE buffer, 10 mM Tris–HCl, 1 mM ethylenediaminetetraacetic acid (EDTA): Add 1 mL of 1 M Tris–HCl (pH 7.5) and 0.2 mL of 500 mM EDTA (pH 8.0) into 99 mL DDW and mix. 1 M Tris (Tris-base): Dissolve 60.57 g in 500 mL DDW and titrate to pH 7.5 using HCl. 0.5 M EDTA: Dissolve 18.6 g in 100 mL DDW and titrate pH to 8.0 using NaOH. EDTA will not dissolve until pH reaches 8.0. Use vigorous stirring and moderate heat. Sterilize by autoclaving and store at RT.
7. 20 % w/v sodium dodecyl sulfate (SDS): Dissolve 100 g of SDS in 400 mL DDW (use mask while handling SDS powder). Heat up to 68 °C and stir to facilitate dissolution. Adjust volume to 500 mL with DDW. Sterilize by autoclaving and store at RT (reheat if necessary to dissolve SDS precipitation).
8. 20 mg/mL proteinase K: Dissolve 100 mg lyophilized Proteinase K in 5 mL DDW. Stored in small single-use aliquots at −20 °C.
9. 5 M NaCl: Dissolve 146.1 g NaCl in 350 mL DDW. Adjust volume to 500 mL. Sterilize by autoclaving.
10. 10 % w/v Hexadecyltrimethylammonium bromide (CTAB)/0.7 M NaCl solution: Dissolve 4.1 g NaCl in 80 mL DDW and slowly add 10 g CTAB while heating and stirring. If necessary, heat to 65 °C to dissolve. Adjust final volume to 100 mL.
11. 70 % v/v ethanol, 70 %: Mix 70 mL 100 % ethanol with 30 mL DDW.
12. 25:24:1 v/v phenol–chloroform–isoamyl Alcohol Solution: Mix 50 mL phenol, 48 mL chloroform, and 2 mL isoamyl alcohol.
13. LongAmp Taq 2X Master Mix.

## 2.6 Validating the Phenotype Using "Clean" Deletions

1. Glycerol stock of *E. coli* IYB5163 and IYB5165 (*see* **Note 1**).
2. 50 mg/mL Kan: *see* Subheading 2.1 **item 2**.
3. 34 mg/mL Cm: *see* Subheading 2.1 **item 3**.
4. 20 % w/v L-arabinose: *see* Subheading 2.1 **item 4**.
5. 1 M IPTG: *see* Subheading 2.1 **item 5**.
6. 10 % v/v glycerol: *see* Subheading 2.1 **item 6**.
7. LB medium and LB agar plates: *see* Subheading 2.1 **items 7** and **9**.
8. 2YT medium: *see* Subheading 2.1 **item 8**.
9. Electroporator.
10. Spectrophotometer.
11. Shaking incubator.

## 2.7 Complementing the Deletion In Trans

1. Glycerol stock of *E. coli* IYB5163 and IYB5165 (*see* **Note 1**).
2. 100 mg/mL Amp: *see* Subheading 2.2 **item 3**.
3. 50 mg/mL Kan: *see* Subheading 2.1 **item 2**.
4. 34 mg/mL Cm: *see* Subheading 2.1 **item 3**.
5. 20 % w/v L-arabinose: *see* Subheading 2.1 **item 4**.
6. 10 % v/v Glycerol: *see* Subheading 2.1 **item 6**.
7. LB medium and LB agar plates: *see* Subheading 2.1 **items 7** and **9**.
8. 2YT medium: *see* Subheading 2.1 **item 8**.

## 3 Methods

### 3.1 Determining the CRISPR-Cas Self Killing Activity by Transformation Efficiency Assays

#### 3.1.1 Electrocompetent Cells Preparation

1. Streak *E. coli hns$^+$* (IYB5237) and *hns$^-$* (IYB5163) from −80 °C glycerol stock onto an LB plate supplemented with 25 µg/mL Kan and grow overnight at 32 °C (*see* **Note 1**).
2. Inoculate an isolated colony into 50 mL plastic tubes containing 5 mL LB with 25 µg/mL Kan. Incubate cultures at 32 °C with shaking at 250 rpm overnight.
3. Inoculate 0.2 mL of the saturated overnight culture into 10 mL LB medium supplemented with 25 µg/mL Kan in 50 mL plastic tubes.
4. Incubate the culture at 32 °C with shaking at 250 rpm until an optical density ($OD_{600}$) of ~0.5 is obtained (~2 h).
5. When the $OD_{600}$ reaches ~0.5, transfer the culture to ice for 15 min (keep the cells cold from this point until after transformation).
6. Centrifuge the cultures for 10 min at $4,400 \times g$ at 4 °C.
7. Decant the supernatant and resuspend the cell pellets in 1 mL of ice-cold DDW and then transfer them into 1.7 mL sterile microcentrifuge tube.
8. Centrifuge the cultures for 1 min at $13,000 \times g$ in a microcentrifuge at RT.
9. Repeat the washing step twice with ice-cold 10 % glycerol and store the tubes at −80 °C.

#### 3.1.2 Transformation and Calculation of Efficiency

1. Thaw 50 µL of electrocompetent cells of IYB5237 and IYB5163 and adjust the volume with ice-cold DDW according to the required number of transformations (*see* **Note 4**).
2. Aliquot 45 µL per sample of the cell suspension into prechilled 2 mm gap electroporation cuvettes.
3. Add 20 ng (5 µL from 4 ng/µL stock plasmid) of purified pWUR477 (encoding control spacers) or pWUR478 (encoding anti-prophage spacers) and mix by pipetting (*see* **Note 2**).

4. Mix the cuvette by flicking and electroporate at 200 Ω, 25 μF, 2.5 kV using Bio-Rad Multiporator.

5. Immediately add 0.2 mL of 2YT medium supplemented with 0.2 % L-arabinose and 0.1 mM IPTG.

6. Transfer the electroporated cells using a 1 mL pipettor into a 1.7 mL sterile microcentrifuge tube and incubate at 32 °C for 1 h with shaking at 1,000 rpm using an Eppendorf Thermomixer.

7. After 1 h, place the samples on ice and prepare serial dilutions of $10^{-1}$, $10^{-2}$, $10^{-3}$, and $10^{-4}$ from the cell suspensions in 2YT medium supplemented with 0.2 % L-arabinose and 0.1 mM IPTG.

8. Plate 50 μL from each dilution on the appropriate selective LB agar plates supplemented with 25 μg/mL Kan, 34 μg/mL Cm, 0.2 % L-arabinose, and 0.1 mM IPTG.

9. Incubate overnight at 32 °C.

10. Count the colonies in each plate and calculate the transformation efficiency by dividing the number of transformants obtained using pWUR478 or pWUR477 by the number obtained using pWUR477. It is expected that transformation efficiency of plasmid pWUR478 will be low in the strain having an active (de-repressed) CRISPR-Cas system (Δ*hns*) whereas it will be normal in the strain having a repressed CRISPR-Cas system (*hns*+).

## 3.2 Transposon Mutagenesis Procedure

### 3.2.1 Recipient Bacteria Preparation

1. Prepare an overnight culture by inoculating RE1093 (*see* **Note 1**) in 2 mL LB medium supplemented with 10 μg/mL Tet and 50 μg/mL Amp in a 10 mL tube and shake at 200 rpm at 32 °C.

2. Dilute the culture by adding 150 μL of the overnight culture into 10 mL LB medium supplemented with 50 μg/mL Amp and shake at 200 rpm at 32 °C in a 100 mL Erlenmeyer flask.

3. Grow to an $OD_{600}$ of 0.8–1.

### 3.2.2 Donor Bacteria Preparation

1. Prepare an overnight culture of WM2672 (*see* **Note 5**) by inoculating it in 2 mL LB supplemented with 12 μg/mL Kan in a 10 mL tube and shake at 200 rpm at 37 °C.

2. Dilute the culture 1:25 by adding 400 μL of the overnight culture into 10 mL LB supplemented with 12 μg/mL Kan in a 50 mL tube and shake at 200 rpm at 37 °C.

3. Grow to an $OD_{600}$ of 0.8–1.

### 3.2.3 Conjugation

1. Concentrate both recipient and donor cultures tenfold (from 10 to 1 mL): Centrifuge cells for 10 min at 4,000 ×*g* at RT and resuspend in 1 mL of LB medium (no antibiotics).

2. In a microcentrifuge tube, mix donor with recipient cells in a 1:10 ratio (50 μL of donor and 500 μL of recipient).

3. Include two controls: (1) 50 µL of donor alone and (2) 500 µL of recipient alone.
4. Incubate for 30 min at 32 °C.
5. Vortex.
6. Spread 20 µL of the cell suspensions on LB agar plates supplemented with 12 µg/mL Kan and 50 µg/mL Amp and incubate overnight at 32 °C (*see* **Note 6**).
7. Add glycerol to a 10 % v/v final concentration to the remaining bacterial suspension in the microcentrifuge tubes from **step 2** and store at −80 °C.
8. Count bacteria grown overnight from **step 6** and determine the transconjugant number (colony-forming units (CFU)/100 µL). Control cultures from **step 3** should yield no CFUs.

*3.2.4 Pooling*

1. After determining the transconjugants number (*see* Subheading 3.2.3 **step 8**), calculate the volume to spread for obtaining ~200 CFU/plate (for example, if there were 120 CFU/10 µL, spread 16 µL) (*see* **Note 6**).
2. Spread the calculated volume on 250 LB agar plates supplemented with 50 µg/mL Amp and 12 µg/mL Kan.
3. Incubate overnight at 32 °C (*see* **Note 6**).
4. Collect all colonies by adding between 0.5 and 1 mL LB medium to ~20 plates and scrape with a spreader. Use the minimal volume possible by transferring the suspension from one plate to another. Approximately 50,000 colonies should be collected (*see* **Note 7**).
5. Add glycerol to give final concentration of 10 % v/v.
6. Store at −80 °C in aliquots of 400 µL.

## 3.3 Selection of Mutants Escaping Self-Killing by the CRISPR-Cas System

### 3.3.1 Growth and Competent Cells Preparation

1. Thaw pooled cultures from Subheading 3.2.4 **step 6**.
2. Add 400 µL of thawed cells into 10 mL LB medium supplemented with 50 µg/mL Amp, 12 µg/mL Kan, and 0.2 % L-arabinose.
3. Shake for 2 h at 200 rpm at 32 °C (*see* **Note 6**).
4. Centrifuge for 10 min at $4,400 \times g$ at 4 °C.
5. Discard the supernatant and resuspend bacteria in 10 mL ice-cold DDW.
6. Centrifuge for 10 min at $4,400 \times g$ at 4 °C.
7. Discard supernatant, resuspend bacteria in 1 mL ice-cold DDW and transfer to a 1.5 mL microcentrifuge tube.
8. Centrifuge for 1 min at $13,000 \times g$ at 4 °C.
9. Resuspend in 1 mL of ice-cold DDW.
10. Centrifuge for 1 min at $13,000 \times g$ at 4 °C.
11. Resuspend in 300 µL of 10 % v/v glycerol solution.

*3.3.2 Electroporation*

1. Place three 0.1 mm electroporation cuvettes on ice.
2. Aliquot 90 µL of competent cells (*see* Subheading 3.3.1 **step 11**) to each cuvette.
3. Add 40 ng pWUR477 to one cuvette, 40 ng pWUR478 to the second cuvette and DDW to the third cuvette (control) (*see* **Note 2**).
4. Pulse in a Bio-Rad Multiporator at 200 Ω, 25 µF, 2.5 kV.
5. Immediately add 0.5 mL LB medium containing 0.2 % w/v L-arabinose and 0.1 mM IPTG.
6. Transfer to a 10 mL tube and shake at 200 rpm at 32 °C for 1 h.

*3.3.3 Plating*

For each one of the three electro-transformations:

1. Plate 100 µL on each LB agar plate supplemented with 10 µg/mL Tet, 100 µg/mL Amp, 12 µg/mL Kan, 35 µg/mL Cm, 1 % w/v L-arabinose, and 1 mM IPTG (*see* **Note 8**). Repeat plating the entire 500 µL from Subheading 3.3.2 **step 6** of pWUR478 transformants.
2. Incubate overnight at 32 °C. Expect about 400 CFU of pWUR478 transformants.

## 3.4 Excluding False Positives

1. Pick colonies from Subheading 3.3.3 **step 2** using toothpicks and streak them on new LB agar plates supplemented with 10 µg/mL Tet, 100 µg/mL Amp, 12 µg/mL Kan, 34 µg/mL Cm, 0.2 % w/v L-arabinose, and 1 mM IPTG.
2. Incubate one plate at 32 °C and the second plate at 42 °C. Grow the plates overnight (*see* **Note 9**).
3. From the plates grown at 32 °C pick colonies gently by using 200 µL pipettor and transfer them into 50 µL DDW (*see* **Note 10**).
4. Prepare PCR master mix which includes the following (per one PCR): 0.125 µL from each primer indicated in **Note 11** (total of 0.625 µL), 1.875 µL DDW, and 3 µL of 2× ready-to-use PCR mix.
5. Aliquot 5.5 µL from the PCR mix into PCR strips tube.
6. Add 0.5 µL from the resuspension prepared in **step 3** to the PCR tube containing the master mix.
7. Program the thermocycler as follows: 94 °C for 5 min, 35 cycles of denaturation at 94 °C for 30 s, annealing at 56 °C for 30 s, extension at 68 °C for 3.5 min, and end by one step of 72 °C for 7 min.
8. Load 5 µL of the PCR on a 1 % agarose gel, separate products by electrophoresis and stain with Ethidium Bromide.
9. Discard all clones that gave positive results in the PCR test and continue mapping the insertion with the others.

### 3.5 Identifying the Transposon Insertion Site by PCR

#### 3.5.1 Genomic DNA Extraction

1. Inoculate the selected bacterial mutant (*see* Subheading 3.4 **step 9**) into 5 mL LB supplemented with 10 μg/mL Tet, 100 μg/mL Amp, 12 μg/mL Kan, and 34 μg/mL Cm.
2. Shake overnight at 200 rpm at 32 °C.
3. Transfer 1.5 mL to a microcentrifuge tube.
4. Centrifuge for 2 min at $13,000 \times g$ at RT (*see* **Note 12**).
5. Discard the supernatant.
6. Resuspend pellet in 567 μL TE buffer by repeated pipetting. Add 30 μL of 10 % w/v SDS (final concentration 0.5 %), 3 μL of 20 mg/mL Proteinase K (final concentration of 100 μg/mL) and 0.6 μL of 10 mg/mL DNase-free RNase.
7. Mix thoroughly and incubate 1 h at 37 °C (*see* **Note 13**).
8. Add 100 μL of 5 M NaCl and mix thoroughly (*see* **Note 14**).
9. Add 80 μL of CTAB/NaCl solution. Mix thoroughly and incubate 10 min at 65 °C.
10. Add an approximately equal volume (0.7–0.8 mL) of Chloroform/Isoamyl Alcohol, mix thoroughly and centrifuge for 4–5 min at $13,000 \times g$ at RT (*see* **Note 15**).
11. Remove aqueous viscous supernatant to a fresh microcentrifuge tube, leaving the interface behind.
12. Add an equal volume of phenol–chloroform–isoamyl alcohol, extract thoroughly, and centrifuge for 5 min at $13,000 \times g$ at RT (*see* **Note 16**).
13. Transfer the supernatant to a fresh tube. Add 0.6 volumes of isopropanol to precipitate the nucleic acids (there is no need to add salt since the NaCl concentration is already high).
14. Shake the tube gently until a stringy white DNA precipitate becomes clearly visible.
15. Transfer the pellet to a fresh tube containing 600 μL of 70 % v/v ethanol by hooking it onto the end of a Pasteur pipette that had been heat-sealed (*see* **Note 17**).
16. Centrifuge for 5 min at $13,000 \times g$ at RT and discard the solution.
17. Add 600 mL of 70 % v/v ethanol (*see* **Note 18**) and recentrifuge for 5 min at $13,000 \times g$ at RT to pellet it again.
18. Carefully remove the supernatant and dry the pellet by leaving the tube open until there are no visible moisture or ethanol smell.
19. Redissolve the pellet in 100 μL of TE buffer (*see* **Note 19**).

3.5.2 PCR for Sequencing

PCR1: 10 μL LongAmp Taq 2× Master Mix, 2 μL primers (*see* **Note 20**): 2.5 μM each, 1 μL template DNA (100 ng/μL): 100 ng DNA, 7 μL DDW, in a 20 μL reaction.
PCR as follows:

| PCR1 | Temp | 94 °C | 94 °C | 42 °C | 70 °C | 94 °C | 65 °C | 70 °C | 4 °C |
|---|---|---|---|---|---|---|---|---|---|
| | Time | 5 min | 30 s | 30 s | 3 min | 30 s | 30 s | 3 min | Forever |
| | Cycles | ×1 | ×6 | | | ×30 | | | |

PCR2: 10 μL LongAmp PCR mix, 2 μL primers (*see* **Note 20**): 2.5 μM each, 6 μL DDW. Template: Dilute PCR1 1:5 and use 2 μL in a 20 μL reaction.
PCR as follows:

| PCR2 | Temp | 94 °C | 94 °C | 65 °C | 70 °C | 70 °C | 4 °C |
|---|---|---|---|---|---|---|---|
| | Time | 5 min | 30 sec | 30 s | 3 min | 3 min | Forever |
| | Cycles | ×1 | ×30 | | | ×1 | |

1. Analyze 5 μL of products of PCR2 by separating by electrophoresis on a 1 % agarose gel and stain with Ethidium Bromide.
2. Extract bands and DNA sequence using primer RE17-2.
3. BLAST-analyze the obtained sequences (http://blast.ncbi.nlm.nih.gov/Blast.cgi) to determine the transposon insertion site.

### 3.6 Validating the Phenotype Using "Clean" Deletions

The following procedure is demonstrated for a specific gene, *htpG*, but can be adjusted to any other gene by P1 transduction:

1. Streak *E. coli hns⁻/htpG⁺* (IYB5163) and *hns⁻/htpG⁻* (IYB5165) from −80 °C glycerol stock onto an LB plate supplemented with 25 μg/mL Kan and grow overnight at 32 °C (*see* **Note 1**).
2. Inoculate an isolated colony into a 50 mL plastic tube containing 5 mL LB supplemented with 25 μg/mL Kan. Incubate cultures at 32 °C with shaking at 250 rpm overnight.
3. Prepare electrocompetent cells as described in Subheading 3.1.1 **step 3**.
4. Thaw 50 μL of electrocompetent cells of IYB5163 and IYB5165 and adjust the volume with DDW according to number of transformations you have (*see* **Note 4**).
5. Continue as described in Subheading 3.1.2 **steps 2–12**.

### 3.7 Complementing the Deletion In Trans

The following procedure is demonstrated for a specific gene, *htpG*, but can be adjusted to any other gene by a P1 transduction and selection of the appropriate plasmid:

1. Prepare electrocompetent cells of IYB5163 and IYB5165 as described in Subheading 3.1.1.

2. Transform the electrocompetent cells with pCA24N (control) and pCA24N-HtpG (*see* **Note 2**).

3. Plate the cells on LB agar plate supplemented with 25 μg/mL Kan and 34 μg/mL Cm and grow overnight at 32 °C.

4. Inoculate an isolated colony into 50 mL plastic tubes containing 5 mL LB supplemented with 25 μg/mL Kan and 34 μg/mL Cm. Incubate cultures at 32 °C with shaking at 250 rpm overnight.

5. Inoculate 0.2 mL of the saturated overnight culture into 10 mL LB medium with 25 μg/mL Kan and 34 μg/mL Cm in 50 mL plastic tubes.

6. Prepare electrocompetent cells as described in Subheading 3.1.1 **step 5**.

7. Thaw 50 μL of the prepared electrocompetent cells and adjust the volume with DDW according to the required number of transformations (*see* **Note 4**).

8. Aliquot 45 μL per sample of the electrocompetent cells into prechilled 2 mm gap electroporation cuvettes.

9. Add 5 μL from 4 ng/μL stock of purified pAC-477 or pAC-478 and mix by pipetting (*see* **Note 2**).

10. Mix the cuvette by flicking and electroporate at 2.5 kV using Bio-Rad Multiporator.

11. Immediately add 0.2 mL of 2YT medium supplemented with 0.2 % w/v L-arabinose and 0.1 mM IPTG.

12. Transfer the electroporated cells using a 1 mL pipettor into a 1.7 mL sterile microcentrifuge tube and incubate at 32 °C for 1 h with shaking at 1,000 rpm using an Eppendorf Thermomixer.

13. After 1 h, place the samples on ice and prepare serial dilutions of $10^{-1}$, $10^{-2}$, $10^{-3}$, and $10^{-4}$ from the cell suspensions in 2YT medium supplemented with 0.2 % w/v L-arabinose and 0.1 mM IPTG.

14. Plate 50 μL from each dilution on the appropriate selective LB agar plates supplemented with 100 μg/mL Amp, 25 μg/mL Kan, 34 μg/mL Cm, and 0.2 % w/v L-arabinose.

15. Count the colonies in each plate and calculate the transformation efficiency by dividing the number of pAC-478 or pAC-477 transformants by the number of pAC-477 transformants.

# 4 Notes

1. The following table summarizes the genotype and properties of the strains that are used in the screen:

| Bacterial strains | Genotype | CRISPR-Cas system | Source |
|---|---|---|---|
| IYB5163 | BW25113 $\Delta hns$ $araB$::T7-RNAp-$tetA$ $\lambda cI857$-Kan | Active | [24] |
| IYB5165 | BW25113 $\Delta hns$ $\Delta htpG$ $araB$::T7-RNAp-$tetA$ $\lambda cI857$-Kan | Not active | [24] |
| IYB5237 | BW25113 $araB$::T7-RNAp-$tetA$ $\lambda cI857$-Kan | Not active | [24] |
| RE1093 | BW25113 $\Delta hns$ $araB$::T7-RNAp-$tetA$ $\lambda cI857$-$bla$ | Active | [24] |

The strains were genetically engineered to carry an arabinose-inducible T7 RNA polymerase (RNAP) to express T7 promoters on plasmids. The T7 promoter is used to drive expression of CRISPR spacers in the pWUR477, pWUR478, pAC-477 and pAC-478 plasmids. H-NS is a transcriptional repressor of the CRISPR-Cas system of *E. coli* and therefore strains lacking *hns* have an active system. The λ prophage ($\lambda cI857$-Kan or $\lambda cI857$-bla) is the target of the spacers encoded by plasmid pWUR478 or pAC-478, and therefore these lysogens cannot be transformed with these plasmids if the CRISPR-Cas system is active.

2. The following table summarizes the properties of the plasmids used:

| Plasmids | Description | Inducer | Source |
|---|---|---|---|
| pWUR477 | pACYCDuet-1 (Novagen) cloned with control spacers under T7 promoter, Cm$^r$ | L-arabinose, IPTG | [3] |
| pWUR478 | pACYCDuet-1 (Novagen) cloned with anti-λ spacers under T7 promoter, Cm$^r$ | L-arabinose, IPTG | [3] |
| pAC-477 | pACYC184-derived, cloned with control spacers under T7 promoter, Amp$^r$ | L-arabinose | [24] |
| pAC-478 | pACYC184-derived, cloned with anti-λ spacers under T7 promoter, Amp$^r$ | L-arabinose | [24] |
| pCA24N-HtpG | pCA24N cloned with *htpG*, Cm$^r$ | IPTG | [26] |
| pCA24N | Empty vector, Cm$^r$ | IPTG | [26] |

3. Glycerol and L-arabinose are easily contaminated. Therefore, it is better to keep them in small aliquots at 4 °C. Also, while sterilizing L-arabinose by autoclaving, the solution might become light to dark brown (due to caramelization). This might occur if the temperature is too high and/or the time of autoclaving is longer than recommended (15 min at 121 °C).

4. Each aliquot of electrocompetent cells can be diluted up to ten times with ice-cold DDW.

5. Strain WM2672 contains the plasmid pRL27. The pRL27 vector encodes a hyperactive transposase gene (*tnp*) under control of the *tetA* promoter. The *tnp* is not encoded on the transposon, and thus is lost after transposition, resulting in stable transposon insertions. OriR6k and the *aph* gene (Kan resistance) are encoded within the transposon. pRL27 is a suicide vector that contains a Pi protein (*pir*) dependent origin of replication, OriR6k, and can only replicate in bacteria encoding the Pi protein. Therefore in non-*pir* strains, Kan resistance will only be conferred when the transposon is inserted into the recipient genome (the plasmid will not be maintained). The plasmid also contains *oriT* (an origin of transfer) that allows the vector to be transferred from the E. coli donor to the recipient strain with high efficiency via conjugation. The transposon insertion site can be cloned from the chromosome and replicated as a plasmid via the *oriR6K* origin of replication in the *pir*+ host like E. coli DH5αλpir.

6. Some mutations might cause temperature sensitivity and therefore plating is at 32 °C. To prevent competition, colonies should be grown on LB agar plates where nutrient scavenging is less competitive than in solution.

7. The E .coli chromosome encodes approximately 5,000 candidate genes. In order to have a good coverage where all the genes are disrupted by the transposon at least once, the chromosome should be covered ten times, i.e., ~50,000 colonies. Repetition of the conjugation procedure might be required to reach this number and to reduce the chance of siblings (identical transposon mutant cells).

8. Tet ensures the presence of T7 RNAP as it is linked to the resistance gene, *tetA*, in the bacterial chromosome (*araB*::T7-RNAp-tetA). Amp ensures the presence of the λ lysogen as the Amp resistance gene, *bla*, is in the prophage genome (λcI857-bla). Kan ensures the presence of the transposon. Cm selects for the plasmid pWUR477 or pWUR478 transformants. L-arabinose induces T7 RNAP, which transcribes the CRISPR spacers from plasmid pWUR477 or pWUR478. IPTG allows the removal of the LacI repressor which is bound to the *lacO* sites upstream of the CRISPR arrays in pWUR477 and pWUR478.

9. The cells are λcI857-*bla* lysogens. At 42 °C, the prophage will be induced and cells will die. Since the λcI857-*bla* prophage is essential for the assay, cells that grow at 42 °C (most likely lost all or part of the prophage) are discarded from further analysis.

10. Resuspend the cells in 50 μL DDW in 96 plates since it is easier to store and handle all the clones.

11. Transposon insertion in any of the *cas* genes (except *cas1* and *cas2*) will abolish CRISPR-Cas activity. The PCR is aimed at detecting transposon insertion into the *cas3* or the *casA-E* region. The presence of a PCR product indicates transposon insertion in one of the *cas* genes.

    PCR was done with a mix of five primers:
    Three primers binding to the *cas* region (P1, P2, P3) and two binding to the Tn5 (KF, KR).

    P1: 5′-TTTGGGATTTGCAGGGATGAC-3′

    P2: 5′-CGCCTCGCCATTATTACGAA-3′

    P3: 5′-GCTACTCCGATGGCCTGCAT-3′

    KR: 5′-GCAATGTAACATCAGAGATTTTGAG-3′

    KF: 5′-ACCTACAACAAAGCTCTCATCAACC-3′

12. It is important to maintain all solutions above 15 °C, as the CTAB will precipitate below this temperature [27].

13. The solution should become viscous as the detergent lyses the bacterial cell walls. There should be no need to predigest the bacterial cell wall with lysozyme [27].

14. This step is very important since a CTAB–nucleic acid precipitate will form if salt concentration drops below about 0.5 M at RT. The aim here is to remove cell wall debris, denatured protein, and polysaccharides complexed to CTAB, while retaining the nucleic acids in solution [27].

15. This extraction removes CTAB–protein/polysaccharide complexes. A white interface should be visible after centrifugation [27].

16. With some bacterial strains the interface formed after chloroform extraction is not compact enough to allow easy removal of the supernatant. In such cases, most of the interface can be fished out with a sterile toothpick before removal of any supernatant. The remaining CTAB precipitate is then removed in the phenol–chloroform extraction [27].

17. If no stringy DNA precipitate forms, this implies that the DNA has sheared into relatively low-molecular-weight pieces. If this is the case, the chromosomal DNA can often still be recovered simply by pelleting the precipitate in a microcentrifuge [27].

18. This step removes residual CTAB [27].

19. This may take some time (up to 1 h) and will require incubation at 37 °C since the DNA is of high-molecular-weight. The typical DNA yield is 5–20 μg DNA per 1 mL starting culture ($10^8$ to $10^9$ cells/mL) [27].

20. In PCR1 two primers were used:

    RE17-1: 5′-AACAAGCCAGGGATGTAACG-3′ (annealing to pRL27-Tn5 [28]).

    RE16: 5′-<u>GGCCACGCGTCGACTAGTAC</u>NNNNNNNNNNGCTGG-3′ (Arbitrary primer annealing to random positions in the genome having the GCTGG pentameric sequence and a few more matching nucleotides for stabilization).

    In PCR2 two primers were used:

    RE17-2: 5′-AGCCCTTAGAGCCTCTCAAAGCAA-3′ (annealing to pRL27-Tn5 [29]).

    RE16a: 5′-<u>GGCCACGCGTCGACTAGTAC</u>-3′ (primer to amplify Tn5 mutagenesis, sequence annealing to the constant portion of the arbitrary primer is underlined in RE16a and in RE16).

21. Use the GFP minus version of the *htpG* from the ASKA collection [26]. It is recommended to verify the identity of the cloned gene from the collection by PCR followed by DNA sequencing.

# Acknowledgements

This research was supported by the Israel Science Foundation grant 611/10 to U.Q., and the Marie Curie International Reintegration Grant PIRG-2010-266717 to R.E.

# References

1. Marraffini LA, Sontheimer EJ (2008) CRISPR interference limits horizontal gene transfer in staphylococci by targeting DNA. Science 322(5909):1843–1845. doi:10.1126/science.1165771

2. Barrangou R, Fremaux C, Deveau H, Richards M, Boyaval P, Moineau S, Romero DA, Horvath P (2007) CRISPR provides acquired resistance against viruses in prokaryotes. Science 315(5819):1709–1712. doi:10.1126/science.1138140

3. Brouns SJ, Jore MM, Lundgren M, Westra ER, Slijkhuis RJ, Snijders AP, Dickman MJ, Makarova KS, Koonin EV, van der Oost J (2008) Small CRISPR RNAs guide antiviral defense in prokaryotes. Science 321(5891):960–964. doi:10.1126/science.1159689

4. Hale CR, Zhao P, Olson S, Duff MO, Graveley BR, Wells L, Terns RM, Terns MP (2009) RNA-guided RNA cleavage by a CRISPR RNA-Cas protein complex. Cell 139(5):945–956. doi:10.1016/j.cell.2009.07.040

5. Goren M, Yosef I, Edgar R, Qimron U (2012) The bacterial CRISPR/Cas system as analog of the mammalian adaptive immune system. RNA Biol 9(5):549–554. doi:10.4161/rna.20177

6. Abedon ST (2012) Bacterial 'immunity' against bacteriophages. Bacteriophage 2(1):50–54. doi:10.4161/bact.18609

7. Bhaya D, Davison M, Barrangou R (2011) CRISPR-Cas systems in bacteria and archaea: versatile small RNAs for adaptive defense and regulation. Annu Rev Genet 45:273–297. doi:10.1146/annurev-genet-110410-132430
8. Wiedenheft B, Sternberg SH, Doudna JA (2012) RNA-guided genetic silencing systems in bacteria and archaea. Nature 482(7385):331–338. doi:10.1038/nature10886
9. Makarova KS, Haft DH, Barrangou R, Brouns SJ, Charpentier E, Horvath P, Moineau S, Mojica FJ, Wolf YI, Yakunin AF, van der Oost J, Koonin EV (2011) Evolution and classification of the CRISPR-Cas systems. Nat Rev Microbiol 9(6):467–477. doi:10.1038/nrmicro2577
10. Pougach K, Semenova E, Bogdanova E, Datsenko KA, Djordjevic M, Wanner BL, Severinov K (2010) Transcription, processing and function of CRISPR cassettes in Escherichia coli. Mol Microbiol 77(6):1367–1379. doi:10.1111/j.1365-2958.2010.07265.x
11. Yosef I, Goren MG, Qimron U (2012) Proteins and DNA elements essential for the CRISPR adaptation process in Escherichia coli. Nucleic Acids Res 40(12):5569–5576. doi:10.1093/nar/gks216
12. Deveau H, Garneau JE, Moineau S (2010) CRISPR/Cas system and its role in phage-bacteria interactions. Annu Rev Microbiol 64:475–493. doi:10.1146/annurev.micro.112408.134123
13. Sorek R, Kunin V, Hugenholtz P (2008) CRISPR—a widespread system that provides acquired resistance against phages in bacteria and archaea. Nat Rev Microbiol 6(3):181–186. doi:10.1038/nrmicro1793
14. Marraffini LA, Sontheimer EJ (2010) CRISPR interference: RNA-directed adaptive immunity in bacteria and archaea. Nat Rev Genet 11(3):181–190. doi:10.1038/nrg2749
15. Gratz SJ, Cummings AM, Nguyen JN, Hamm DC, Donohue LK, Harrison MM, Wildonger J, O'Connor-Giles KM (2013) Genome engineering of Drosophila with the CRISPR RNA-guided Cas9 nuclease. Genetics 194(4):1029–1035. doi:10.1534/genetics.113.152710
16. Gaj T, Gersbach CA, Barbas CF III (2013) ZFN, TALEN, and CRISPR/Cas-based methods for genome engineering. Trends Biotechnol 31(7):397–405. doi:10.1016/j.tibtech.2013.04.004
17. Wang H, Yang H, Shivalila CS, Dawlaty MM, Cheng AW, Zhang F, Jaenisch R (2013) One-step generation of mice carrying mutations in multiple genes by CRISPR/Cas-mediated genome engineering. Cell 153(4):910–918. doi:10.1016/j.cell.2013.04.025
18. Blackburn PR, Campbell JM, Clark KJ, Ekker SC (2013) The CRISPR system—keeping zebrafish gene targeting fresh. Zebrafish 10(1):116–118. doi:10.1089/zeb.2013.9999
19. DiCarlo JE, Norville JE, Mali P, Rios X, Aach J, Church GM (2013) Genome engineering in Saccharomyces cerevisiae using CRISPR-Cas systems. Nucleic Acids Res 41(7):4336–4343. doi:10.1093/nar/gkt135
20. Ramalingam S, Annaluru N, Chandrasegaran S (2013) A CRISPR way to engineer the human genome. Genome Biol 14(2):107. doi:10.1186/gb-2013-14-2-107
21. Jiang W, Bikard D, Cox D, Zhang F, Marraffini LA (2013) RNA-guided editing of bacterial genomes using CRISPR-Cas systems. Nat Biotechnol 31(3):233–239. doi:10.1038/nbt.2508
22. Cong L, Ran FA, Cox D, Lin S, Barretto R, Habib N, Hsu PD, Wu X, Jiang W, Marraffini LA, Zhang F (2013) Multiplex genome engineering using CRISPR/Cas systems. Science 339(6121):819–823. doi:10.1126/science.1231143
23. Edgar R, Qimron U (2010) The Escherichia coli CRISPR system protects from lambda lysogenization, lysogens, and prophage induction. J Bacteriol 192(23):6291–6294. doi:10.1128/JB.00644-10
24. Yosef I, Goren MG, Kiro R, Edgar R, Qimron U (2011) High-temperature protein G is essential for activity of the Escherichia coli clustered regularly interspaced short palindromic repeats (CRISPR)/Cas system. Proc Natl Acad Sci U S A 108(50):20136–20141. doi:10.1073/pnas.1113519108
25. Vercoe RB, Chang JT, Dy RL, Taylor C, Gristwood T, Clulow JS, Richter C, Przybilski R, Pitman AR, Fineran PC (2013) Cytotoxic chromosomal targeting by CRISPR/Cas systems can reshape bacterial genomes and expel or remodel pathogenicity islands. PLoS Genet 9(4):e1003454. doi:10.1371/journal.pgen.1003454
26. Kitagawa M, Ara T, Arifuzzaman M, Ioka-Nakamichi T, Inamoto E, Toyonaga H, Mori H (2005) Complete set of ORF clones of Escherichia coli ASKA library (a complete set of E. coli K-12 ORF archive): unique resources for biological research. DNA Res 12(5):291–299. doi:10.1093/dnares/dsi012
27. Wilson K (1994) Preparation of genomic DNA from bacteria, p. 2.4. 1-2.4. 5. InIn FA Ausubel, R. Brent, RE Kingston, DD Moore, JG Seidman, JA Smith, and K. Struhl. Current protocols in molecular biology John Wiley & Sons, Inc, New York, NY

28. Larsen RA, Wilson MM, Guss AM, Metcalf WW (2002) Genetic analysis of pigment biosynthesis in Xanthobacter autotrophicus Py2 using a new, highly efficient transposon mutagenesis system that is functional in a wide variety of bacteria. Arch Microbiol 178(3):193–201. doi:10.1007/s00203-002-0442-2

29. Walker CB, Stolyar S, Chivian D, Pinel N, Gabster JA, Dehal PS, He Z, Yang ZK, Yen HC, Zhou J, Wall JD, Hazen TC, Arkin AP, Stahl DA (2009) Contribution of mobile genetic elements to Desulfovibrio vulgaris genome plasticity. Environ Microbiol 11(9):2244–2252. doi:10.1111/j.1462-2920.2009.01946.x

# Chapter 16

# Analysis of Nuclease Activity of Cas1 Proteins Against Complex DNA Substrates

Natalia Beloglazova, Sofia Lemak, Robert Flick, and Alexander F. Yakunin

## Abstract

Cas1 genes encode the signature protein of the CRISPR/Cas system, which is present in all CRISPR-containing organisms. Recently, Cas1 proteins (together with Cas2) have been shown to be essential for the formation of new spacers in *Escherichia coli*, and purified Cas1 proteins from *Pseudomonas aeruginosa* and *E. coli* have been shown to possess a metal-dependent endonuclease activity. Here we describe the protocols for the analysis of nuclease activity of purified Cas1 proteins against various DNA substrates including Holliday junctions and other intermediates of DNA recombination and repair.

**Key words** CRISPR, Cas1, Nuclease assay, DNA cleavage, Holliday junction

## 1 Introduction

Clustered Regularly Interspaced Short Palindromic Repeats (CRISPR) and associated proteins (Cas) represent a novel prokaryotic immune system, which protects microbial cells from invading viruses and plasmids [1]. During the first step of the CRISPR mechanism (adaptation), Cas proteins recognize foreign DNA and insert short sequences of this DNA into the CRISPR loci on the host chromosome as new spacers. In the processing step of the CRISPR mechanism, these viral-specific spacers are transcribed and processed by other Cas proteins to produce CRISPR guide RNAs (crRNAs), which direct different Cas nucleases to specifically recognize and destroy foreign DNA or RNA during the interference stage [2]. Presently, at least 65 different Cas proteins have been identified in microbial genomes, and can be classified into up to 45 protein families and three major CRISPR types: I, II, and III [1, 3, 4]. The proteins from six Cas families (Cas1-Cas6) represent the core group of CRISPR-associated proteins with Cas1 and Cas2 proteins found in all CRISPR-containing genomes. Both Cas1 and Cas2 proteins have been shown to be essential for efficient addition of new spacers to the CRISPR array in *Escherichia coli* [5]. Cas1 proteins from *Pseudomonas aeruginosa* and *E. coli* exhibit metal-dependent endonuclease activity against single stranded (ss), double stranded (ds), and branched DNA substrates, whereas the Cas1 protein SSO1450 from *Sulfolobus solfataricus* appears not to possess nuclease activity [6–8]. Therefore, further biochemical studies of Cas1 proteins from diverse organisms are required to delineate the functional versatility of this protein family and their role in the CRISPR mechanism [9]. In this chapter, we present the protocols for the analysis of nuclease activity of purified Cas1 proteins against various DNA substrates.

## 2 Materials

All solutions should be prepared using ultrapure water (e.g., Milli-Q) and molecular biology grade reagents. Unless stated otherwise, all chemicals can be purchased from Sigma (St. Louis, MO).

### 2.1 Protein Purification

1. Pelleted *E. coli* cells from 1 L of culture over-expressing CRISPR protein of interest in a 50 mL conical tube.
2. Binding buffer/wash buffer/elution buffer: 50 mM HEPES (pH 7.5), 0.25 M NaCl, 5/30/250 mM imidazole, 5 % glycerol. Add about 250 mL water to a 500 mL beaker (*see* **Note 1**). Weigh 5.96 g HEPES, 7.31 g NaCl, 0.17 g/1.02 g/8.51 g imidazole and transfer to beaker. Add 25 mL of 100 % glycerol to beaker. Mix and adjust pH with NaOH and/or HCl. Adjust volume to 500 mL using a graduated cylinder and store at 4 °C.
3. Equilibrated Ni-NTA resin: Spin 4 mL of Ni-NTA resin at $1,000 \times g$ for 1 min, decant supernatant and resuspend to 4 mL with Binding Buffer (*see* **Note 2**). Repeat this three times and resuspend resin to 4 mL with Binding Buffer.
4. 125 mL stainless steel beakers.
5. Sonicator equipped with standard probes.
6. 50 mL high-speed centrifuge tubes.
7. High-speed centrifuge.
8. Mini shaker.
9. Chromatography columns, $1.5 \times 15$ cm (Bio-Rad or other supplier).
10. Two-way stopcock.
11. Flow Adaptor. 1.5 cm diameter (Bio-Rad or other supplier).
12. Peristaltic pump.
13. Hot plate stirrer and magnetic stir bar.
14. Glass bottle with rubber septum.

15. Three needles (two long and one short).
16. Bradford reagent concentrate.

## 2.2 SDS PAGE

1. Separating Buffer: 1.5 M Tris–HCl (pH 8.8). To make 100 mL dissolve 18.21 g of Trizma base in water in a beaker, adjust to pH 8.8 using HCl. Adjust volume to 100 mL using a graduated cylinder. Store at 4 °C.
2. 30 % acrylamide–bis-acrylamide (29:1). Store at 4 °C.
3. 10 % SDS: Dissolve 10 g of SDS in water, and bring to 100 mL using a graduated cylinder. Store at room temperature.
4. 10 % APS: 1 g of ammonium persulfate in water, and bring to 10 mL using a graduated cylinder. Store at 4 °C for 1 week.
5. $N,N,N,N'$-tetramethyl-ethylenediamine (TEMED). Store at 4 °C.
6. Water-saturated butanol.
7. Stacking buffer: 0.5 M Tris–HCl (pH 6.8). To prepare 100 mL, dissolve 6.07 g of Trizma base in water in a beaker, adjust to pH 6.8 using HCl. Adjust volume to 100 mL using a graduated cylinder. Store at 4 °C.
8. 5× SDS gel loading buffer: In a 15 mL tube combine: 3.55 mL water, 1.25 mL stacking buffer, 2.5 mL glycerol, 2 mL 10 % SDS, 500 µL of β-mercaptoethanol and 0.2 mL 0.5 % (w/v) bromophenol blue. Store at −20 °C.
9. Dry Block Heating and Cooling Shaker.
10. Microcentrifuge.
11. Empty Gel Cassettes, Mini, 1.0 mm.
12. Mini cell for 1D vertical gel electrophoresis.
13. Protein Ladder.
14. SDS Running Buffer: Mix 28.8 g glycine, 6.04 g Trizma base, 2 g of SDS, and water in a beaker. Bring to 2 L using a graduated cylinder. Store at room temperature.
15. SimplyBlue™ SafeStain (Invitrogen).
16. Mini Shaker.

## 2.3 Nuclease Assay

1. Oligonucleotides: Oligodeoxy- and oligoribo-nucleotides obtained commercially (Integrated DNA Technology).
2. ATP, (γ-32P)- 3,000 Ci/mmol 10 mCi/mL, 1 mCi.
3. T4 Polynucleotide Kinase (PNK) and 10× PNK Buffer.
4. Annealing Buffer: 10 mM HEPES (pH 7.5), 20 mM sodium chloride. Store at 4 °C.
5. Dry Block Heating and Cooling Shaker.
6. Formamide loading buffer: 80 % formamide, 0.025 % bromophenol blue, 0.025 % xylene cyanol, and 10 mM EDTA (pH 8.0).

7. Glycerol gel loading buffer: 10 % glycerol, 0.025 % bromophenol blue, 0.025 % xylene cyanol.
8. Mini cell for 1-D vertical gel electrophoresis.
9. 0.75 mm × 10 cm × 8 cm notched glass plate set and comb.
10. Universal Power Supply.
11. Autoradiography Cassette.
12. Biomax MS X-Ray Film (Kodak).
13. Dark Room.
14. Tabletop X-Ray Film Processor.
15. GBX developer/replenisher Liquid concentrate.
16. GBX fixer/replenisher Liquid concentrate.
17. Microcentrifuge.
18. Razor blade.
19. Acetone.
20. 2 % lithium perchlorate: Dissolve 0.8 g of lithium perchlorate in 40 mL of acetone. Store at room temperature.
21. 500 mM Tris–HCl (pH 8 and 8.5): Dissolve 6.05 g of Trizma base in water in a beaker, adjust to desired pH using HCl. Adjust to 100 mL using a graduated cylinder then filter through a 0.2 μm membrane. Store at 4 °C.
22. 1 M potassium chloride (KCl): Dissolve 7.45 g of KCl in water in a beaker with stirring in about 85 mL nuclease-free water. Adjust the final volume to 100 mL using a graduated cylinder and filter through a 0.2 μm membrane. Store at room temperature.
23. 100 mM magnesium chloride ($MgCl_2$): Dissolve 952 mg of $MgCl_2$ powder in water in a beaker, adjust to 100 mL using a graduated cylinder and filter through a 0.2 μm membrane. Store at room temperature.
24. 20 mM dithiothreitol (DTT): Mix 308.5 mg of DTT in water in a beaker and bring to 100 mL using a graduated cylinder. Store at −20 °C.

*2.4 Nucleic Acids Gel*

1. 24 % acrylamide mixture (19:1): Dissolve 480.48 g of Urea in a large beaker containing 700 mL of water (*see* **Note 3**). Add 240 g acrylamide, and 12 g bis-acrylamide, bring up to 1 L using a graduated cylinder. Filter through a 0.45 μm membrane. Store at room temperature.
2. 40 % acrylamide Mixture (29:1): Dissolve 380 g acrylamide and 12.6 g bis-acrylamide in a large beaker containing water, bring up to 1 L using a graduated cylinder. Filter through a 0.45 μm membrane. Store at room temperature.

3. 8 M urea: Dissolve 480.48 g of Urea in a large beaker of water (*see* **Note 3**), and bring to 1 L using a graduated cylinder. Filter through a 0.2 μm membrane. Store at room temperature.

4. 10× TBE Buffer: Dissolve 108.7 g Tris, 55.3 g boric acid, and 43.8 g EDTA in a large beaker of water, bring to 1 L using a graduated cylinder. Store at room temperature.

5. 10 % APS: 1 g of Ammonium persulfate in 10 mL of water. Store at 4 °C.

6. *N,N,N,N'*-tetramethyl-ethylenediamine (TEMED). Store at 4 °C.

7. Mini cell for 1-D vertical gel electrophoresis.

8. Vertical Electrophoresis Systems, Large Format for DNA/RNA sequencing.

9. 10 cm × 8 cm notched glass plate set, 1 mm spacers and combs, 0.75 mm spacers and combs.

10. 33 cm × 42 cm Notched Plate Set, 0.4 mm spacers and combs.

11. Universal Power Supply.

12. Grade 1 Filter Paper, Whatman®.

13. Saran wrap.

14. Gel Dryer with Vacuum Pump.

## 2.5 Imaging

1. Autoradiography Cassette.

2. Storage Phosphor Screen.

3. Phosphorimager and imaging software.

---

# 3 Methods

## 3.1 Gel Preparation

### 3.1.1 SDS Gel (for Proteins)

1. In a small (50 mL) beaker combine 6.46 mL distilled water, 5 mL Separating Buffer, 8.54 mL 30 % acrylamide–bisacrylamide, 200 μL 10 % SDS, 100 μL 10 % APS, and 20 μL TEMED. Mix well and quickly pour (or transfer using pipette) into 1 mm gel cassettes (can be purchased from Invitrogen) about 2/3 of the way. This recipe makes 15 mL of gel solution, which is enough for three 12 % gels. Let polymerize for ~20 min (*see* **Note 4**).

2. In a small (50 mL) beaker combine 5.7 mL distilled water, 2.5 mL Stacking Buffer, 1.7 mL 30 % acrylamide–bisacrylamide, 100 μL 10 % SDS, 50 μL 10 % APS, and 10 μL TEMED. Mix well and quickly pour (or transfer using pipette) to fill the remaining volume in the three cassettes. Insert comb and let polymerize for ~20 min.

### 3.1.2 Native Gel (for Nucleic Acids)

1. For making 10 mL of gel solution combine 2.5 mL of 40 % acrylamide with 1 mL of 10× TBE buffer and add distilled

water to 10 mL. Mix gently, and add 100 μL of 10 % APS and 10 μL of TEMED. Mix gently and quickly pour (or transfer using pipette) into 0.75 mm cassette, trying to avoid bubbles and insert a 0.75 mm comb immediately. Allow gel to polymerize for ~15 min.

2. Remove comb and wash wells with 1× TBE buffer (100 mL 10× TBE plus 900 mL of $H_2O$) using a syringe. Place gel in electrophoresis cell filled with 1× TBE buffer (*see* **Note 5**). Load entire sample into each lane and run at low voltage.

3. After electrophoresis, remove gel from cassette and place on filter paper (*see* **Note 6**). Cover with plastic wrap and dry using the gel dryer and vacuum pump.

### 3.1.3 Denaturing gel (for Nucleic Acids)

1. Combine 5.41 mL of 24 % Acrylamide mixture with 1 mL of 10× TBE buffer and add 8 M Urea to 10 mL. Mix gently and add 100 μL of 10 % APS and 10 μL of TEMED. Mix gently. Cast gel in a 0.75 mm × 10 cm × 8 cm cassette trying to avoid bubbles and insert a 0.75 mm comb immediately. Allow gel to set for approximately 10 min.

2. Remove comb and wash wells with 1× TBE buffer using a syringe. Place gel in electrophoresis cell filled at least halfway with 1× TBE buffer. Rinse the urea from the wells just before loading the samples. Load 1/3 of the sample amount and run for ~45 min at about 20 V/cm gel length (for a 10 cm long gel this will be about 200 V).

3. After electrophoresis, remove gel from cassette and place on filter paper. Cover with plastic wrap and dry using the gel dryer and vacuum pump.

### 3.1.4 Sequencing gel (for Nucleic Acids)

1. The following recipe makes 60 mL of gel solution, which is enough for one 42 × 30 cm × 0.4 mm gel. Combine 37.5 mL of 24 % acrylamide mixture with 6 mL of 10× TBE buffer and add 8 M urea to 60 mL. Add 300 μL of 10 % APS and 35 μL of TEMED. Mix gently (*see* **Note 7**). Cast gel in 33 cm × 43 cm glass plates with 0.4 mm spacers on either side trying to avoid bubbles and insert a large 0.4 mm comb immediately (*see* **Note 8**). Allow gel to polymerize for ~20 min.

2. Remove comb and wash wells with 1× TBE buffer using a syringe. Place gel in Electrophoresis cell filled with 1× TBE buffer. Load 1/2 of the sample amount and run at 65 W for ~2 h (*see* **Note 9**).

3. After electrophoresis, remove gel from cassette and place on filter paper. Cover with plastic wrap and dry using the gel dryer and vacuum pump.

### 3.2 Gel Images

1. Place dried gels face-up inside an autoradiography cassette (*see* **Note 10**).

2. Place a storage phosphor screen on top of the gels and lock cassette overnight (*see* **Note 11**).

3. Scan the phosphor screen using a system and software capable of imaging $^{32}$P.

**3.3 Protein Purification**

1. Resuspend cell pellet with 25 mL of Binding Buffer by vortexing in 50 mL conical tube (*see* **Note 12**). Decant resuspended cells into 150 mL metal beaker and place on ice.

2. Sonicate resuspended cells for a total of 25 min on ice with the following cycle settings: 10 s at 100 amplitude, followed by 4 s at 0 amplitude (*see* **Note 13**). Take a 20 μL aliquot as a representative of the "whole cell lysate" for SDS-PAGE.

3. Transfer lysate into high-speed centrifuge tubes and spin at 61,000 × $g$ for 30 min. Transfer supernatant to a 50 mL conical tube. Take a 20 μL aliquot as a sample of the "cleared cell lysate" for SDS-PAGE.

4. Add 4 mL of equilibrated Ni-NTA resin to cleared lysate and mix by rotation for 30 min at 4 °C. Decant into 15 cm × 1.5 cm (L × d) glass Econo-Column (Bio-Rad) fitted with a stopcock in the closed position and allow resin to settle for 10 min (Fig. 1). Open stopcock and keep a 20 μL aliquot of the flow through as a sample of "unbound protein" for SDS-PAGE.

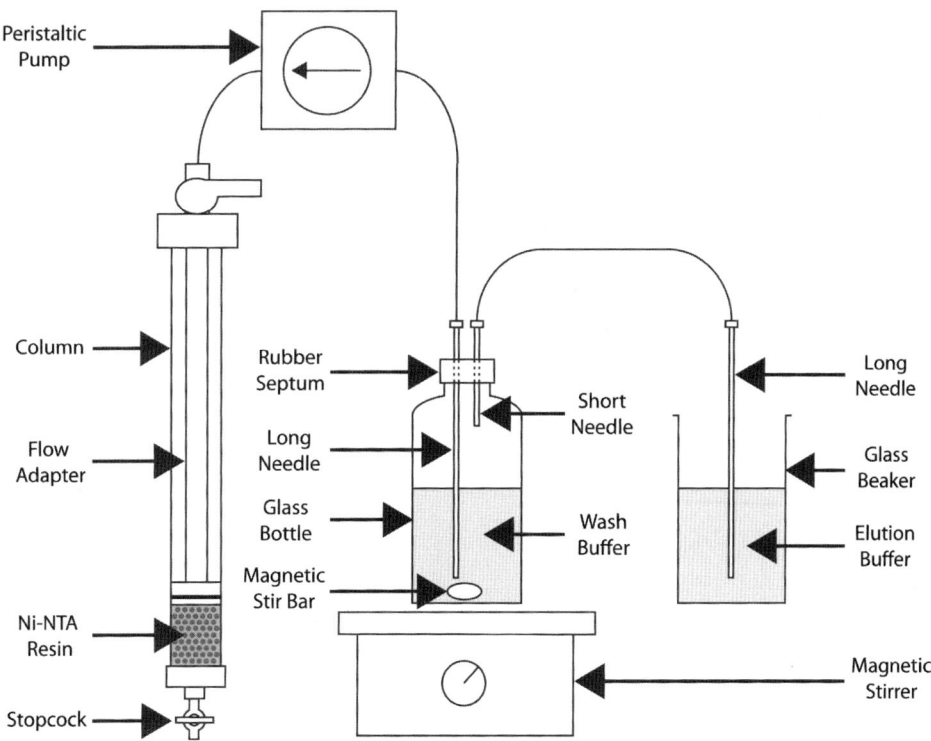

**Fig. 1** Diagram of the protein purification apparatus

5. Attach a 1.5 cm diameter Flow adaptor (Bio-Rad) connected to a peristaltic pump, primed with Wash Buffer to the glass column and pump 50 mL of Wash Buffer through the resin via the peristaltic pump at a flow rate of 2 mL/min with the stopcock in the open position (Fig. 1) (*see* **Note 14**). Stop pump, close stopcock, and take a 20 μL aliquot of the flow through as a sample of the "wash fraction" for SDS-PAGE.

6. Add 75 mL of Wash Buffer to a 125 mL glass bottle containing a magnetic stir bar, seal with a rubber septum and place on a magnetic stir plate. Use a peristaltic pump equipped with a needle to evacuate all air from the bottle (*see* **Note 15**). Attach a long needle to the peristaltic pump line connected to the Flow adaptor and insert needle through the rubber septum and submerge needle in Wash buffer (Fig. 1).

7. Connect a short needle to a length of tubing and a long needle to the other end and submerge long needle in 200 mL of Elution buffer in a glass beaker. Insert short needle through rubber septum of 125 mL glass bottle from previous step, open stopcock on column and begin pumping at 2 mL/min (Fig. 1). Collect 2 mL fractions from column.

8. Monitor protein elution periodically by adding 5 μL of eluate to 100 μL of Bradford Reagent and stop collecting fractions when Bradford Reagent no longer changes to a blue color (*see* **Note 16**).

## 3.4 SDS-PAGE Analysis of Purified Proteins

1. Prepare samples by adding 5× SDS loading buffer to a final concentration of 1×, incubating for 10 min at 95 °C, and centrifuging at ~13,000 rpm for 30 s.

2. Remove comb and white sticker from cassette, lock in the X-Cell gel box, and fill at least 1/3 of the way with SDS Running Buffer.

3. Load samples (*see* **Note 17**) alongside 4 μL of Protein Ladder, and run at 120 V.

4. Remove gel from the running box and remove from the cassette by breaking open with a spatula. Add staining solution (*see* **Note 18**) and let sit for ~30 min, gently rotating. Destain gel by replacing dye with water and gently rotate overnight.

## 3.5 Substrate Preparation for Nuclease Assays

### 3.5.1 5′-End Labeling of Oligonucleotides

1. Prepare reaction mixture containing 20 pmol ($\gamma$-32P)-ATP, 10–50 pmol of oligonucleotide, 10 U of T4 polynucleotide kinase (PNK) in T4 PNK buffer, mix thoroughly, spin briefly, and incubate at 37 °C for 30 min (*see* **Notes 19–23**).

2. At the end of incubation add an equal volume of Formamide loading buffer, and separate labeled DNA from unincorporated ATP by electrophoresis in a 10–15 % PAAG under denaturing conditions (*see* **Note 24**).

3. After separation, visualize gel by exposing on an X-ray film for 1 min (*see* **Note 25**).

4. Excise a slice of the gel containing labeled oligonucleotide, put it in a 1.5 mL Eppendorf tube and add 200 μL of nuclease-free water. Incubate for 2 h at 37 °C with constant shaking. Collect the supernatant and precipitate the radiolabeled oligonucleotide by the addition of 10 volumes of a 2 % solution of lithium perchlorate in acetone.

5. Incubate at −20 °C for 1 h and collect the pellet by centrifugation at 13,000 rpm for 15 min. Aspirate the acetone, wash the precipitate with acetone twice, and dry at 30 °C.

6. Dissolve the oligonucleotide in Milli-Q water and store in aliquots at −20 °C.

*3.5.2 Preparation of Branched DNA Substrates*

Holliday Junction (HJ) substrates and other branched DNA substrates such as replication forks, Y-junctions, splayed arms, and 3′- and 5′-flaps are formed in all organisms during DNA integration, recombination, and recombinational DNA repair, as well as the regression or restart of a replication fork. Branched DNA substrates are formed by partially complementary oligonucleotides, one of which is (5′-$^{32}$P)-labeled.

1. Prepare a mixture of partially complementary oligonucleotides, one of which is (5′-$^{32}$P)-labeled in Annealing buffer.

2. Incubate the mixture at 90 °C for 5 min, and then slowly cool down to room temperature.

3. Analyze branched DNA substrates by electrophoresis in a 10–12 % native PAAG, followed by phosphorimaging or exposure on X-ray film.

4. Store substrates in aliquots at +4 °C for 1–2 weeks.

## 3.6 Nuclease Activity of the E. Coli Cas1 Protein YgbT

*3.6.1 Cas1 Nuclease Assays with Linear Substrates*

1. Add 100 ng of purified YgbT to reaction mixture containing 50 mM 5′-$^{32}$P)-labeled ssDNA, 50 mM Tris–HCl (pH 8.0), 100 mM KCl, 10 mM $MgCl_2$, 1 mM dithiothreitol, mix and incubate at 37 °C.

2. Remove 10 μL aliquots after 15, 30, 45, and 60 min of incubation, add an equal volume of Formamide loading buffer, mix and store the samples on ice until the gel is loaded.

3. Before loading, incubate all samples at 50 °C for 5 min, then load them on a 12–15 % denaturing PAAG of 30–40 cm length running at 40 V/cm (*see* **Note 26**).

4. After electrophoresis, dry the gel on a gel dryer and visualize it by exposing to X-ray film or phosphorimaging (*see* **Note 11**). Representative images showing activity of YgbT against ssDNA or ssRNA are presented in Fig. 2.

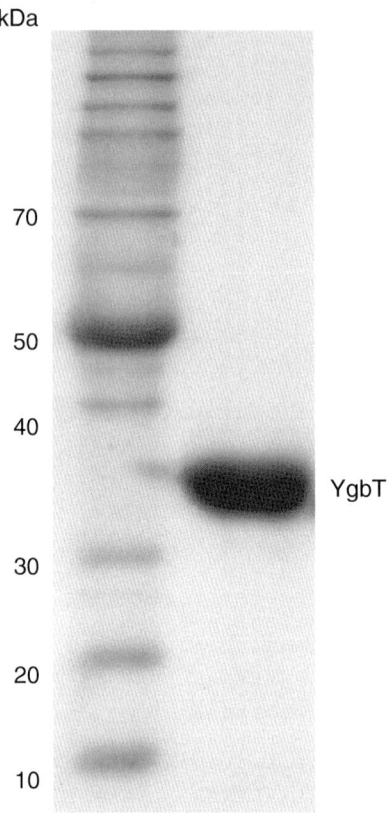

**Fig. 2** SDS-PAGE analysis of the purified Cas1 protein YgbT from *E. coli*

*3.6.2 Cas1 Nuclease Assays with HJ*

1. Prepare reaction mixtures containing 20 nM (5′-$^{32}$P)-labeled substrate in 50 mM Tris–HCl (pH 8.5), 100 mM KCl, 10 mM $MgCl_2$, 1 mM DTT.
2. Add YgbT to 100 µg/mL, mix and incubate at 37 °C for 30 min.
3. Stop the reaction by adding Glycerol gel loading buffer and resolve the products on a native 8 % polyacrylamide gel (Fig. 3).

*3.6.3 Cas1 Nuclease Assays with Cruciform-Like Substrate*

1. Prepare reaction mixtures containing 20 nM (5′-$^{32}$P)-labeled substrate in 50 mM Tris–HCl (pH 8.5), 100 mM KCl, 10 mM $MgCl_2$, 1 mM DTT.
2. Add YgbT to 100 µg/mL, mix and incubate at 37 °C for 30 min.
3. After 13, 30, 60, and 90 min of incubation remove 10 µL aliquots and quench reactions by adding an equal volume of Formamide loading buffer and resolve the reaction products by electrophoresis in 12–15 % polyacrylamide/8 M urea gels using TBE as a running buffer (Fig. 4).

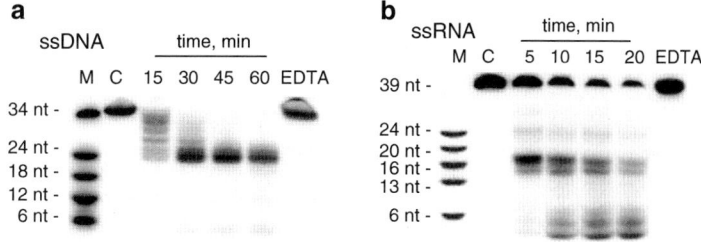

**Fig. 3** Nuclease activity of the *E. coli* Cas1 protein YgbT against linear DNA (**a**) and RNA (**b**) substrates. The 5'-($^{32}$P)-labeled ssDNA (34 nt) or ssRNA (37 nt) was incubated without protein (C), with YgbT for the indicated times (37 °C), or in the presence of 10 mM EDTA for 60 min (EDTA). The lane M shows the nucleotide markers

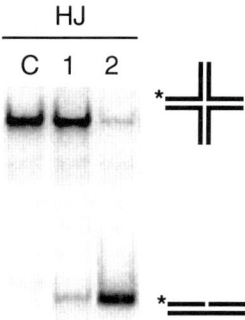

**Fig. 4** Cleavage of HJ by the *E. coli* YgbT (native PAGE/autoradiography). The 5'-($^{32}$P)-labeled HJ substrate was incubated at 37 °C for 45 min without protein (C) or with YgbT (1) or the *E. coli* RuvC (2)

## 4 Notes

1. Water at the bottom of the beaker helps dissolve HEPES, NaCl and imidazole quicker. Also, having the magnetic stir bar stirring before adding any additional chemicals helps prevent the stir bar from becoming stuck in undissolved powder.

2. Take care when decanting supernatant as to not lose any of the Ni-NTA resin. Retaining a small amount of supernatant is fine as the resin does not form a tightly packed pellet.

3. Low heat will help the mixture to dissolve.

4. Fill up the rest of the cassette with water-saturated butanol to level gel. Pour it out after gel has polymerized and rinse with distilled water. Use a Kimwipes to dry and remove all excess liquid from the gel.

5. Make sure that sufficient volume of buffer is added to the cell as this helps keep the native gels cooler, preventing substrate denaturing.

6. Make sure to note the orientation of the lanes to prevent mixing up your samples.

7. Using a 250 mL beaker for mixing the gel components will ensure better mixing as well easier casting of the gel.

8. Thorough washing of the glass plates with soap and water, followed by rinsing with ethanol will ensure a smooth surface for the gel mixture and reduce formation of bubbles. Slightly elevating the glass plates will help the gel mixture to spread quicker along the large plate surface.

9. Pre-electrophoresis prior to sample loading is recommended because it will create a uniform gel temperature and bring the gel temperature to the recommended run temperature. This will help eliminate any "smile" patterns from developing early in the run. Run sequencing gels at 50 °C for best results.

10. Removal of excess paper will make it easier to determine the exact position of your gel.

11. Substrates labeled with recently acquired isotope can produce a bright image after 2 h of exposure.

12. Using cell pellets that were frozen at −20 °C for at least one night helps with cell lysis, but are difficult to resuspend. If using frozen cell pellets, the pellets should be thawed in room temperature water for approximately 15 min before adding buffer. This allows for easier resuspension of the pellet as frozen pellets do not readily resuspend even with vortexing.

13. Make sure that water is added to the ice, creating an ice bath, before sonication to allow for better thermal transfer between beaker and ice. Failure to do so may result in overheating of cell pellets and lysate, denaturing the proteins.

14. Priming the peristaltic pump before attaching the Flow adaptor to the column prevents damaging the Ni-NTA resin through the introduction of air. Also make sure that there is no gap between the flow adaptor and the Ni-NTA resin. This prevents the formation of a mixing chamber that may disrupt the packing of the Ni-NTA resin.

15. Purging air from the bottle generates a negative pressure environment. This allows for the generation of the imidazole gradient by drawing fluid from the beaker containing elution buffer mentioned in Subheading 3.3 **step** 7 (Fig. 1).

16. Monitoring elution of protein by Bradford Reagent prevents dilution of the protein, removing a need for further concentration of protein following purification.

17. Loading ~5 μg of protein/well will give the best understanding of purity (Fig. 5).

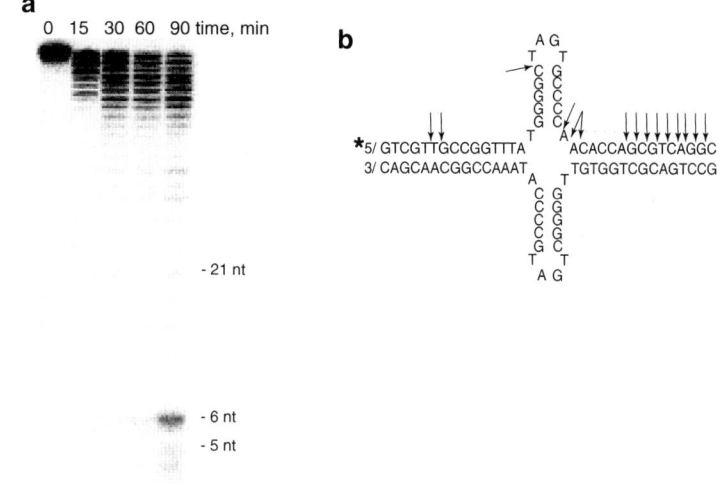

**Fig. 5** Cleavage of the CRISPR cruciform-like substrate by YgbT. (**a**), denaturing PAGE analysis; (**b**), the sequence of the cruciform-like substrate and the YgbT cleavage positions

18. For best results submerge the gel in water and heat to boiling point (~30 s in a microwave). Swap for fresh water and repeat. Pour out the water, submerge gel in SimplyBlue stain and heat as above.
19. Always wear gloves and maintain a nuclease-free bench area for RNA and DNA manipulations.
20. When working with radioactive material, strictly follow the safety rules established at your institution.
21. All reagents should be centrifuged briefly before opening to prevent loss and/or contamination of material that may be present around the rim of the tube.
22. Mix ingredients thoroughly. Gently flick the tube or pipette the mixture up and down gently, then microfuge tube briefly to collect the reaction mixture at the bottom of the tube.
23. The DNA or RNA substrates should not be precipitated with, or dissolved in buffers containing ammonium salts prior to PNK reactions because these compounds are inhibitory to T4 Polynucleotide Kinase (PNK).
24. There are a number of ways to remove free nucleotides from the PNK reaction, including gel filtration, purification on a denaturing polyacrylamide gel, ammonium acetate/ethanol or lithium chloride precipitations. RNA and DNA substrates should be electrophoretically homogeneous, therefore we recommend using denaturing (8 M urea) polyacrylamide gels with the percentage of polyacrylamide appropriate for the size

of the nucleic acid being purified as a reliable and simple method to purify the labeled nucleic acids away from free ($\gamma$-$^{32}$P) ATP and longer and shorter fragments.

25. When recovering the labeled oligonucleotide from the denaturing gel, fluorescent stickers are the easiest way to orient the film with the gel to cut out the band.

26. For most DNA/RNA sequencing or nucleic acid separations, a 19:1 acrylamide–bis-acrylamide solution is required. A 1× TBE (Tris, boric acid, and EDTA) solution is the preferred electrophoresis buffer. High-quality reagents for gel and buffer preparations are preferable because this affects reproducibility.

## Acknowledgements

This work was supported by the Government of Canada through Genome Canada and Ontario Genomics Institute (2009-OGI-ABC-1405), Ontario Research Fund (ORF-GL2-01-004), and Natural Science and Engineering Research Council of Canada.

## References

1. Makarova KS, Haft DH, Barrangou R, Brouns SJ, Charpentier E, Horvath P, Moineau S, Mojica FJ, Wolf YI, Yakunin AF, van der Oost J, Koonin EV (2011) Evolution and classification of the CRISPR-Cas systems. Nat Rev Microbiol 6:467–477
2. Westra ER, Swarts DC, Staals RH, Jore MM, Brouns SJ, van der Oost J (2012) The CRISPRs, they are a-changin': how prokaryotes generate adaptive immunity. Annu Rev Genet 46:311–339
3. Haft DH, Selengut J, Mongodin EF, Nelson KE (2005) A guild of 45 CRISPR-associated (Cas) protein families and multiple CRISPR/Cas subtypes exist in prokaryotic genomes. PLoS Comput Biol 1:e60
4. Makarova KS, Grishin NV, Shabalina SA, Wolf YI, Koonin EV (2006) A putative RNA-interference-based immune system in prokaryotes: computational analysis of the predicted enzymatic machinery, functional analogies with eukaryotic RNAi, and hypothetical mechanisms of action. Biol Direct 1:7
5. Yosef I, Goren MG, Qimron U (2012) Proteins and DNA elements essential for the CRISPR adaptation process in Escherichia coli. Nucleic Acids Res 12:5569–5576
6. Han D, Lehmann K, Krauss G (2009) SSO1450–a CAS1 protein from Sulfolobus solfataricus P2 with high affinity for RNA and DNA. FEBS Lett 583:1928–1932
7. Wiedenheft B, Zhou K, Jinek M, Coyle SM, Ma W, Doudna JA (2009) Structural basis for DNase activity of a conserved protein implicated in CRISPR-mediated genome defense. Structure 17:904–912
8. Babu M, Beloglazova N, Flick R, Graham C, Skarina T, Nocek B, Gagarinova A, Pogoutse O, Brown G, Binkowski A, Phanse S, Joachimiak A, Koonin EV, Savchenko A, Emili A, Greenblatt J, Edwards AM, Yakunin AF (2011) A dual function of the CRISPR-Cas system in bacterial antivirus immunity and DNA repair. Mol Microbiol 79(2): 484–502
9. Nunez JK, Lee ASY, Engelman A, Doudna JA (2015) Integrase-mediated spacer aquisition during CRISPR-Cas adaptive immunity. Nature 519(7542):193–198

# Chapter 17

# Characterizing Metal-Dependent Nucleases of CRISPR-Cas Prokaryotic Adaptive Immunity Systems

## Ki H. Nam, Matthew P. DeLisa, and Ailong Ke

### Abstract

CRISPRs (clustered regularly interspaced short palindromic repeats), together with the nearby CRISPR-associated (*cas*) operon, constitute a prokaryotic RNA-based adaptive immune system against exogenous genetic elements. Here, we describe nuclease assays that are useful for characterizing the substrate-specific function of CRISPR-associated protein Cas2. We also provide methods for characterizing the stoichiometry and affinity between Cas2 and divalent metal ions using isothermal titration calorimetry (ITC).

**Key words** Cas2, CRISPR/Cas, DNase, Isothermal titration calorimetry (ITC), Microbial adaptation, Nuclease

## 1 Introduction

CRISPRs (Clustered regularly interspaced short palindromic repeats), together with an operon of CRISPR-associated (Cas) proteins, form an RNA-based prokaryotic immune system against exogenous genetic elements [1–5]. The CRISPR-Cas pathway can be subdivided into three distinct molecular events: CRISPR adaptation, CRISPR RNA (crRNA) expression and processing, and crRNA-mediated nucleic acid degradation [1–5]. During the CRISPR adaptation step, new spacers derived from foreign genetic elements are inserted into the CRISPR locus. Two universally conserved *cas* genes, namely, *cas1* and *cas2* that are present in all CRISPR-Cas subtypes, are required in this step [6–10].

While the Cas1 protein has been established as a $Mn^{2+}$-dependent nuclease, the function and biochemical activity of Cas2 remains is still debated. Cas2 is a small size protein (~100 amino acids) sharing certain sequence homology with the VapD family of proteins that are functionally linked to the VapBC toxin–antitoxin system [6, 11, 12]. The Cas2 monomer contains an N-terminal ferredoxin fold composed of two α-helices packed against the four-stranded antiparallel β-sheet. The β5-strand at the C-terminal region mediates Cas2 dimerization via interactions with the β4-strand in the partner protomer to extend the existing anti-parallel β-sheet. Several crystal structures of Cas2 have been reported. While they all similarly maintain an overall dimeric structure [12–14], closer biochemical characterization revealed that the substrate specificity among the Cas2 family of proteins is quite distinct [12–14]. *Sulfolobus solfataricus* Cas2 was characterized as a metal-dependent single-stranded (ss) endoribonuclease with preference for uracil-rich ssRNA [12]; however, this RNase activity was not observed in the Cas2 enzymes from *Desulfovibrio vulgaris*, *Bacillus halodurans*, and *Thermus thermophilus* [13, 14]. Instead, metal-dependent double-stranded (ds) DNase activity are found in the *B. halodurans* and *T. thermophilus* Cas2 proteins, and this distinct preference for DNA versus RNA substrates can be rationalized in part from structural and sequence comparisons [13]. The DNase activity in Cas2 agrees well with its essential function in the process of new spacer acquisition, which could involve the dicing of foreign dsDNA into short fragments (proto-spacers) and the subsequent insertion of these fragments into CRISPR *loci* as new spacers. This process minimally requires Cas1 and Cas2 proteins [8]. Neither protein produces dsDNA fragments as short as found in the CRISPR spacers, therefore it is likely that these two proteins function in a concerted fashion to generate the proper sized proto-spacers. This scenario agrees well with a recent study showing that the combined action of Cas1 and Cas2 extends the CRISPR region in *Escherichia coli* [8].

We previously demonstrated the presence of DNase activity in a subset of Cas2 proteins and revealed that the activity is critically dependent on pH, metal ion, and monovalent ion concentration [13]. Here, we describe in detail the methods used to characterize the nuclease activity of Cas2. In addition, we provide our protocols for measuring the stoichiometry and affinity between Cas2 and divalent metal ions using isothermal titration calorimetry (ITC). ITC is a method to measure of the heat exchange generated when ligand is mixed with the binding partner in a series of injections. It has becoming a popular method to derive binding stoichiometry, thermodynamic parameters, and even the kinetic parameters, of a binding process.

## 2 Materials

Prepare all solutions using ultrapure $H_2O$ (prepared by purifying deionized $H_2O$ to attain a sensitivity of 18 M Ω cm at 25 °C) and analytical grade reagents. Degas and store all purification buffers at 4 °C. Protein and DNA reagents are stored frozen. They are brought to room temperature for nuclease assay and ITC before the start of the experiment.

## 2.1 Protein Expression and Purification

The bacterial host and expression plasmid are items that can be selected by the researcher.

1. Expression vector with T7 promoter, e.g. pET28b plasmid DNA (Novagen) (*see* **Note 1**).
2. Expression host BL21 Star™ (DE3) *E. coli* cells (Novagen).
3. Luria–Bertani (LB) broth.
4. Isopropyl β-D-1-thiogalactopyranoside (IPTG).
5. Thrombin (2–3 U/μl): solution in 20 mM Tris–HCl, pH 8.4, 150 mM NaCl, and 2.5 mM $CaCl_2$.
6. Ni-NTA agarose resin.
7. Open column or glass column (1.5 × 5 cm), e.g. Econo-Column (Bio-Rad).
8. Ion exchange column or heparin column, e.g. Mono-Q column (GE Healthcare) or heparin column (GE Healthcare).
9. Size exclusion chromatography media, e.g. Superdex 200 10/30 (GE Healthcare).
10. Chromatography system.
11. Bradford assay reagents.
12. 15 % SDS-PAGE gel

## 2.2 Purification Buffers (See Note 2)

All purification buffers should be stored at 4 °C.

1. Lysis buffer: 50 mM Tris–HCl, pH 8.0, 0.1 M NaCl, 2 mM β-mercaptoethanol, and 0.2 mM phenylmethylsulfonyl fluoride.
2. Wash buffer (W1): 50 mM Tris–HCl, pH 8.0, 50 mM NaCl, 2 mM β-mercaptoethanol, and 20 mM imidazole.
3. Elution buffer (E1): 50 mM Tris–HCl, pH 8.0, 50 mM NaCl, 2 mM β-mercaptoethanol, and 300 mM imidazole.
4. Wash buffer 2 (W2): 20 mM Tris–HCl, pH 8.0, 2 mM DTT, and 0.1 M NaCl.
5. Elution buffer 2 (E2): 20 mM Tris–HCl, pH 8.0, 2 mM DTT, and 0.5 M NaCl.
6. Size-exclusion buffer (SE): 10 mM Tris–HCl, pH 8.0, 2 mM DTT, and 0.2 M NaCl.

## 2.3 Nuclease Assays

1. dsDNAs (*see* **Note 3**).
2. 2× reaction buffer: 20 mM HEPES, pH 7.5 and 50 mM NaCl (*see* **Note 4**).
3. pH-dependency nuclease assay (PDNA) buffer: 200 mM Sodium citrate (pH 3.0–5.0), MES (pH 6.0), Tris–HCl (pH 7.0–8.0), and CAPS (pH 9.0–11.0) solution. Sodium

citrate and MES adjust pH with NaOH. Tris and CAPS adjust pH with HCl. Store at 4 °C (*see* **Note 5**).

4. Metal-dependent nuclease assay (MDNA) buffer: 20 mM MgSO$_4$, MnSO$_4$, CuSO$_4$, NiSO$_4$, ZnSO$_4$, FeSO$_4$, and EDTA solution. Store at 4 °C (*see* **Note 6**).

5. Physiological environment-dependent nuclease assay (PEDNA) buffer: 50, 100, 200, and 400 mM NaCl and KCl solution in H$_2$O (*see* **Note 7**).

6. Stop reaction buffer: 20 mM Tris–HCl, pH 8.0 and 20 mM EDTA (*see* **Note 8**).

7. Proteinase K (0.02 U/μl) in 50 mM Tris–HCl, pH 8.0 and 1 mM CaCl$_2$ (*see* **Note 9**).

### 2.4 Protein Electrophoresis

1. Electrophoresis system.
2. 6× loading dye: 30 % (v/v) glycerol, 0.25 % (w/v) bromophenol blue, and 0.25 % (w/v) xylene cyanol FF.
3. 1.0 % (w/v) agarose gel (*see* **Note 10**).
4. Ethidium bromide (EtBr) solution: Add 5 μl ethidium bromide stock (10 mg/ml) per 100 ml water or 1× TBE buffer.
5. 5× TBE buffer: weigh 54 g Tris-base, 27.5 g boric acid, 20 ml of 0.5 M EDTA (pH 8.0), then bring final volume to 1,000 ml.

### 2.5 ITC Assays

1. Dialysis buffer A: 10 mM Tris–HCl, pH 8.0, 100 mM NaCl and 2 mM EDTA (*see* **Note 11**).
2. Dialysis buffer B: 10 mM Tris–HCl, pH 8.0 and 100 mM NaCl (*see* **Note 12**).
3. Metal solution: 20 mM Mg$^{2+}$ (or optimized metal ions) in dialysis buffer B (*see* **Note 13**).

### 2.6 Additional Equipment

1. Thermocycler or incubator.
2. Microcentrifuge tubes.
3. Microcentrifuge.
4. UV spectrophotometer (*see* **Note 14**).
5. UV transilluminator.
6. Nano-ITC instrument (*see* **Note 15**).

## 3 Methods

### 3.1 Expression and Purification of Cas2

All purification steps should be performed at 4 °C.

1. Grow *E. coli* BL21(DE3) carrying plasmid encoding Cas2 in 1 L LB media at 37 °C.
2. When the culture reaches an optical density at 600 nm (OD$_{600}$) of ~0.4–0.8 (*see* **Note 16**), induce expression of

plasmid-encoded Cas2 with IPTG (or other appropriate inducer) at a final concentration of 0.1 mM.

3. Grow cells at 18 °C for 18 h (*see* **Note 17**).
4. Harvest cells by centrifugation at 3,500 × *g* for 20 min.
5. Resuspend cell pellet in ice-cold lysis buffer and disrupt by sonication (*see* **Note 18**).
6. Remove the cell debris by centrifugation at 18,000 × *g* for 30 min.
7. Pull the 5 ml of Ni-NTA agarose resin into an Econo-Column.
8. Wash a Ni-NTA column with 5 CV of water.
9. Wash a Ni-NTA column with 10 CV of buffer W1.
10. Load the supernatant onto a Ni-NTA column.
11. Wash the resin using 10 column volumes (CV) of buffer W1.
12. Elute the protein using 5 CV of buffer E1.
13. Measure protein concentration using Bradford Assay and/or UV spectrophotometer.
14. Add the thrombin of 10 μl for 1 mg Cas2 protein for removing the N- or C- terminal His6-tag.
15. Pipette and incubate at 4 °C overnight.
16. Analyze the protein on SDS-PAGE gel (*see* **Note 19**).
17. Load the proteins onto a mono-Q or heparin column on a chromatography system.
18. Wash the column with W2 buffer until the signal of the 280 nm returns to baseline.
19. Elute the protein with E2 buffer by salt gradient elution.
20. Analyze the peak fractions using 15 % SDS-PAGE gel
21. Concentrate the protein fraction.
22. Load the proteins onto a Superdex 200 column on a chromatography system.
23. Elute the protein with SE buffer at flow rate of 0.5 ml/min.
24. Analyze the peak fractions using 15 % SDS-PAG gel.

### 3.2 Cas2 Nuclease Activity Assay

All nuclease experiments should be performed at 25 °C. Experimental setup and expected results are depicted in Fig. 1a.

1. Add the following materials to a microcentrifuge tube: 2 μl dsDNA (1.5–3.0 μg); 10 μl 2× reaction buffer; 8 μl nuclease-free $H_2O$; and 2 μl purified Cas2 (0.1–20 μM).
2. Resuspend gently and centrifuge for a few seconds.
3. Incubate at incubator or thermal cycler at 25 °C for 0–10 h.

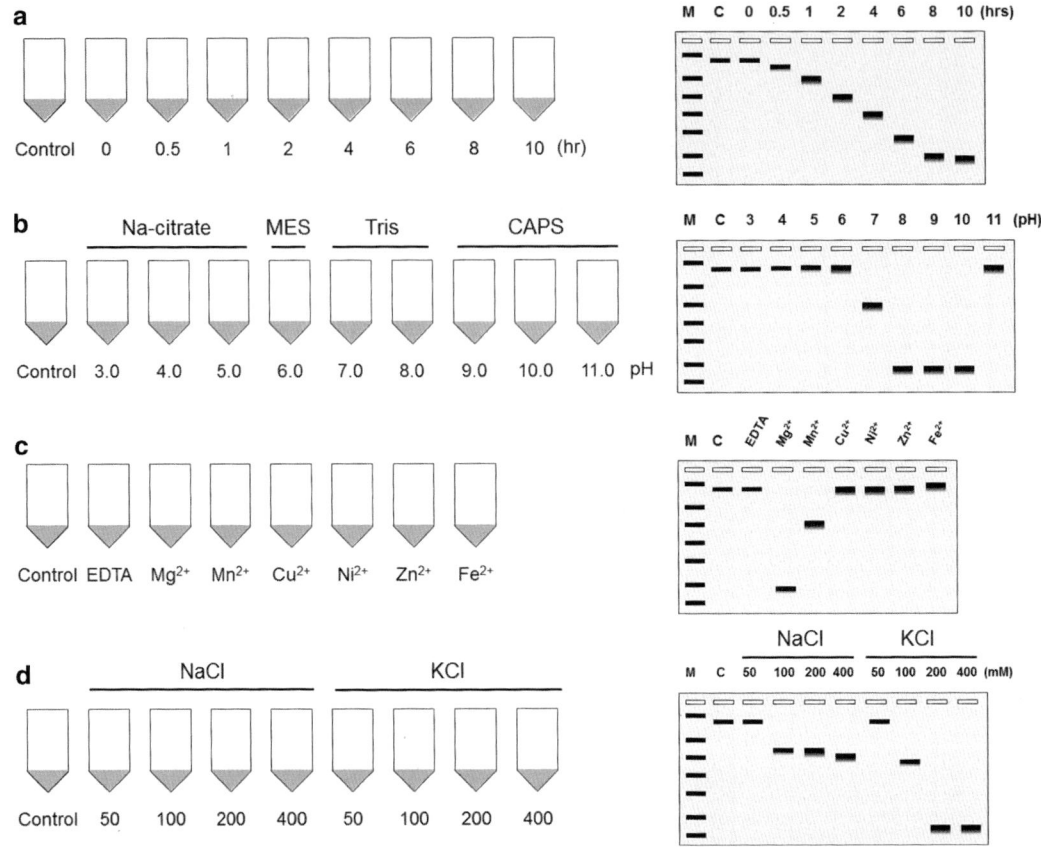

**Fig. 1** Nuclease activity assays for Cas2 protein. (**a**) Nuclease assay. (**b**) pH-dependent nuclease activity assay. (**c**) Metal-dependent nuclease activity assay. (**d**) Physiological environment-dependent nuclease activity assay. Right panels are expected results based on the nuclease activity for *B. halodurans* Cas2 [13]. Marker (M) and substrate-only control (C) are also shown

4. Add 2 μl of stop buffer, mix immediately, and centrifuge for a few seconds.
5. Add 1 μl of Proteinase K (0.02 U) and incubate at 37 °C for 10 min.
6. Centrifuge at $15,000 \times g$ for 5 min.
7. Transfer 20 μl of supernatant to microcentrifuge tube.
8. Add 5 μl of loading dye, mix, and centrifuge for a few seconds.
9. Load the reaction products onto a 1.0 % (w/v) agarose gel.
10. Transfer agarose gel into 50 ml of EtBr solution in tray.
11. Shake the tray at room temperature for 10 min.
12. Transfer agarose gel into 100 ml of water in tray (*see* **Note 20**).
13. Shake the tray at room temperature for 5 min.
14. Analyze the products using a UV transilluminator.

### 3.2.1 pH-Dependent Nuclease Activity Assay

All nuclease experiments should be performed at 25 °C. Experimental setup and expected results are depicted in Fig. 1b.

1. Add the following materials to a microcentrifuge tube: 2 μl dsDNA (1.5–3.0 μg); 10 μl each PDNA buffer (pH 3–11); 8 μl nuclease-free $H_2O$; and 2 μl purified Cas2 (0.1–20 μM).
2. Resuspend gently and centrifuge for a few seconds.
3. Incubate in thermocycler or incubator at 25 °C for 30–60 min.
4. Add 2 μl of stop buffer, mix immediately, and centrifuge for a few seconds.
5. Add 1 μl of Proteinase K (0.02 U) and incubate at 37 °C for 10 min.
6. Centrifuge at $15,000 \times g$ for 5 min.
7. Transfer 20 μl of supernatant to the microcentrifuge tube.
8. Add 5 μl of loading dye, mix, and centrifuge for a few seconds.
9. Load the reaction products onto a 1.0 % (w/v) agarose gel.
10. Transfer agarose gel into 50 ml of EtBr solution in tray.
11. Shake the tray at room temperature for 10 min.
12. Transfer agarose gel into 100 ml of water in tray (*see* **Note 20**).
13. Shake the tray at room temperature for 5 min.
14. Analyze the products using a UV transilluminator.

### 3.2.2 Metal-Dependent Nuclease Activity Assay

All nuclease experiments should be performed at 25 °C. Experimental setup and expected results are depicted in Fig. 1c.

1. Add the following materials to a microcentrifuge tube: 2 μl dsDNA (1.5–3.0 μg); 10 μl 2× reaction buffer; 2 μl each MDNA buffer (final 2 mM of $Mg^{2+}$, $Mn^{2+}$, $Cu^{2+}$, $Ni^{2+}$, $Zn^{2+}$, and EDTA); 6 μl nuclease-free $H_2O$; and 2 μl purified Cas2 (0.1–20 μM).
2. Resuspend gently and centrifuge for a few seconds.
3. Incubate in thermocycler or incubator at 25 °C for 30–60 min.
4. Add 2 μl of stop buffer, mix immediately, and centrifuge for a few seconds.
5. Add 1 μl of Proteinase K (0.02 U) and incubate at 37 °C for 10 min.
6. Centrifuge at $15,000 \times g$ for 5 min.
7. Transfer 20 μl of supernatant to the microcentrifuge tube.
8. Add 5 μl of loading dye, mix, and centrifuge for a few seconds.

9. Load the reaction products onto a 1.0 % (w/v) agarose gel.
10. Transfer agarose gel into 50 ml of EtBr solution in tray.
11. Shake the tray at room temperature for 10 min.
12. Transfer agarose gel into 100 ml of water in tray (*see* **Note 20**).
13. Shake the tray at room temperature for 5 min.
14. Analyze the products using a UV transilluminator.

*3.2.3 Physiological Environment-Dependent Nuclease Activity Assay*

All nuclease experiments should be performed at 25 °C. Experimental setup and expected results are depicted in Fig. 1d.

1. Add the following materials to a microcentrifuge tube: 2 μl dsDNA (1.5–3.0 μg); 10 μl 2× reaction buffer; 2 μl PEDNA buffer (final 0, 100, 200, 300, 400, 500 mM of KCl or NaCl); 6 μl nuclease-free $H_2O$; and 2 μl purified Cas2 (0.1–20 μM).
2. Resuspend gently and centrifuge for a few seconds.
3. Incubate in thermocycler or incubator at 25 °C for 30–60 min.
4. Add 2 μl of stop buffer, mix immediately, and centrifuge for a few seconds.
5. Add 1 μl of Proteinase K (0.02 U) and incubate at 37 °C for 10 min.
6. Centrifuge at $15,000 \times g$ or 5 min.
7. Transfer 20 μl of supernatant to the microcentrifuge tube.
8. Add 5 μl of loading dye, mix, and centrifuge for a few seconds.
9. Load the reaction products onto a 1.0 % (w/v) agarose gel.
10. Transfer agarose gel into 50 ml of EtBr solution in tray.
11. Shake the tray at room temperature for 10 min.
12. Transfer agarose gel into 100 ml of water in tray (*see* **Note 20**).
13. Shake the tray at room temperature for 5 min.
14. Analyze the products using a UV transilluminator.

## 3.3 Identification of Metal Binding Using Isothermal Thermal Calorimetry (ITC)

Dialysis and ITC experiments should be performed at 4 °C and 25 °C, respectively. Experimental setup and expected results are depicted in Fig. 2.

1. Wash the dialysis tubing (< MW 10,000) using dialysis buffer.
2. Pipette the Cas2 protein into the dialysis tubing and close off.
3. Insert the dialysis tubing into 1,000 ml of dialysis buffer A in a beaker.
4. Stir the buffer slowly with a magnetic stir plate for at least 2 h at 4 °C.

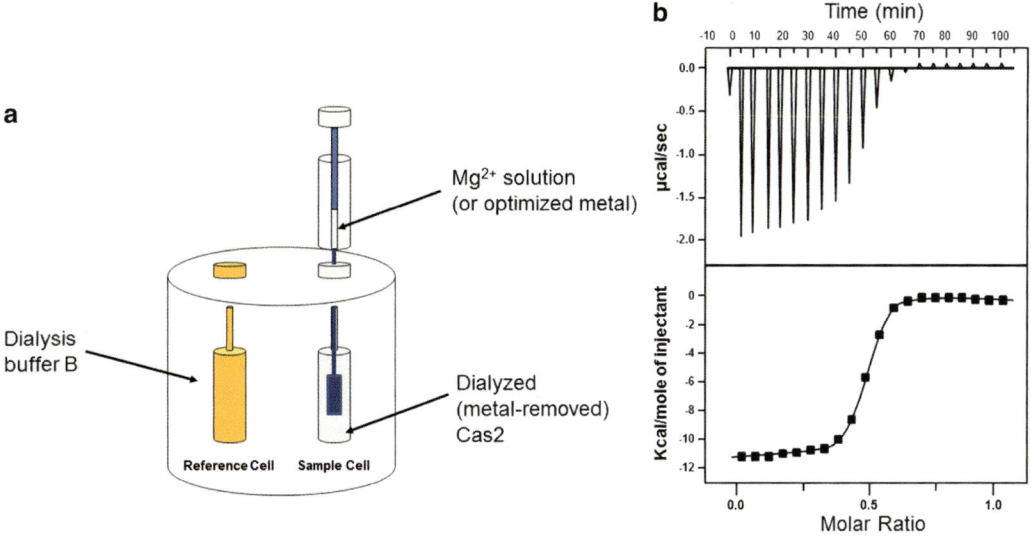

**Fig. 2** Use of ITC to identify stoichiometry between Cas2 and metal ions. (**a**) Experimental setup for ITC. Concentration of metal ion solution is 20 times higher than protein solution. (**b**) Expected results based on ITC analysis of *B. halodurans* Cas2 [13]

5. Discard the dialysis buffer A and replace with the same volume of fresh dialysis buffer.
6. Dialyze for an additional 2 h.
7. Discard the dialysis buffer B and replace with the same volume of fresh dialysis buffer.
8. Dialyze for an additional 2 h.
9. Discard the dialysis buffer B and replace with the same volume of fresh dialysis buffer.
10. Dialyze for an additional 2 h.
11. Pipette the dialyzed protein solution into tubes.
12. Degas the protein, dialysis and metal solution.
13. Add the dialysis buffer B into reference cell in nano-ITC instrument.
14. Add the protein (0.8–1 mM) solution of 190 μl into sample cell in nano-ITC instrument (*see* **Note 21**).
15. Add the metal solution (20 mM) of 50 μl into syringe in nano-ITC instrument (*see* **Note 22**).
16. Stir the syringe with 300 rpm in nano-ITC instrument until consistent temperature.
17. Inject the 2 μl of metal solution by 20 injections with 240 s interval with continuous stirring at 300 rpm (*see* **Note 23**).
18. Measure the binding constant, enthalpy and stoichiometry using nano-ITC instrument (*see* **Note 24**).

## 4 Notes

1. We used the pET28b vector with T7 promoter for protein expression. This expression vector can be selected by other desired vector and promoter depends on Lab source. The nuclease activity of Cas2 is not inhibited when short epitope tags such as $His_6$ or $His_{10}$, are used. However, we have observed inhibition of Cas2 DNase activity when larger epitope tags (e.g., SUMO tag) are used [13]. All Cas2 proteins form stable dimers in solution and in crystal structures [12–14]; based on modeling, the N-terminal SUMO fusion may interfere with the hinge motion in Cas2 and cause an allosteric inhibition of the dsDNase activity [13]. Thus, we strongly recommend the use of the short epitope tags.

2. Purification buffer can be selected based on desired type of purification. However, we have observed inhibition of Cas2 DNase activity at acidic pH [13]. We recommend using buffers at basic pH.

3. We use the pET-28a or pUC19 plasmids as substrate dsDNA. For preparing linear dsDNA, appropriate restriction enzymes can be used.

4. This buffer condition is not optimized for the nuclease activity. If DNase activity is not observed, proceed to pH- or metal-dependent nuclease activity assays.

5. We find that Cas2 nuclease activity is highly pH-dependent. In the case of both *B. halodurans* and *T. thermophilus* Cas2 enzymes, high DNase activity is observed at neutral or basic pH whereas nuclease activity was inhibited at acidic pH. To optimize the DNase activity of Cas2, pH screening may be required.

6. We find that Cas2 nuclease activity is highly metal-dependent. In the case of *B. halodurans* and *T. thermophilus* Cas2 enzymes, nuclease activity is inhibited in the presence of EDTA whereas nuclease activity is increased in the presence of $Mg^{2+}$ or $Mn^{2+}$.

7. We find that the nuclease activity of Cas2 is increased in the presence of high concentrations of NaCl or KCl.

8. Cas2 proteins are metal-dependent nucleases. Stop solution containing EDTA can be used to effectively stop the nuclease reaction.

9. We observed that Cas2 proteins bind to dsDNA after incubation, generating a smear for the product bands when run on argarose gel and visualized. In contrast, distinct bands on agarose gel are observed when reaction products are treated with Proteinase K prior to electrophoresis.

10. Gel percentage is not critical, but we suggest using ~1 % (w/v) agarose gel to observe the products.

11. This buffer is used to remove the metal ions in Cas2 proteins.
12. This buffer is used to remove the EDTA solution for ITC analysis.
13. The metal is selected after metal-dependent nuclease assay.
14. UV spectrophotometer is used to measure DNA concentration.
15. This instrument must be cleaned and pre-equilibrated at desired experimental temperature.
16. The optical density of cells and IPTG concentration can be selected based on optimal condition for protein overexpression.
17. This temperature can be selected depend on the optimal temperature for protein overexpression.
18. The cell lysis method by sonication can be changed into other methods (*i.e.* French press and microfluodizer)
19. This step is required for analyzing the $His_6$-tag cleaving. If not completely cleaved, increase the incubation time.
20. This step for washing of agarose gel using water can remove the EtBr background on gel by EtBr staining.
21. This protein concentration was optimized to observe the metal binding heat of *B. haldourans* Cas2. If the binding heat signal is high, then decrease the protein concentration. In contrast, increase the protein concentration, if weak heat signal.
22. The concentration of metal solution is depending on concentration of protein solution. Based on our experience, we recommend use the 10 to 20-fold higher concentration of metal solution than protein solution, in case of one-metal binding to dimer protein such as Cas2.
23. The injection volume and time of metal solution, interval time of injection, and speed of stirring were optimized for *B. haldourans* Cas2. If you do not have any reference, we recommend use our protocol first, then optimizes.
24. We observed excellent signal for metal binding affinity of *B. haldourans* Cas2 at pH 8.0; however, there is no signal for metal binding of *B. haldourans* Cas2 at pH 6.0 [13]. We recommend using buffers at basic pH.

# Acknowledgements

This work was supported by the National Institutes of Health Grant GM-102543 (to A.K.).

## References

1. Sorek R, Kunin V, Hugenholtz P (2008) CRISPR–a widespread system that provides acquired resistance against phages in bacteria and archaea. Nat Rev Microbiol 6:181–186
2. Hale CR, Zhao P, Olson S, Duff MO, Graveley BR, Wells L, Terns RM, Terns MP (2009) RNA-guided RNA cleavage by a CRISPR RNA-Cas protein complex. Cell 139:945–956
3. Koonin EV, Makarova KS (2009) CRISPR-Cas: an adaptive immunity system in prokaryotes. *F1000 Biol Reports* 1, 95
4. Sorek R, Lawrence CM, Wiedenheft B (2013) CRISPR-mediated Adaptive Immune Systems in Bacteria and Archaea. Annu Rev Biochem 82:237–266
5. Barrangou R (2013) CRISPR-Cas systems and RNA-guided interference. Wiley Interdiscip Rev RNA 4:267–278
6. Haft DH, Selengut J, Mongodin EF, Nelson KE (2005) A guild of 45 CRISPR-associated (Cas) protein families and multiple CRISPR/Cas subtypes exist in prokaryotic genomes. PLoS Comput Biol 1:e60
7. Fineran PC, Charpentier E (2012) Memory of viral infections by CRISPR-Cas adaptive immune systems: acquisition of new information. Virology 434:202–209
8. Yosef I, Goren MG, Qimron U (2012) Proteins and DNA elements essential for the CRISPR adaptation process in Escherichia coli. Nucleic Acids Res 40:5569–5576
9. Wiedenheft B, Zhou K, Jinek M, Coyle SM, Ma W, Doudna JA (2009) Structural basis for DNase activity of a conserved protein implicated in CRISPR-mediated genome defense. Structure 17:904–912
10. Babu M, Beloglazova N, Flick R, Graham C, Skarina T, Nocek B, Gagarinova A, Pogoutse O, Brown G, Binkowski A, Phanse S, Joachimiak A, Koonin EV, Savchenko A, Emili A, Greenblatt J, Edwards AM, Yakunin AF (2011) A dual function of the CRISPR-Cas system in bacterial antivirus immunity and DNA repair. Mol Microbiol 79:484–502
11. Kwon AR, Kim JH, Park SJ, Lee KY, Min YH, Im H, Lee I, Lee KY, Lee BJ (2012) Structural and biochemical characterization of HP0315 from Helicobacter pylori as a VapD protein with an endoribonuclease activity. Nucleic Acids Res 40:4216–4228
12. Beloglazova N, Brown G, Zimmerman MD, Proudfoot M, Makarova KS, Kudritska M, Kochinyan S, Wang S, Chruszcz M, Minor W, Koonin EV, Edwards AM, Savchenko A, Yakunin AF (2008) A novel family of sequence-specific endoribonucleases associated with the clustered regularly interspaced short palindromic repeats. J Biol Chem 283:20361–20371
13. Nam KH, Ding F, Haitjema C, Huang Q, DeLisa MP, Ke A (2012) Double-stranded endonuclease activity in Bacillus halodurans clustered regularly interspaced short palindromic repeats (CRISPR)-associated Cas2 protein. J Biol Chem 287:35943–35952
14. Samai P, Smith P, Shuman S (2010) Structure of a CRISPR-associated protein Cas2 from Desulfovibrio vulgaris. Acta Crystallogr Sect F: Struct Biol Cryst Commun 66:1552–1556

# Chapter 18

## Cas3 Nuclease–Helicase Activity Assays

Tomas Sinkunas, Giedrius Gasiunas, and Virginijus Siksnys

### Abstract

Cas3 is a signature protein of the type I CRISPR-Cas systems and typically contains HD phosphohydrolase and Superfamily 2 (SF2) helicase domains. In the type I CRISPR-Cas systems Cas3 functions as a slicer that provides foreign DNA degradation. Biochemical analysis indicate that Cas3 of the *Streptococcus thermophilus* DGCC7710 (St-Cas3) CRISPR4 system is a single-stranded DNA nuclease which also possesses a single-stranded DNA-stimulated ATPase activity, which is coupled to unwinding of DNA/DNA and RNA/DNA duplexes in 3′ to 5′ direction. The interplay between the nuclease and ATPase/helicase activities of St-Cas3 results in DNA degradation. Here, we describe assays for monitoring of St-Cas3 nuclease, ATPase and helicase activities in a stand-alone form and in the presence of the Cascade ribonucleoprotein complex. These assays can be easily adapted for biochemical analysis of Cas3 proteins from different microorganisms.

**Key words** Cas3, CRISPR, Cascade, *Streptococcus thermophilus*

## 1 Introduction

CRISPR-Cas is a recently discovered bacterial and archaeal adaptive immune system which provides acquired resistance against invading viral or plasmid DNA [1]. CRISPR locus consists of an array of short conserved repeat sequences interspaced by unique DNA sequences of similar size called spacers, which often originate from phage or plasmid DNA. Spacer-repeat array is accompanied by a cluster of *cas* (CRISPR-associated) genes. CRISPR-Cas systems function by incorporating fragments of the invading nucleic acid as spacers into a host genome and later use these spacers as templates to generate small RNA molecules (crRNA) that are combined with Cas proteins into an effector complex which silences foreign nucleic acids in the subsequent rounds of infection [1–4]. CRISPR-Cas systems have been categorized into three main types, based on core elements content and sequences [5]. Each type is specified by the so-called signature protein, which is conserved in a particular type, accordingly, Cas3 in type I, Cas9 in type II, and Cas10 in type III.

*Streptococcus thermophilus* DGCC7710 strain contains four different CRISPR systems which belong to three different type [2]. The CRISPR4-Cas system of *S. thermophilus* DGCC7710 (St-CRISPR4-Cas) is orthologous to the I–E CRISPR-Cas system of *Escherichia coli* [2, 6]. In the St-CRISPR4-Cas system, five *cas* genes are arranged into a cluster (cse1-cse2-cas7-cas5-cas6e) and form an St-Cascade complex that binds to the target DNA guided by the crRNA. St-Cascade binding to the matching sequence in the target DNA requires a PAM sequence located in the vicinity of a protospacer. In the St-CRISPR4 system, the PAM identified in the in vitro binding assay is extremely promiscuous and limited to a single A(-1) or T(-1) nucleotide [7]. If the correct PAM is present, St-Cascade guided by the crRNA generates an R-loop. Cascade binding to the matching sequence in the invading DNA does not trigger silencing; degradation of the foreign DNA requires an accessory Cas3 protein which acts a slicer in the type I CRISPR-Cas systems.

Cas3 typically contains HD phosphohydrolase and Superfamily 2 (SF2) helicase domains arranged into a single subunit protein, however sometimes HD- and helicase domains are encoded as individual Cas3′ and Cas3″ subunits, respectively [5, 8, 9]. Furthermore, in some CRISPR systems the single chain Cas3 or separate Cas3 domains are fused to other Cas proteins (Cas2-Cas3, Cas3-Cse1) [5, 10]. Single chain Cas3 variants from four different bacteria/archaea strains and Cas3 domains (subunits) have been purified and biochemically characterized [4].

Cas3 of *S. thermophilus* DGCC7710 (St-Cas3) is a large (926 aa) protein that contains an HD-nuclease domain at the N-terminus and the ATPase motif at the C-terminus. Biochemical analysis revealed that St-Cas3 is a single-stranded DNA (ssDNA) nuclease and ATP-dependent helicase which unwinds of DNA/DNA and RNA/DNA duplexes in 3′ to 5′ direction [6]. St-Cas3 alone neither binds nor cleaves double-stranded DNA (dsDNA) and has no ATPase or helicase activity [6]. To unleash St-Cas3 full catalytic activity, an ssDNA is required and St-Cascade contributes to its formation. St-Cascade binding to the dsDNA guided by crRNA creates an R-loop where the non-target strand of a protospacer is expelled as ssDNA [7]. This displaced DNA strand in the St-Cascade-target DNA complex serves as a platform for the St-Cas3 binding and triggers ATPase and nuclease activities. The nuclease function of Cas3 located in the HD-domain seems to be coupled to ATP hydrolysis in the C-terminal helicase domain. Indeed, if ATP is missing, the single-stranded DNA nuclease (DNase) activity of the HD domain is weak and cleavage is limited to the expelled non-target DNA strand. On the other hand, in the presence of ATP, the HD-nuclease of Cas3 produces multiple cuts beyond the protospacer region in the 3′ to 5′ direction presumably due to the unwinding of DNA duplex by the C-terminal helicase [7]. The interplay between the nuclease and ATPase/helicase activities of St-Cas3 results in DNA degradation.

Here, we describe assays for monitoring of the *S. thermophilus* Cas3 nuclease, ATPase and helicase activities in a stand-alone form and in the presence of the Cascade ribonucleoprotein complex.

## 2 Materials

Solutions and buffers: prepare all solutions and buffers using deionized water and analytical grade reagents. Store all reaction buffers and solutions at −20 °C if not stated otherwise.

Proteins: Use St-Cas3 and St-Cascade preparations purified to near homogeneity [6, 7] in all assays described below (*see* **Note 1**). Store St-Cas3 and St-Cascade stock solution in the storage buffer containing 50 % glycerol (*see* **Note 2**) at −20 or −70 °C for prolonged time intervals (*see* **Note 3**).

### 2.1 Nuclease Assay

1. NB1 nuclease buffer: 20 mM Tris–HCl (pH 7.5 at 25 °C), 20 mM $MgCl_2$, 0.2 mg/ml BSA (*see* **Note 4**).

2. Substrate DNA solution: prepare 20 mM single-stranded M13mp18 (ssM13mp18) DNA solution in $H_2O$.

3. Cas3 storage buffer: 10 mM Tris–HCl (pH 7.5 at 25 °C), 300 mM KCl, 50 % glycerol, 1 mM DTT (*see* **Note 5**).

4. 3× stop solution: 67.5 mM EDTA, 27 % (v/v) glycerol, 0.3 % (w/v) SDS (*see* **Note 6**).

5. 10× NBE buffer: 1 M sodium borate (pH 8.3 at 25 °C), 20 mM EDTA.

6. 0.8 % agarose gels prepared in 1× NBE.

7. 0.5 mg/L ethidium bromide solution prepared in 1× NBE buffer (*see* **Note 7**).

8. Equipment for a horizontal agarose gel electrophoresis.

9. UV transilluminator for ethidium bromide stained DNA visualization.

10. NB2 nuclease buffer: 40 mM Tris–HCl (pH 7.5 at 25 °C), 6 mM $MgCl_2$, 0.4 mM $NiCl_2$, 8 mM ATP, 0.4 mg/ml BSA (*see* **Note 8**).

11. pSP plasmid solution: prepare 20 nM of pSP plasmid containing a protospacer solution in $H_2O$ (*see* **Note 9**).

12. CDB buffer: 10 mM Tris–HCl (pH 7.5 at 25 °C), 200 mM NaCl.

13. Seq-stop solution: 95 % (v/v) formamide, 25 mM EDTA, pH 7.5, 0.1 % (w/v) bromophenol blue.

14. Vertical gel electrophoresis system (see **Note 10**).
15. 10× TBE buffer: 0.89 M Tris, 0.89 M boric acid, 20 mM EDTA (pH 8.5 at 25 °C).
16. Synthetic oligodeoxyribonucleotides, HPLC purified (see **Note 11**):

    oligodeoxyribonucleotide-1: 5′-GACCACCCTTTTTGATA TAATATACCTATATCAAT GCCTCCCACGCATAAGC GCAGATACGTTCTGAGGGAA-3′;

    oligodeoxyribonucleotide-2: 5′-TTCCCTCAGAACGTATC TGCGCTTATGCGTGGGAG GCCATTGATATAGGTA TATTATATCAAAAAGGGTGGTC-3′.

    Protospacer sequence is underlined and PAM sequence is shown in bold.

## 2.2 ATPase Assays

### 2.2.1 Thin Layer Chromatography (TLC) Analysis of ($\alpha^{32}$P)ATP Hydrolysis

1. AB1 buffer: 40 mM Tris–HCl (pH 7.5 at 25 °C), 0.4 mg/ml BSA (see **Note 12**).
2. AM1 substrate solution: 2 mM ATP, 4 nM ($\alpha^{32}$P)ATP, 8 mM $MgCl_2$ in water (see **Notes 13 and 14**).
3. Polyethyleneimine-cellulose TLC plates.
4. TLC buffer: 0.325 M $KH_2PO_4$ (pH 3.5).
5. Imaging plates and phosphorimager.

### 2.2.2 Monitoring of ATP Hydrolysis by Malachite Green Orthophosphate Detection

1. AM2 substrate solution: 2 mM ATP, 8 mM $MgCl_2$ (see **Note 15**).
2. Malachite green phosphate assay kit.
3. ATPase stop solution: 60 mM EDTA solution in water.
4. Flat bottom 96 well plates for visible light absorbance measurements.
5. 96 Well plate absorbance reader with 620 nm light filter.
6. $KH_2PO_4$ standards (0, 4, 8, 12, 16, 24, 32, 40 μM of $KH_2PO_4$ in water).
7. AB2 solution: 80 mM Tris–HCl (pH 7.5 at 25 °C), 0.8 mg/ml BSA (see **Note 16**).
8. pSP plasmid: 40 nM double-stranded pSP plasmid DNA solution in $H_2O$ (see **Note 9**).
9. CDB buffer: 10 mM Tris–HCl (pH 7.5 at 25 °C), 200 mM NaCl.
10. AM3 solution: 8 mM ATP, 6 mM $MgCl_2$ (see **Note 17**).

## 2.3 Helicase Assays

### 2.3.1 Substrate Assembly

1. Synthetic oligodeoxyribonucleotides, HPLC purified:

    oligodeoxyribonucleotide-3: 5′-CCTGCAGGTCGACT CTAGAG-3′ (see **Note 18**);

    oligodeoxyribonucleotide-4: 5′-GGGGGGGGGGGTAGTTG AGAA-3′;

oligodeoxyribonucleotide-5: 5′-CCCGCGCGCGTCGTCAT GCG-3′;

oligodeoxyribonucleotide-6: 5′- <u>GGGCGCGCGCAGCAGT ACGC</u>TAGTACTGTTCCCATGT CTAAGGAGGGGTT GCG**TTCTCAACTACCCCCCCCC**-3′. Sequences of oligodeoxyribonucleotide-6 complementary for oligodeoxyribonucleotide-5 and oligodeoxyribonucleotide-4 are underlined and shown in bold, respectively (*see* **Note 19**).

2. Single-stranded DNA: 250 nM ssM13mp18 DNA solution in H$_2$O.

3. ($\gamma^{33}$P)ATP (*see* **Note 14**).

4. T4 Polynucleotide Kinase (PNK) 10 U/μL and 10× PNK reaction buffer.

5. Tris buffer: 10 mM Tris–HCl solution in H$_2$O (pH 8 at 25 °C).

### 2.3.2 Helicase Activity and Polarity Assay

1. HB buffer: 20 mM Tris–HCl (pH 7.5), 0.2 mg/ml BSA, 2 mM MgCl$_2$, 4 mM ATP.

2. 40 % acrylamide–bis-acrylamide (29:1) solution.

3. 10 % Ammonium persulfate in water.

4. $N,N,N′,N′$-Tetramethylethylenediamine (TEMED).

5. 50×TAE buffer: 2 M Tris, 1 M acetic acid, 50 mM EDTA, pH 8 at 25 °C.

6. Equipment for a vertical polyacrylamide gel electrophoresis.

7. Imaging plate and phosphorimager.

## 3 Methods

### 3.1 Nuclease Assay

Cas3 alone degrades single-stranded DNA using HD nuclease domain but does not hydrolyse double-stranded DNA. Hydrolysis of dsDNA by Cas3 is triggered by Cascade binding to the target sequence. Here we provide assays for monitoring of Cas3 nuclease activity on the ssDNA and dsDNA (in the presence of Cascade).

### 3.1.1 Single-Stranded DNA Degradation by the Stand-Alone Cas3 Protein

1. Mix 70 μl of NB1 buffer with 35 μl of 20 nM ssM13mp18 DNA solution (*see* **Note 20**).

2. Sample 15 μl aliquots of the reaction mixture into six empty 1.5-ml tubes.

3. Prepare 10 μl serial dilutions of the Cas3 protein (e.g., 0, 20, 200, 400, 1,000, 2,000 nM). Serially dilute protein into Cas3 storage buffer briefly vortexing between dilutions (*see* **Notes 21 and 22**).

4. Prewarm the Cas3 solution and other reaction components by incubating for a few minutes at 37 °C.

5. Start reactions by adding 5 μl of Cas3 solution to the reaction mixture containing other components.

6. Incubate the reaction mixture for 2 h at 37 °C and terminate the reaction by adding 10 μl of 3× stop solution.

7. Separate cleavage products by electrophoresis through 0.8 % (w/v) agarose gel prepared in 1× NBE buffer (*see* **Note 23**).

8. Stain agarose gel with ethidium bromide solution and UV visualize reaction products (*see* **Note 24**).

*3.1.2 Double-Stranded DNA Degradation by Cas3 in the Presence of Cascade*

1. Dilute Cascade complex stock into CDB buffer to prepare 40 μl of 80 nM Cascade solution (*see* **Notes 22** and **25**).

2. Mix 35 μl of NB2 buffer with 35 μl of 20 nM of pSP and 35 μl of Cascade solutions (*see* **Note 26**).

3. Sample 15 μl aliquots of the reaction mixture into 6 empty 1.5-ml tubes.

4. Prepare 10 μl serial dilutions of the Cas3 protein (e.g., 0, 20, 200, 400, 1,000, 2,000 nM). Serially dilute protein into Cas3 storage buffer briefly vortexing between dilutions (*see* **Notes 21** and **22**).

5. Prewarm the Cas3 solutions and other reaction components by incubating for a few minutes at 37 °C.

6. Start reactions by adding 5 μl of the Cas3 protein solution to the reaction mixture containing other components.

7. Incubate the reaction mixture for 10 min at 37 °C and terminate by adding 10 μl of 3× stop solution.

8. Separate cleavage products by electrophoresis through 0.8 % (w/v) agarose gel in 1× NBE buffer (*see* **Note 23**).

9. Stain agarose gel with ethidium bromide solution and UV visualize reaction products (*see* **Note 24**).

## 3.2 Mapping of Cas3 Cleavage Positions in the Presence of Cascade

Cascade binds to the protospacer forming an R-loop which serves as the signal for Cas3 loading and cleavage [7]. To map the cleavage positions of Cas3 in R-loop we employed oligodeoxyribonucleotide duplex substrates.

*3.2.1 Assembly of the Radioactively Labelled Oligoduplexes*

Here we describe assembly of oligodeoxyribonucleotide duplexes radioactively labelled at the 5′-end of one strand. These oligoduplexes contain a protospacer sequence and PAM sequence required for Cascade binding.

1. Prepare PNK reaction mixtures by adding 1 μl of 5 μM of oligodeoxyribonucleotide-1 and 1 μl of 5 μM oligodeoxyribo-

nucleotide-2 in separate tubes, then add 1 μl PNK reaction buffer, 6.5 μl H$_2$O, and 1 μl PNK into each tube.

2. Add 0.5 μl ($\gamma^{33}$P)ATP to the PNK reaction mixtures and incubate for 20 min at 37 °C (*see* **Note 14**).

3. Stop reaction by heating at 95 °C for 5 min.

4. Prepare two separate annealing reaction mixtures by mixing 2 μl of labelled oligodeoxyribonucleotide-1 with 2 μl of 0.75 μM unlabelled oligodeoxyribonucleotide-2, and 2 μl of labelled oligodeoxyribonucleotide-2 with 2 μl of 0.75 μM unlabelled oligodeoxyribonucleotide-1, respectively (*see* **Note 27**).

5. Add 20 μl of Tris buffer and 76 μl of H$_2$O to both mixtures.

6. To obtain radioactively labelled oligoduplexes SP1 and SP2 containing $^{33}$P label at the 5′-end of one strand, place tubes containing reaction mixtures described in stage 4 and 5, into a boiling water bath and leave overnight for slow cool down (*see* **Note 28**).

### 3.2.2 Mapping of the Cas3 Cleavage Products

1. Dilute Cascade complex stock into CDB to prepare 20 μl of 40 nM Cascade solution (*see* **Notes 22** and **25**).

2. Sample 5 μl aliquots of SP1 and SP2 oligoduplexes into separate 1.5-ml tubes.

3. Add 5 μl of NB2 solution and 5 μl of Cascade solution into these tubes and vortex briefly.

4. Prepare 20 μl of 400 nM Cas3 protein solution in the Cas3 storage buffer (*see* **Note 22**).

5. Prewarm the Cas3 solution and other reaction components by incubating for a few minutes at 37 °C.

6. Start reactions by adding 5 μl of the Cas3 protein solution to the reaction mixture, incubate for 10 min at 37 °C and terminate the reaction by adding 20 μl of seq-stop solution.

7. Separate cleavage products and DNA marker by vertical electrophoresis through 20 % (w/v) denaturing polyacrylamide gel (PAAG) in 1× TBE buffer (*see* **Note 29**).

8. Visualize reaction products using a phosphorimager.

9. Map cleavage products on DNA size marker sequence.

### 3.3 ATPase Assays

In the presence of a single-stranded DNA or a double-stranded DNA in complex with Cascade, Cas3 hydrolyses ATP using the ATPase/helicase domain [6, 7]. To monitor ATPase activity we employed (1) thin layer chromatography for detection of the reaction products of ($\alpha^{32}$P)ATP hydrolysis, (2) detection of the released orthophosphate by the malachite green.

### 3.3.1 Thin Layer Chromatography (TLC) Analysis of ($\alpha^{32}$P)ATP Hydrolysis by the Stand-Alone Cas3 Protein

ATP hydrolysis by Cas3 can be visualized by TLC using radioactively labelled ATP. The method provides high sensitivity, however requires special precautions and permissions working with radioactive material (*see* **Note 14**).

1. Prepare 15 μl of 1 μM Cas3 protein solution by diluting protein stock into the Cas3 storage buffer (*see* **Note 22**).
2. Mix 10 μl of AB1 buffer with 10 μl of 20 nM ssM13mp18 DNA solution and then add 10 μl of the diluted Cas3 protein solution (*see* **Note 30**).
3. Prewarm AM1 solution and mixture of other reaction components by incubating for a few minutes at 37 °C in separate tubes (*see* **Note 14**).
4. Start reactions by adding 10 μl of AM1 solution to reaction mix with other components.
5. Sample 6 μl aliquots at fixed time intervals (e.g., after 1, 5, 10, 30, 60 min) and stop reaction by mixing with 2 μl of 3× stop solution (*see* **Note 31**).
6. Spot 0.5 μl of each sample onto a polyethyleneimine-cellulose thin layer plate (*see* **Note 32**).
7. Wait for a few minutes until spots dry up.
8. Put the plate vertically into container with TLC buffer (*see* **Note 33**).
9. Wait until front of the TLC buffer reaches top of the plate.
10. Pull out the plate, put it horizontally and let it dry up.
11. Visualize reaction products using a phosphorimager.

### 3.3.2 Monitoring of ATP Hydrolysis by a Stand-Alone Cas3 Protein Using Malachite Green Orthophosphate Detection

Phosphate released from Cas3 hydrolysed ATP can be detected spectrophotometrically using malachite green. This method is less sensitive than ($\alpha^{32}$P)ATP assay; however, it allows for analyzing a large number of samples in parallel and does not require radioactive materials.

1. Prepare 15 μl of 1 μM Cas3 protein solution by diluting protein into the Cas3 storage buffer (*see* **Note 22**).
2. Mix 15 μl of AB1 with 15 μl of 20 nM ssM13mp18 DNA solution then add 15 μl of diluted Cas3 protein solution (*see* **Note 30**).
3. Prewarm the AM2 substrate solution and other reaction components by incubating for a few minutes at 37 °C.
4. Start reactions by adding 15 μl of AM2 solution to the reaction mix containing other components.
5. Sample 10 μl aliquots at fixed time intervals (e.g., after 1, 5, 10, 30, 60 min) and stop reaction by rapidly mixing with 90 μl of ATPase stop solution (*see* **Note 31**).

6. Sample 80 μl aliquots of the stopped reaction mixes and KH$_2$PO$_4$ standards into the flat bottom 96 well plate and add 20 μl of the phosphate detection reagent into each well (*see* **Note 34**).

7. Incubate for 30 min at room temperature and measure absorbance at 620 nm.

8. Plot absorbance values against phosphate standard concentrations and use the linear dependence to determine the concentration of liberated phosphate in the ATP samples hydrolysed by Cas3.

9. Plot the liberated phosphate concentration versus the reaction time and calculate the ATP cleavage rate.

*3.3.3 Monitoring of ATP Hydrolysis by the Cas3 Protein in the Presence of Cascade Using Malachite Green Orthophosphate Detection*

1. Prepare 15 μl of 400 nM Cas3 protein solution by diluting protein into the Cas3 storage buffer (*see* **Note 22**).

2. Dilute Cascade complex stock into CDB to prepare 15 μl of 80 nM Cascade solution (*see* **Notes 22** and **25**).

3. Mix 5 μl of AB2 solution with 5 μl of 40 nM pSP then add 10 μl of diluted Cascade and Cas3 protein solutions (*see* **Note 26**).

4. Prewarm AM3 substrate solution and mixture of other reaction components by incubating for a few minutes at 37 °C.

5. Start reactions by adding 10 μl of AM3 solution to reaction mix with other components.

6. Sample 5 μl aliquots at fixed time intervals (e.g., after 1, 5, 10, 30, 60 min) and stop the reaction by rapidly mixing with 95 μl of the ATPase stop solution (*see* **Note 31**).

7. Mix 80 μl of the stopped reaction samples and KH$_2$PO$_4$ standards with 20 μl of the phosphate detection reagent in flat bottom 96 well plate (*see* **Note 34**).

8. Incubate for 30 min at room temperature and measure absorbance at 620 nm.

9. Plot absorbance values against phosphate standard concentrations and use the linear dependence to determine the concentration of liberated phosphate in the ATP samples hydrolysed by Cas3.

10. Plot the liberated phosphate concentration versus the reaction time and calculate the ATP cleavage rate.

## 3.4 Cas3 Helicase Activity Assay Monitoring the Radioactively Labelled Oligodeoxyribonucleotide Displacement

Cas3 binds to a single-stranded DNA and unwinds DNA duplex in an ATP-dependent manner. To monitor DNA unwinding activity of Cas3 we assembled a partial DNA duplex annealing a 20 nt 5′-end labelled oligodeoxyribonucleotide to the complementary sequence in the circular ssM13mp18 DNA. Cas3 binds to the single-stranded DNA fragment, translocates through it and unwinds this duplex region releasing the labelled oligodeoxyribonucleotide into solution.

### 3.4.1 Assembly of a Partial Duplex on a Circular Single-Stranded DNA

1. Mix 1 μl of 5 μM oligodeoxyribonucleotide-3 with 1 μl of 10× PNK reaction buffer and 6.5 μl H$_2$O, then add 1 μl PNK.
2. Add 0.5 μl ($\gamma^{33}$P)ATP to the PNK reaction mixture and incubate for 20 min at 37 °C (*see* **Note 14**).
3. Stop reaction by heating at 95 °C for 5 min.
4. Mix 15 μl of 250 nM ssM13mp18 DNA solution with 10 μl of Tris buffer and 20 μl H$_2$O, then add 5 μl of 5′-end labelled oligodeoxyribonucleotide (*see* **Note 27**).
5. To obtain duplex between radioactively labelled oligodeoxyribonucleotide-3 and ssM13mp18 DNA, place tube into a boiling water bath for 5 min and leave overnight for slow cool down (*see* **Note 28**).
6. Dilute 5 μl of annealed substrate in 120 μl of H$_2$O to achieve 2 nM final concentration of the helicase substrate.

### 3.4.2 Helicase Activity Assay

1. Mix 70 μl of HB buffer with 35 μl of helicase substrate (Subheading 3.4.1).
2. Sample 15 μl aliquots of the reaction mixture into 6 empty 1.5-ml tubes.
3. Prepare 10 μl serial dilutions of the Cas3 protein (e.g., 0, 20, 200, 400, 1,000, 2,000 nM). Serially dilute protein into Cas3 storage buffer briefly vortexing between dilutions (*see* **Notes 21 and 22**).
4. Prewarm the Cas3 solutions and other reaction components by incubating for a few minutes at 37 °C.
5. Start reactions by adding 5 μl of Cas3 protein solution to reaction mixture containing other components.
6. Incubate the reaction mixture for 60 min at 37 °C and terminate the reaction by adding 10 μl of 3× stop solution.
7. Separate unwound products by electrophoresis through 8 % (w/v) non-denaturing PAAG in 1× TAE buffer.
8. Visualize reaction products using a phosphorimager (*see* **Note 35**).

### 3.5 Cas3 Helicase Polarity Assay

To monitor the polarity of the Cas3 helicase we used a pair of partial duplexes containing a single-stranded DNA region flanked by the duplex regions at the 5′-or 3′-ends (5′- or 3′-substrates, respectively) (Fig. 1). The oligodeoxyribonucleotide either at the 5′-or 3′-end is radioactively labelled to enable the detection of the Cas3 unwinding reaction products in the gel.

### 3.5.1 Assembly of Partial Oligoduplexes

1. Sample 1 μl of 5 μM oligodeoxyribonucleotide-4 and oligodeoxyribonucleotide-5 in separate tubes then add 1 μl PNK reaction buffer, 6.5 μl H$_2$O and 1 μl PNK in both tubes.

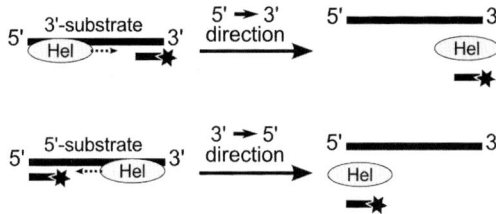

**Fig. 1** Schematic representation of the helicase polarity assay. Helicase (Hel) binds to a single-stranded part of a partial DNA duplex. Helicase polarity is 3′→5′, if a radioactively labelled oligodeoxyribonucleotide at the 5′-end is displaced. Conversely, if helicase polarity is 5′→3′, a radioactively labelled oligodeoxyribonucleotide at the 3′-end is displaced

2. Add 0.5 μl ($\gamma^{33}$P)ATP to the PNK reaction mixtures and incubate for 20 min at 37 °C (*see* **Note 14**).

3. Stop reaction by heating at 95 °C for 5 min.

4. Sample 5 μl of each labelled substrate into different tubes.

5. Mix 12.5 μl of 0.75 μM oligodeoxyribonucleotide-6 solution with 25 μl of Tris buffer and 75 μl H$_2$O, then aliquot 45 μl of this mixture into each tube with labelled oligodeoxyribonucleotide-4 or oligodeoxyribonucleotide-5.

6. To obtain radioactively labelled 5′- and 3′-substrates (Fig. 1) place tubes containing reactions mixtures described in stage 4 and 5, into a boiling water bath and leave overnight to cool down slowly (*see* **Notes 19** and **28**).

7. Dilute 5 μl of annealed substrate in 120 μl of H$_2$O to achieve 2 nM final concentration of the 3′- or 5′-substrates.

*3.5.2 Helicase Polarity Assay*

1. Mix 70 μl of HB buffer with 35 μl of the 5′-substrate in one tube and 3′-substrate in other.

2. Prepare 80 μl of 2 μM Cas3 protein solution by diluting Cas3 stock protein in the Cas3 storage buffer (*see* **Note 22**).

3. Prewarm the protein and solution with other reaction components by incubating for a few minutes at 37 °C.

4. Start reactions by adding 35 μl of Cas3 protein solution to reaction mixtures containing other components.

5. Sample 20 μl aliquots at fixed time intervals (e.g., after 1, 5, 10, 30, 60 min) and stop reaction by rapidly mixing with 10 μl of 3× stop solution (*see* **Note 31**).

6. Separate unwound products by electrophoresis through 8 % (w/v) non-denaturing PAAG in 1× TAE buffer.

7. Visualize reaction products using a phosphorimager (*see* **Note 36**).

## 4 Notes

1. Traces of nuclease or ATPase impurities present in the preparations may interfere with the assays.

2. Storage buffer for St-Cas3 is 10 mM Tris–HCl (pH 7.5 at 25 °C), 300 mM KCl, 50 % glycerol, 1 mM DTT. Storage buffer for St-Cascade is 10 mM Tris–HCl (pH 8 at 25 °C), 500 mM NaCl, 50 % glycerol.

3. Split Cas3 protein stock into smaller aliquots to avoid frequent thawing and freezing of the stock solution if the protein will be kept at −70 °C.

4. NB1 buffer is used for Cas3 nuclease activity assay in the absence of Cascade.

5. Prepare 2× buffer containing 20 mM Tris–HCl and 600 mM KCl. Adjust pH to 7.8 with concentrated HCl, then add glycerol and slowly adjust pH to 7.5. Store buffer at +4 °C. Prepare 1 M DTT stock solution in water, split it into smaller aliquots and store at −20 °C. Keep in mind that DTT is sensitive to freeze/thaw cycles. Add DTT just before using buffer.

6. Add 0.01 % (w/v) bromophenol blue or orange G stain for easier visualization of the sample when applying it to wells in the agarose gel.

7. Ethidium bromide is toxic. Do not inhale ethidium bromide dust when weighing and avoid contact with the skin. Wear gloves and protective clothing.

8. NB2 buffer is used for Cas3 nuclease activity assay in the presence of Cascade. Nickel ions are better cofactor than magnesium ions for St-Cas3 and *Thermus thermophilus* Cas3 [7, 11] nucleases.

9. pSP plasmid is described in [7]. Protospacer should be complementary for crRNA in the Cascade complex and contain a proper PAM sequence (e.g., 5′-AA-3′ in vicinity of the protospacer in the case of St-Cascade). Protospacer sequence and PAM can be inserted into any plasmid by ligating an oligoduplex as described in [7].

10. Vertical gel electrophoresis system should allow one nucleotide resolution to precisely map cleavage positions (e.g., Sequi-Gen GT System, Bio-Rad, Hercules, CA, USA).

11. Complementary oligodeoxyribonucleotides-1 and -2 contain a crRNA matching protospacer sequence in the middle (underlined), AA PAM at the 5′-end (bold) and flanking 18 nt sequences on both sides.

12. AB1 buffer is used in Cas3 ATPase activity detection using radioactively labelled ATP assay.

13. AM1 substrate solution is used to start ATPase reactions. Hydrolysis of radioactively labelled ATP was detected by thin layer chromatography. AM1 solution contains radioactive ($\alpha^{32}$P)ATP.

14. $^{32}$P and $^{33}$P emits high-energy beta radiation. Take all precautions to reduce radiation when working with radioactive materials. Refer to the rules established by your local radioactivity control agency for handling and proper disposal of radioactive materials and waste.

15. AM2 substrate solution is used to start ATPase reactions. Liberated phosphate is detected by malachite green phosphate assay.

16. AB2 buffer is used to monitor Cas3 ATPase activity in the presence of Cascade by malachite green assay.

17. AM3 substrate solution is used to start ATPase reactions in the presence of Cascade. Liberated phosphate is detected by malachite green phosphate assay.

18. Any oligodeoxyribonucleotide with a sequence complementary to a fragment in a ssM13mp18 or other circular ssDNA can be used.

19. Sequences of oligodeoxyribonucleotide-6 complementary for oligodeoxyribonucleotide-5 and oligodeoxyribonucleotide-4 are underlined and shown in bold, respectively. oligodeoxyribonucleotide-4 annealed to oligodeoxyribonucleotide-6 forms a partial duplex with a 20 bp duplex region at the 3′-end and 52 nt single stranded DNA. Oligodeoxyribonucleotide-5 annealed to oligodeoxyribonucleotide-6 forms a partial duplex with a 20 bp duplex region at the 5′-end and 52 nt single stranded DNA. Different sets of oligodeoxyribonucleotides forming partial DNA duplexes similar to oligodeoxyribonucleotides-4, -5 and -6 can be used. Keep in mind that for the Cas3 helicase loading a single-stranded DNA region in the partial oligoduplex should be sufficiently long (at least 50 nt).

20. If you see low nuclease activity in the presence of Mg$^{2+}$ ions, it is worthwhile to try other divalent ions such as Ni$^{2+}$ which stimulate nuclease activity of different Cas3 proteins [7, 11].

21. The simplest way to prepare solutions of different protein concentrations is to make serial dilutions into storage buffer. In this way you change protein concentrations keeping constant concentration of other components such as salts or glycerol. Serial dilution is made by stepwise decreasing protein concentration. Firstly, stock protein is diluted into a storage buffer getting dilution of maximal protein concentration. Then part of the resultant aliquot is diluted into the next tube with a

storage buffer, etc. In this way you obtain a set of dilutions with different protein concentrations.

22. Keep protein samples on ice while performing dilutions.

23. 1× TAE or TBE buffers might be used instead of 1× NBE.

24. Degraded DNA smear should be visible at higher Cas3 concentrations.

25. For the St-Cas3 reactions in the presence of Cascade, we use saturating St-Cascade concentration at which all target DNA is bound in the gel shift assay [7]. Keep in mind that in the case of Cascade complexes from different bacteria this concentration could differ and therefore should be established experimentally.

26. Plasmid DNA (e.g., pUC18) lacking a protospacer is used instead pSP as negative control.

27. Use 1:1.5 ratio of labelled vs unlabelled DNA to ensure that all labelled oligodeoxyribonucleotide is in the duplex form.

28. Use screw cap tubes for annealing. Alternatively, annealing could be performed in PCR machine decreasing temperature from 95 to 25 °C at 0.5 °C/min increments.

29. Sanger sequencing reactions of oligodeoxyribonucleotide-1 and oligodeoxyribonucleotide-2 can be used as DNA size markers [12, 13]. Alternatively, 5′-end labelled synthetic oligodeoxyribonucleotides with a 1 or 2 nt increment truncations from 3′-end of oligodeoxyribonucleotide-1 and oligodeoxyribonucleotide-2 can be used as DNA size markers.

30. Single-stranded DNA activates ATPase activity of Cas3. Cas3 alone or in the presence of a double-stranded DNA, for example pUC19 plasmid, show traces of ATPase activity.

31. Prepare tubes with stop solution before starting reactions.

32. Samples should be spotted on a start line sufficiently wide not to merge with each other. It is useful to draw a line with marker on TLC plate's backside 2 cm from the bottom and mark spot positions on the line every 1 cm. This should ease spotting of the samples.

33. Sample spots should be in the bottom edge of the plate but the surface of the TLC buffer should be always lower than sample spots. For example, if samples are spotted on the line located 2 cm away from the bottom edge of the TLC plate then the buffer surface should be approximately 1 cm below the spots line.

34. Carefully tip off samples into plate wells. Do not bubble the samples because it may misrepresent absorbance values.

35. Band corresponding to the free single-stranded labelled oligodeoxyribonucleotide-3 should appear in the presence of Cas3.

36. Helicase polarity is $3' \to 5'$ if $5'$-substrate is unwound, while $3'$-substrate stays intact. Conversely, if helicase polarity is $5' \to 3'$, $3'$-substrate is unwound (Fig. 1).

## Acknowledgement

We thank Rodolphe Barrangou and Philippe Horvath for discussions. The work on CRISPR systems in Siksnys's laboratory was funded by the European Social Fund under Global Grant measures.

## References

1. Barrangou R, Fremaux C, Deveau H et al (2007) CRISPR provides acquired resistance against viruses in prokaryotes. Science 315(5819):1709–1712. doi:10.1126/science.1138140
2. Horvath P, Barrangou R (2010) CRISPR/Cas, the immune system of bacteria and archaea. Science 327:167–170, 10.1126/science.1179555
3. Wiedenheft B, Sternberg SH, Doudna JA (2012) RNA-guided genetic silencing systems in bacteria and archaea. Nature 482:331–338. doi:10.1038/nature10886
4. Gasiunas G, Sinkunas T, Siksnys V (2013) Molecular mechanisms of CRISPR-mediated microbial immunity. Cell Mol Life Sci. doi:10.1007/s00018-013-1438-6
5. Makarova KS, Haft DH, Barrangou R et al (2011) Evolution and classification of the CRISPR-Cas systems. Nat Rev Microbiol 9:467–477. doi:10.1038/nrmicro2577
6. Sinkunas T, Gasiunas G, Fremaux C et al (2011) Cas3 is a single-stranded DNA nuclease and ATP-dependent helicase in the CRISPR/Cas immune system. EMBO J 30:1335–1342. doi:10.1038/emboj.2011.41
7. Sinkunas T, Gasiunas G, Waghmare SP et al (2013) In vitro reconstitution of Cascade-mediated CRISPR immunity in Streptococcus thermophilus. EMBO J 32:385–394. doi:10.1038/emboj.2012.352
8. Haft DH, Selengut J, Mongodin EF, Nelson KE (2005) A guild of 45 CRISPR-associated (Cas) protein families and multiple CRISPR/Cas subtypes exist in prokaryotic genomes. PLoS Comput Biol 1:e60. doi:10.1371/journal.pcbi.0010060
9. Makarova KS, Grishin NV, Shabalina SA et al (2006) A putative RNA-interference-based immune system in prokaryotes: computational analysis of the predicted enzymatic machinery, functional analogies with eukaryotic RNAi, and hypothetical mechanisms of action. Biol Direct 1:7. doi:10.1186/1745-6150-1-7
10. Westra ER, van Erp PBG, Künne T et al (2012) CRISPR immunity relies on the consecutive binding and degradation of negatively supercoiled invader DNA by Cascade and Cas3. Mol Cell 46:595–605. doi:10.1016/j.molcel.2012.03.018
11. Mulepati S, Bailey S (2011) Structural and biochemical analysis of nuclease domain of clustered regularly interspaced short palindromic repeat (CRISPR)-associated protein 3 (Cas3). J Biol Chem 286:31896–31903. doi:10.1074/jbc.M111.270017
12. Maxam AM, Gilbert W (1977) A new method for sequencing DNA. Proc Natl Acad Sci U S A 74:560–564
13. Maxam AM, Gilbert W (1980) Sequencing end-labeled DNA with base-specific chemical cleavages. Methods Enzymol 65:499–560

# Chapter 19

# Chemical and Enzymatic Footprint Analyses of R-Loop Formation by Cascade-crRNA Complex

Ümit Pul

## Abstract

Cascade-crRNA complexes mediate the identification of the invading foreign DNA and initiate its neutralization by formation of an R-loop (RNA-induced DNA-loop) at the crRNA-complementary sequence (protospacer). After initial unspecific binding to the double-stranded DNA, Cascade-crRNA complex slides along the DNA to find the protospacer. Once the target site is detected, the crRNA hybridizes to the complementary strand with subsequent displacement of the non-complementary strand to form an R-loop structure. Here, we describe how Cascade-DNA complexes and the Cascade-induced strand separation can be characterized in detail by combining chemical and enzymatic footprint analyses. Selective modification of unpaired thymines by permanganate ($KMnO_4$) and the specific cleavage of single-stranded DNA by Nuclease P1 can be used to probe an R-loop formation by Cascade. Localization of the Cascade-crRNA complex on the DNA can be achieved by an Exonuclease III protection assay.

**Key words** DNA footprint, Structural probing, Potassium permanganate, $KMnO_4$, Nuclease P1, Exonuclease III, Cascade, CRISPR, R-loop

## 1 Introduction

Footprinting techniques are valuable tools to characterize complexes of proteins with nucleic acids at single-nucleotide resolution in a fast and easy way. Originally developed for the simple examination of specific binding of proteins to their cognate sites on DNA [1], today several variations of the method are available allowing not only the mapping of DNA binding sites but also the probing of structural deformation of the DNA helix upon protein binding. In general, footprinting methods are based on the investigation of the accessibility of the DNA to nucleolytic enzymes or chemical probes in the presence or absence of a binding factor. The DNA of interest must be labeled at one end of one strand only (Fig. 1a). Addition of limited amounts of a DNA cleaving enzyme (e.g., DNase I) or chemical probe (e.g., hydroxyl radicals) to unbound DNA or DNA–protein complexes produces strand breaks

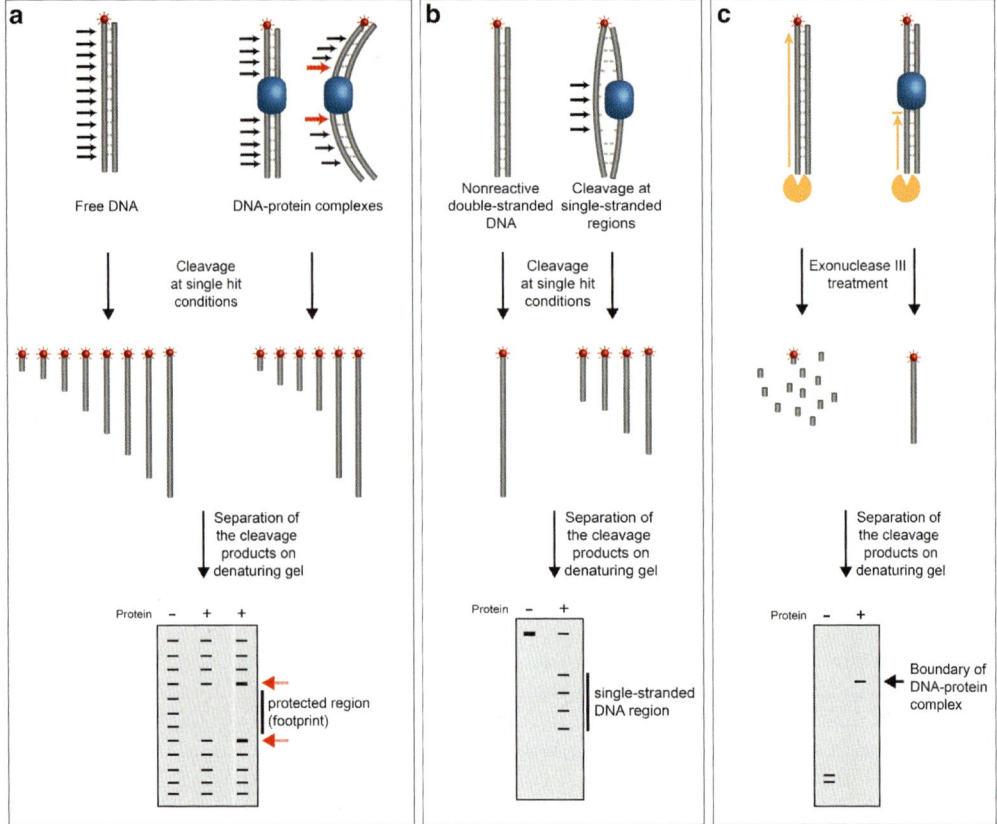

**Fig. 1** Principles of different DNA footprinting methods are shown schematically. (**a**) Protection assay using endonucleolytic enzymes (e.g., DNase I) or chemical reagents (e.g., hydroxyl radicals). (**b**) Probing of DNA-melting with single-strand specific enzymes (e.g., Nuclease P1 or S1) or chemical reagents (e.g., $KMnO_4$ or DEPC). (**c**) Protection assay with double-strand specific 3′-5-exonuclease assay for determination of boundaries of the contact region. *Black arrows* indicate accessible positions of unbound DNA or DNA–protein complexes to cleavage. *Red arrows* point to hyperreactive sites, arising from protein-induced bending of the DNA (**a**). The reactions are shown only for the labeled DNA strand (marked with *red stars*). The same analyses have to be repeated with the DNA labeled on the opposite strand

at each susceptible position. A critical step in the experimental procedure is that the enzymatic or chemical probe has to be added in small quantities in order to achieve a "single hit" condition at which, on average, each DNA molecule in the sample is randomly hydrolyzed at a single site (Fig. 1a). The visualization of the accessible sites occurs through separation of the cleavage products on denaturing polyacrylamide gels according to their size. The protein-bound regions are protected from hydrolysis, which results in the disappearance of the corresponding product bands in form of a "footprint" on the gel (Fig. 1a). A co-separation of DNA sequencing reaction enables a precise determination of the protected or reactive regions at single-nucleotide resolution.

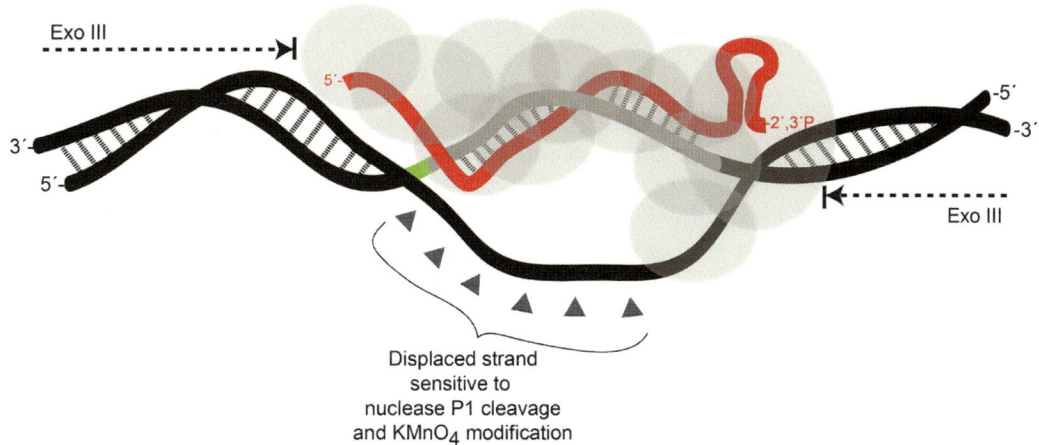

**Fig. 2** Model of the crRNA-Cascade:DNA complex, obtained from chemical and enzymatic footprint assays [3], is shown. Cascade complex is indicated by *transparent colored circles*, crRNA is shown in *red*; the protospacer region is marked in *grey* and the protospacer adjacent motif in *green*. *Arrow heads* indicate the positions within the displaced strand, which are accessible to permanganate modification and Nuclease P1 cleavage. The borders of the protospacer-bound Cascade complex is determined by Exo III protection assay, as indicated

While local protein-bound regions are protected, structural deformation of the DNA can lead to an increased sensitivity at positions flanking the protected region (Fig. 1a). Therefore, beside the determination of protein binding sites, footprint studies are also helpful to obtain information on the structures of DNA–protein complexes, e.g., DNA-bending, -wrapping, or -melting (Fig. 1b). The latter occurs for instance during the formation of RNA polymerase–promoter open complexes or R-loop structures in DNA–RNA hybrids [2]. An R-loop structure is also formed by binding of Cascade-crRNA to its target site (protospacer) to initiate the hydrolysis of the foreign DNA [3, 4]. Here, we show that the Cascade-induced R-loop formation can be demonstrated by probing of the Cascade-DNA complexes with single-strand specific potassium permanganate ($KMnO_4$) and Nuclease P1. The borders of the Cascade-DNA complexes can be determined by Exonuclease III protection assay (Fig. 2).

Potassium permanganate oxidizes the 5,6-double bond of unpaired thymines with high selectivity [5]. The single-strand selectivity of the oxidation reaction is based on the attack of permanganate to the 5,6-double bond of thymine perpendicular to the aromatic ring, which is impeded in double-stranded DNA due to the stacking of the bases [6]. Owing to the single-strand selectivity of the reaction, permanganate modification is well suited to probe the appearance of single-stranded regions within double-stranded DNA (Fig. 1b). Since the permanganate modification does not cause a hydrolysis of the sugar-phosphate backbone by itself, the detection of the susceptible positions is achieved by treatment of the modified DNA with piperidine at high temperatures. Piperidine is a strong base and induces the cleavage of the phosphodiester bond at the 3′-side of permanganate modified positions [7].

The single-strand specific endonucleases S1 and P1 are often used as enzymatic probes to map unwound DNA regions or structural distortions within double-stranded DNA [8]. Nuclease P1 (EC 3.1.30.1) from *Penicillium citrinum* cleaves phosphodiester bonds of single-stranded RNA or DNA leaving a 5′-phophate at the cleavage site. Compared to the S1 enzyme, Nuclease P1 is active at neutral pH and thus more suitable for studying protein-DNA complexes [9]. Structural probing of Cascade-induced R-loop with permanganate and Nuclease P1 complements each other in terms of specificity and accessibility: while permanganate can also reach and modify unpaired thymines that are in close contact to the protein (and thus likely be protected from enzymatic cleavage), Nuclease P1 can only cleave single-stranded regions that are not shielded by the protein. Therefore, P1 susceptible sites provide additional information about single-stranded regions that are exposed in the protein-DNA complex.

Exonuclease III (EC 3.1.11.2) exhibits a double-strand DNA specific 3′-5′-exonuclease [10]. The exonucleolytic digestion of double-stranded DNA (with blunt or 5′-overhang ends) occurs from its 3′-ends in a nonprocessive way by release of mononucleotides. If the DNA is stably bound by a protein, Exo III stops upon reaching the protein, enabling thus the determination of the extremities of protein-bound regions on a 5′-end labeled DNA (Fig. 1c) [11]. Owing to the complete degradation of unbound DNA, Exo III protection assay is applicable even under conditions in which the binding of the protein to the DNA is not saturated. However, the method is only limited applicable on protein-DNA complexes with rather short half-lifes or weakly-bound proteins, which can be displaced by Exo III from their DNA-binding sites [12].

## 2 Materials

### 2.1 crRNA-Loaded Cascade Complex and Double-Stranded Target DNA

1. Cascade-crRNA complexes can be purified as described [3, 13] (*see* **Note 1**).

2. Double-stranded DNA, which contains the protospacer sequence (crRNA-matching sequence): prepare two 10 nM stock solutions, one with the DNA labeled at the 5′-end of the targeted strand (contains the crRNA-complementary sequence), and another labeled at the 5′end of the nontargeted strand (*see* **Note 2**).

## 2.2 Components for KMnO₄, Nuclease P1 and Exonuclease III Footprint Assays

1. $KMnO_4$ stock solution: to prepare a 100 ml of a 370 mM $KMnO_4$ solution, dissolve 6 g $KMnO_4$ in 110 ml distilled water and boil the solution until it reaches a volume of 100 ml. If not using a dark bottle for storage, wrap the bottle with aluminum foil. The stock solution can be stored for several months.

2. Nuclease P1 from *Penicillium citrinum* (can be obtained in form of lyophilized powder with ~300 U/mg protein from for example Sigma). To prepare a 0.3 U/μl stock solution, dissolve 1 mg protein in 1 ml of 30 mM sodium acetate, pH 5.3. The stock solution can be stored at −20 °C for several months.

3. Exonuclease III (200 U/μl is available from for example Promega). Can be stored at −20 °C for several months.

4. 500 mM EDTA, pH 8.0.

5. Phenol–chloroform (1:1) (*see* **Note 3**).

6. Piperidine (99 %). Can be stored at 4 °C for several months.

7. 1 μg/μl glycogen.

8. 3 M sodium acetate, pH 5.3.

9. 14.3 M β-mercaptoethanol.

10. Formamide loading buffer: 50 mM Tris-borate, pH 8.3, 95 % (v/v) deionized formamide, 20 mM EDTA, pH 8.0, 0.025 % (w/v) xylene cyanol, 0.025 % (w/v) bromophenol blue.

## 2.3 Components for Gel Electrophoresis

1. 40 % (w/v) 19:1 acrylamide–bisacrylamide solution.

2. 10× TBE: 0.89 M Tris-borate, pH 8.3 and 25 mM EDTA.

3. $N,N,N',N'$- tetramethylenediamine (TEMED).

4. 10 % (w/v) ammonium persulfate (APS).

5. Urea, ultra-pure.

6. Prepare 10 % sequencing gel: Weigh 42 g of urea. Add 25 ml of 40 % (w/v) 19:1 acrylamide–bisacrylamide solution (be careful, acrylamide is neurotoxic). Add 10 ml of 10× TBE solution. Add distilled water to a final volume of 100 ml. Agitate the mixture at room temperature with magnetic stirring until urea is completely dissolved. Degas the solution. Add 80 μl of TEMED. Add 800 μl of 10 % (w/v) APS. Mix thoroughly and cast the gel immediately.

## 2.4 Other Items

1. Gel electrophoresis system.

2. Vacuum concentrator.

3. X-ray film or phosphorimaging system.

4. Intensifier screen (optional).

5. Liquid scintillation counter.

## 3 Methods

To establish the optimal condition for the formation of Cascade-DNA complexes it is recommended to perform electrophoretic mobility shift assays (EMSA) prior starting with the footprint studies. In contrast to some other footprint assays (e.g., DNase I), all three footprint methods described here, have the advantage that the DNA must not be saturated by the protein, because the unbound DNA fraction in the sample is either unreactive (to $KMnO_4$ or Nuclease P1) or completely degraded (by Exonuclease III) and, therefore, does not produces background signals. However, there are some important recent findings on the DNA-binding activity of Cascade complexes, which should be considered in order to obtain specific and stable Cascade-DNA complexes and clear footprints (*see* **Notes 2** and **4**).

### 3.1 $KMnO_4$ Probing

1. Prepare a 30 μl binding mixture by adding appropriate amount of Cascade-crRNA to 0.5 to 1 nM of 5′-end labeled target DNA (10,000–15,000 cpm/sample). Incubate the mixture for 30 min at 30 °C to allow complex formation (*see* **Note 5**).

2. Dilute the binding samples 1:1 by adding 30 μl distilled water.

3. Start the modification reaction by adding of 4 μl of 160 mM $KMnO_4$ solution (freshly diluted in distilled water; corresponds to a final concentration of 10 mM). Mix promptly but gently with a micropipette.

4. Allow modification for 2 min at 30 °C (*see* **Note 6**).

5. Stop the reaction by adding 4.8 μl β-mercaptoethanol, shake the samples briefly on vortex and put them on ice to cool.

6. Add 5.3 μl of 500 mM EDTA, pH 8.0.

7. Extract the samples with phenol–chloroform: add 100 μl phenol–chloroform and vortex the samples for 1 min vigorously and centrifuge 5 min at 14,000 rpm. Transfer the aqueous phase (upper layer) in new reaction tubes. Repeat the phenol–chloroform extraction once. Finally, extract with 100 μl chloroform as above and transfer the aqueous layer to new reaction tubes (*see* **Note 7**).

8. Precipitate the modified DNA with 2.5 volumes of ice-cold absolute ethanol. Shake on vortex and incubate at −20 °C for at least 30 min. Centrifuge the samples for 30 min at 14,000 rpm. Carefully decant the supernatant and wash the pellet with 100 μl of ice-cold 80 % (v/v) ethanol and centrifuge for 10 min at 14,000 rpm. Remove the supernatant with micropipette and dry the pellets in vacuum concentrator.

   *Hot piperidine cleavage reaction*

9. Dissolve the pellets in 50 μl of freshly diluted 10 % (v/v) piperidine and incubate at 90 °C for 30 min. Place the tubes on ice to cool. Centrifuge briefly to collect the samples on the bottom of the tubes (*see* **Note 8**).

10. Dry the samples in a vacuum concentrator.

11. Wash the pellet by adding 30 μl distilled water, gently mix with micropipette and dry the samples again in a vacuum concentrator. Repeat the wash step once (*see* **Note 9**).

12. Dissolve the pellets in 50 μl of 0.3 M sodium acetate, pH 5.3.

13. Add 2 μl glycogen (1 μg/μl) (to increase the recovery of the cleavage products during the precipitation).

14. Precipitate the samples with 2.5 volumes of ice-cold absolute ethanol as described in **step 8**.

15. Dissolve the pellets in 10 μl of formamide loading buffer.

### 3.2 Nuclease P1 Probing

1. Dilute the 30 μl binding mixture (from **step 1** of Subheading 3.1) with 30 μl of 60 mM sodium acetate, pH 5.3.

2. Add 3 μl (0.09 U) of freshly made dilution of Nuclease P1 (0.3 U/μl stock solution diluted 1:10 in 30 mM sodium acetate, pH 5.3) (*see* **Note 10**).

3. Incubate at 37 °C for 30 min.

4. Stop the reaction by the addition of 100 μl phenol–chloroform. Extract the samples with phenol–chloroform and precipitate as described in **steps 7** and **8** of Subheading 3.1 (*see* **Note 11**).

5. Dissolve the pellets in 10 μl formamide loading buffer.

### 3.3 Exonuclease III Assay

1. Dilute the 30 μl binding mixture (from **step 1** of Subheading 3.1) with 30 μl distilled water.

2. Add 3 μl (60 U) of freshly made dilution of Exonuclease III (200 U/μl stock solution diluted 1:10 in 10× reaction buffer provided by for example Promega: 660 mM Tris, pH 8.0, 6.6 mM MgCl$_2$). Mix briefly and gently with a micropipette (*see* **Note 12**).

3. Incubate at 37 °C for 20 min.

4. Stop the reaction by adding 1.2 μl of 500 mM EDTA pH 8.0. Shake the samples on vortex.

5. Add 1/10 volume of 3 M sodium acetate, pH 5.3 (*see* **Note 13**).

6. Extract the samples with phenol–chloroform and precipitate as described in steps 7 and **8** of Subheading 3.1.

7. Dissolve the pellet in 10 μl of formamide loading buffer.

## 3.4 Separation of the Cleavage Products by Denaturing Gel Electrophoresis

1. Determine the total recovery of the cleavage products by counting Cherenkov radiation of the samples (suspended in 10 μl formamide loading buffer) using a scintillation counter (*see* **Note 14**).

2. Pre-warm the gel: run the gel in 1× TBE at constant 25 W for 10 min and at 50 W for another 10 min.

3. Denature the cleavage products by incubation of the samples at 95 °C for 3 min. Put the samples immediately on ice for 5 min.

4. Rinse the gel pockets with TBE buffer and apply equal amounts of radioactivity (about 3,000–8,000 cpm/lane) onto pre-warmed gel. Load also a length standard onto the gel (Maxam–Gilbert sequencing reaction [14] of the 5′-end labeled DNA probes would be the best way to map the exact positions).

5. Run the gel at constant 75 W for 2–3 h.

6. Disassemble the gel apparatus and transfer the gel into an exposure cassette (*see* **Note 15**).

7. Expose the gel to X-ray film at −80 °C or a phosphorimager screen at room temperature for 18–36 h (*see* **Note 16**).

## 3.5 Evaluation of the Footprint Results

Examples of autoradiograms from the footprint assays with type I-E Cascade and an 85 bp target DNA are shown in Fig. 3 [3]. Probing studies with DNA labeled on the nontargeted strand revealed that several thymines within the protospacer region of this strand were susceptible to permanganate modification and Nuclease P1 cleavage upon binding of Cascade-R44 (loaded with matching crRNA) (Fig. 3a). In addition to thymines, one adenine (at position 33, Fig 3d) and one guanine (at position 36) were also modified. Note that beside the preferential modification of single-stranded thymines, permanganate can also react to a much lesser extent with unpaired cytosines, adenines and guanines [7]. The targeted strand was almost unreactive to modification, consistent with a base-pairing of the protospacer region with the crRNA, with exception of a single thymine residue (T33). The susceptibility of T33 suggests that the base at position 6 of the protospacer is flipped out of the crRNA-DNA hybrid [3]. Actually, recent study has shown that position 6 within the seed region of the crRNA is not involved in target recognition [15], which highlights the sensitivity of the permanganate probing. The same analyses with a Cascade complex loaded with nontargeting crRNA did not yield any signals, demonstrating that the permanganate and Nuclease P1 sensitivity is not caused by a binding of Cascade *per se* but requires the hybridization of the crRNA to the targeted strand. Within the targeted strand, a region immediately upstream of protospacer was sensitive to P1 cleavage, indicating a structural deformation of the DNA at the PAM region upon binding of the Cascade. Such a

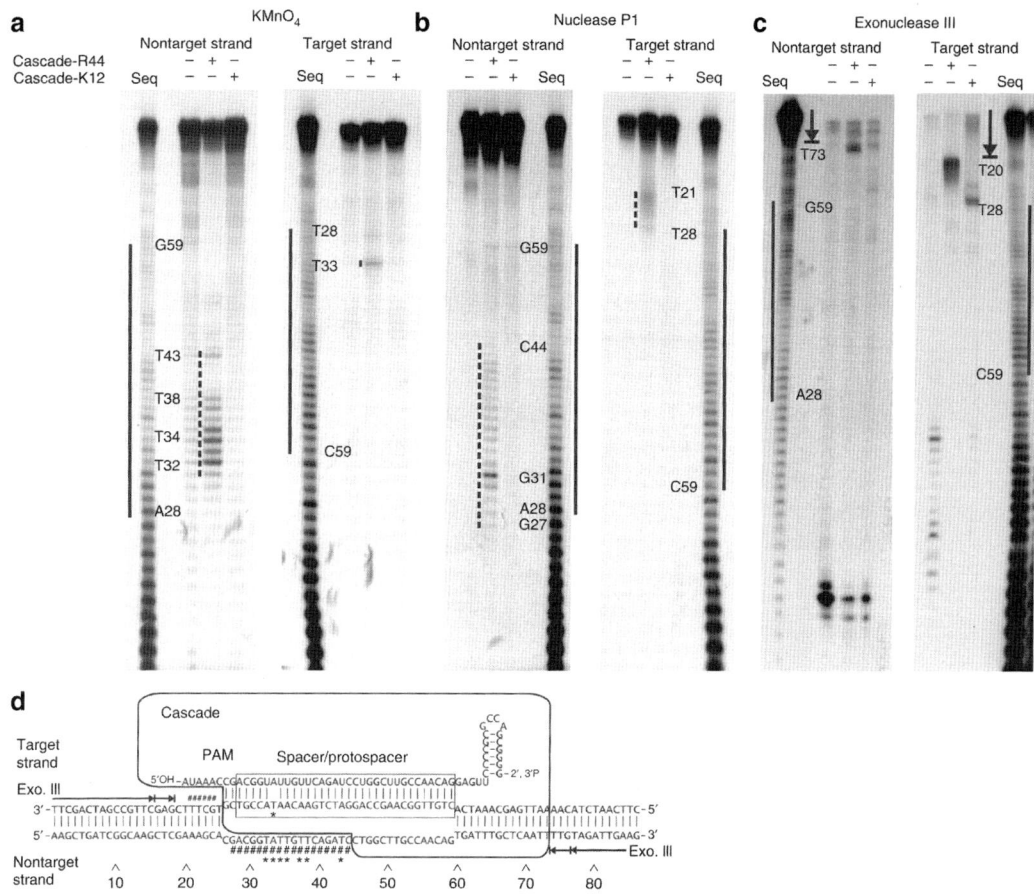

**Fig. 3** Autoradiograms of the KMnO$_4$ (**a**) and Nuclease P1 (**b**) probing and Exo III protection assays (**c**) with 85 bp double-stranded target DNA labeled on the 5'-end of either nontarget or target strand, respectively. Cascade-R44 contained targeting crRNA. Cascade-K12 loaded with nontargeting crRNAs was used as negative control. The protospacer region is indicated by *solid lines* next to the sequencing lanes. Permanganate and Nuclease P1 susceptible sites are marked with *dashed lines* (**a**, **b**). Exonuclease III stop sites are indicated with *arrows* (**c**). In (**d**) the results are summarized schematically (*hash*: Nuclease P1 sensitive sites; *asterisk*: permanganate sensitive sites) (reproduced from [3] with kind permission from Nature Publishing Group)

distortion of the DNA at the PAM region is suggested to be caused by the Cse1 subunit of Cascade in order to initiate the base-pairing of the crRNA with the targeted strand [16]. The Exonuclease III protection assay yielded precise information on the upstream and downstream boundaries of the Cascade protected region (Fig. 3c). In presence of the targeting Cascade (R44) the cleavage products had a length of about 73 bp (nontargeted strand) or 65 bp (targeted strand), consistent with a protected DNA region of about 60 base pairs (Fig. 3d).

## 4 Notes

1. Cascade is able to bind unspecific to double-stranded DNA (in order to scan the DNA for the protospacer sequence [3, 16], therefore, it is highly recommended to prepare also a Cascade complex loaded with nontargeting crRNA to use it as negative control in the footprint assays. The formation of an R-loop structure by melting of the protospacer DNA occurs only with the targeting Cascade complex. Thus any footprint signals also obtained with the nontargeting Cascade should be regarded as unspecific.

2. There are several possible ways to prepare double-stranded DNA, which is labeled at one end of one strand only. Cascade-crRNA complex from *E. coli* has been shown to bind short linear DNA fragments (65 bp or 85 bp) [3, 15]. Therefore, an easy way to prepare short DNA fragments is to anneal gel-purified complementary oligonucleotides, one of which is labeled on its 5′-end. The 5′-end labeled oligonucleotide should be hybridized with two to threefold molar excess of its unlabeled complementary strand to ensure complete saturation of the labeled strand. Otherwise the labeled single-stranded oligonucleotide would cause high background in the permanganate and Nuclease P1 assays. The 5′-end labeling of the oligonucleotides can be achieved by using T4 polynucleotide kinase and $\gamma$-$^{32}$P-ATP [17].

    The following points should be considered for the design of the target DNA:

    (a) The DNA must contain the protospacer DNA, which is complementary to the spacer-part of the crRNA in the Cascade complex.

    (b) The protospacer sequence should be flanked by at least 25 bp sequences on both sides (this will also help to obtain clear Exo III stop products).

    (c) The protospacer region of the DNA should contain the appropriate protospacer adjacent motif (PAM) [18]: while DNA fragments without PAM can be in principle bound by Cascade-crRNA [3], the binding affinity seems to be 10 to 50-fold higher in presence of an accurate PAM [15, 19].

    (d) According to Westra et al. [19], the binding affinity of Cascade-crRNA appears to be reduced with increasing length of the linear DNA fragment (likely due to the requirement of higher energy for strand separation during the R-loop formation in large DNA fragments). The use of 65–85 bp DNA fragments seems to work well [3, 15].

The preparation of the target DNA labeled at only one strand can also be achieved by standard PCR amplification using 5′-end labeled primer. Moreover, in the permanganate and Nuclease P1 assays, one can also use 3′-end labeled DNA fragment obtained by "fill-in" reaction at a 5′-protruding end using Klenow fragment and α-$^{32}$P-dNTP. A prerequisite for the Exo III protection assays is the use of 5′-end labeled DNA fragments, however.

3. Phenol and chloroform are both hazardous and require handling with extreme caution in a laboratory fume hood. Prepare a 1:1 mixture of phenol–chloroform by adding 100 ml chloroform to 100 ml of TE-saturated phenol. Shake the solution and leave standing until the phenol phase is clear. Discard the aqueous layer. Add 0.1 % 8-hydroxychinolin, which avoids oxidation of phenol. Wrap the bottle with aluminum foil.

4. Cascade-crRNA complex has been shown to bind with much higher affinity when using negatively supercoiled target DNA. Negative supercoiling facilitates the melting of the protospacer DNA during the formation of the R-loop [19]. It could be beneficial, therefore, to use circular plasmid DNA for probing with $KMnO_4$ and Nuclease P1. In that case, the detection of the permanganate or Nuclease P1 susceptible sites can be achieved by primer extension reaction with labeled oligonucleotides [7]. Exonuclease III footprints, however, require the use of linear DNA fragments.

5. In case that quite excess of Cascade is needed to complex more than half of the input DNA, it is advisable to reduce the DNA amounts (e.g., to a final concentration of about 0.05 nM; 3,000–5,000 cpm) in the $KMnO_4$ samples, instead further increasing the Cascade concentration (high protein concentration can quench permanganate reaction).

   Short DNA fragments with high AT-content could be intrinsically sensitive to permanganate modification. Therefore, one should include control sample with DNA in absence of protein for each reaction.

6. Optimal $KMnO_4$ concentration and incubation time should be determined in pilot experiments (e.g., by titration of 5–20 mM $KMnO_4$). In our laboratory, permanganate modification for 2 min with 10–15 mM $KMnO_4$ worked also well to localize RNA polymerase open complexes [20, 21].

7. Alternatively, proteinase K and SDS treatment can be performed to remove Cascade from the sample. However, extraction with phenol–chloroform is recommended if high concentration of Cascade (e.g., >1 μM) has been used for complex formation (note that 1 μM of type I-E Cascade corresponds to 11 μM protein in total). We made the experience

that the DNA bands are significantly sharper when DNA/protein mixtures are extracted with phenol–chloroform.

8. Piperidine is toxic and should be handled in a laboratory fume hood. An opening of the lids during heating should be prevented (e.g., weigh down the lids by placing a lead onto the tubes or use reaction tubes with screw cap).

9. These washing steps are necessary to remove residual piperidine, which interferes with the migration of the DNA in the gel. In order to obtain sharp bands, one should wash the pellets after the cleavage at least twice with distilled water as described.

10. It is advisable to establish the optimal concentration of Nuclease P1 in pilot experiments: e.g., by adding of 3 µl of undiluted and 1:10, 1:100, 1:1,000 dilutions of 0.3 U/µl Nuclease P1 stock solution.

11. The radioactivity levels of the phenol and aqueous phase can be monitored using a Geiger counter. In case that the phenol phase contains most of the radioactivity, a re-extraction of the phenol layer could be helpful to increase the recovery of DNA cleavage products: add 100 µl of TE buffer to the phenol phase, shake on vortex for 1 min and centrifuge at 14,000 rpm for 5 min. Transfer aqueous layer into fresh tubes and extract once with chloroform. Pool the aqueous layer with that from the first extraction step prior to precipitation with ethanol.

12. Optimal concentration of Exo III and the incubation time needed will depend on the DNA and stability of the complexes. The experiments should be done under conditions, which ensure the complete degradation of the unbound DNA.

13. In our experience, adding sodium acetate before extraction with phenol–chloroform increases the yield of recovered DNA significantly.

14. If a scintillation counter is unavailable, a Geiger counter can be used here.

15. Our laboratory uses developed (old) X-ray films to peel off the gel from the glass plate to transfer it into exposition cassette: lift off one glass plate using a spatula and gently overlay the gel with a used X-ray film. Trapped bubbles between the gel and the overlying X-ray film should be removed by gently wiping over the X-ray film with gloved hand. Carefully peel off the gel by holding the top corners of the X-ray film, place the X-ray film with the attached gel into the exposition cassette (gel side up) and cover the gel with plastic wrap. Overlay the gel with an X-ray film or a phosphorimager screen in the dark room.

16. If necessary, use intensifier screen. It may require several exposures to obtain appropriate autoradiograms.

## References

1. Galas DJ, Schmitz A (1978) DNAse footprinting: a simple method for the detection of protein-DNA binding specificity. Nucleic Acids Res 5(9):3157–3170
2. Aguilera A, Garcia-Muse T (2012) R loops: from transcription byproducts to threats to genome stability. Mol Cell 46(2):115–124
3. Jore MM, Lundgren M, van Duijn E, Bultema JB, Westra ER, Waghmare SP, Wiedenheft B, Pul Ü, Wurm R, Wagner R, Beijer MR, Barendregt A, Zhou K, Snijders AP, Dickman MJ, Doudna JA, Boekema EJ, Heck AJ, van der Oost J, Brouns SJ (2011) Structural basis for CRISPR RNA-guided DNA recognition by Cascade. Nat Struct Mol Biol 18(5):529–536
4. Sinkunas T, Gasiunas G, Waghmare SP, Dickman MJ, Barrangou R, Horvath P, Siksnys V (2013) In vitro reconstitution of Cascade-mediated CRISPR immunity in *Streptococcus thermophilus*. EMBO J 32(3):385–394
5. Hayatsu H, Ukita T (1967) The selective degradation of pyrimidines in nucleic acids by permanganate oxidation. Biochem Biophys Res Commun 29(4):556–561
6. McCarthy JG, Williams LD, Rich A (1990) Chemical reactivity of potassium permanganate and diethyl pyrocarbonate with B DNA: specific reactivity with short A-tracts. Biochemistry 29(25):6071–6081
7. Kahl BF, Paule MR (2001) The use of diethyl pyrocarbonate and potassium permanganate as probes for strand separation and structural distortions in DNA. Methods Mol Biol 148:63–75
8. Bramhill D, Kornberg A (1988) Duplex opening by dnaA protein at novel sequences in initiation of replication at the origin of the *E. coli* chromosome. Cell 52(5):743–755
9. Rao BJ, Dutreix M, Radding CM (1991) Stable three-stranded DNA made by RecA protein. Proc Natl Acad Sci U S A 88(8):2984–2988
10. Rogers SG, Weiss B (1980) Exonuclease III of *Escherichia coli* K-12, an AP endonuclease. Methods Enzymol 65(1):201–211
11. Siebenlist U, Simpson RB, Gilbert W (1980) *E. coli* RNA polymerase interacts homologously with two different promoters. Cell 20(2):269–281
12. Metzger W, Heumann H (2001) Footprinting with exonuclease III. Methods Mol Biol 148:39–47
13. Brouns SJ, Jore MM, Lundgren M, Westra ER, Slijkhuis RJ, Snijders AP, Dickman MJ, Makarova KS, Koonin EV, van der Oost J (2008) Small CRISPR RNAs guide antiviral defense in prokaryotes. Science 321(5891):960–964
14. Maxam AM, Gilbert W (1980) Sequencing end-labeled DNA with base-specific chemical cleavages. Methods Enzymol 65(1):499–560
15. Semenova E, Jore MM, Datsenko KA, Semenova A, Westra ER, Wanner B, van der Oost J, Brouns SJ, Severinov K (2011) Interference by clustered regularly interspaced short palindromic repeat (CRISPR) RNA is governed by a seed sequence. Proc Natl Acad Sci U S A 108(25):10098–10103
16. Sashital DG, Wiedenheft B, Doudna JA (2012) Mechanism of foreign DNA selection in a bacterial adaptive immune system. Mol Cell 46(5):606–615
17. Maniatis T, Fritsch EF, Sambrock J (1982) Molecular cloning, a laboratory manual. Cold Spring Harbor Laboratory, Cold Spring Harbor, NY
18. Mojica FJ, Diez-Villasenor C, Garcia-Martinez J, Almendros C (2009) Short motif sequences determine the targets of the prokaryotic CRISPR defence system. Microbiology 155(Pt 3):733–740
19. Westra ER, van Erp PB, Kunne T, Wong SP, Staals RH, Seegers CL, Bollen S, Jore MM, Semenova E, Severinov K, de Vos WM, Dame RT, de Vries R, Brouns SJ, van der Oost J (2012) CRISPR immunity relies on the consecutive binding and degradation of negatively supercoiled invader DNA by Cascade and Cas3. Mol Cell 46(5):595–605
20. Pul Ü, Wurm R, Arslan Z, Geissen R, Hofmann N, Wagner R (2010) Identification and characterization of *E. coli* CRISPR-*cas* promoters and their silencing by H-NS. Mol Microbiol 75(6):1495–1512
21. Stratmann T, Pul Ü, Wurm R, Wagner R, Schnetz K (2012) RcsB-BglJ activates the *Escherichia coli leuO* gene, encoding an H-NS antagonist and pleiotropic regulator of virulence determinants. Mol Microbiol 83(6):1109–1123

# Chapter 20

# Creation and Analysis of a Virome: Using CRISPR Spacers

## Michelle Davison and Devaki Bhaya

### Abstract

Advances in sequencing technology have allowed for the study of complex and previously unexplored microbial and viral populations; however, linking host–phage partners using in silico techniques has been challenging. Here, we describe the flow-through for creation of a virome, and its subsequent analysis with the viral assembly and analysis module "Viritas," which we have recently developed. This module allows for binning of contigs based on tetranucleotide frequencies, putative phage-host partner identification by CRISPR spacer matching, and identification of ORFs.

**Key words** CRISPR-Cas, Next-generation sequencing, Viral genome assembly, Metagenomics, Phage annotation, Bioinformatics

## 1 Introduction

Pioneering work by several groups [1–4] using high-throughput sequencing technologies has provided a glimpse of microbial and viral diversity in a range of different environments. However, the lack of fully assembled viral and corresponding host genomes has limited the analysis of such data. Despite recent progress in handling regions of uneven coverage [5–7], tackling complex populations [8–10], development of comprehensive metagenomic analysis pipelines [11, 12], as well as ORF identification and calling in viromes [13], robust assembly and analysis of viral sequence data remains a significant obstacle.

For viruses to initiate successful infections, they must mutate to evade the host CRISPR defense system, which relies on a close match between acquired spacer and incoming viral sequence [13, 14]. Because new spacers are being acquired into host CRISPR arrays they are useful markers, both for the analysis of host–phage relationships, and to provide a time-line of past viral infections [15]. The spacers of the CRISPR-Cas adaptive immunity system, therefore, can also be a tool that can be utilized to "bin" novel viral sequences. [13]. We recently investigated viral populations in the hot spring polymicrobial mat community of Yellowstone National Park. By using CRISPR spacers identified in full genome sequences, as well as from metagenomic reads, we were able to successfully bin and identify potential viral-host partners [13, 16]. The processes to perform these analyses are described in this chapter.

## 2 Materials

1. 1× Tris–EDTA: 10 mM Tris–HCl, 1 mM EDTA, pH 8.0.
2. Nitrocellulose filters, 0.45 μm and 0.2 μm (Nalgene, Thermo Scientific).
3. Microfuge Tube Polyallomer 1.5 mL ultramicrocentrifuge tubes.
4. Illustra GenomiPhi V2 kit, GE.
5. Beckman TL-100 Ultracentrifuge, TLA 100.3.
6. 10× PCR buffer (Qiagen).

## 3 Methods

### 3.1 Creation of a Virome

#### 3.1.1 Enrichment of Viral Particles from an Environmental Sample of Interest

Resuspend microbial mat sample in 50 mL 1× Tris–EDTA by vigorous vortexing (*see* **Note 1**). Pellet down intact cells and cellular debris at 6,000 rpm for 10 min in a Sorvall GS5C. Aspirate the supernatant, and pass sequentially through 0.45 μm and 0.2 μm filters to remove any remaining cells and/or cellular debris. Aliquot filtered supernatant into 1.5 mL microfuge tubes and ultracentrifuge at 50,000 rpm or 1 h for concentrate viral particles. Carefully remove all but 5–10 μL of supernatant. Use 1 μL of enrichment as template for amplification reaction with $\phi$29 polymerase as per kit instructions (Illustra GenomiPhi V2 kit; *see* **Note 2**) (Fig. 1).

#### 3.1.2 Qualitative Determination of Bacterial and Viral Ratios in Amplified Viral Sample

Assemble 25 μL PCRs with the following 8× Mastermix recipe: 20 μL 10× PCR Buffer, 111.0 μL of double distilled water, 10.0 μL of DMSO, 16.0 μL dNTP (2.5 mM stock), 20.0 μL forward primer, and 20.0 μL reverse universal bacterial 16S primers [27], 1.0 μL of $\phi$29 amplified viral template reaction and 2.0 μL of Taq Polymerase. Visualize PCRs by gel electrophoresis in 0.8 % agarose with ethidium bromide (0.0002 μg/μL) run at 100 mV for 20 min.

#### 3.1.3 Sequencing of Viral Enrichment

Pool duplicate $\phi$29 amplified viral templates and carry out appropriate Next-Generation Sequencing (e.g., 454 sequencing, Illumina, PacBio) based on read length needs, acceptable error rates, and coverage depth requirements (*see* **Note 3**).

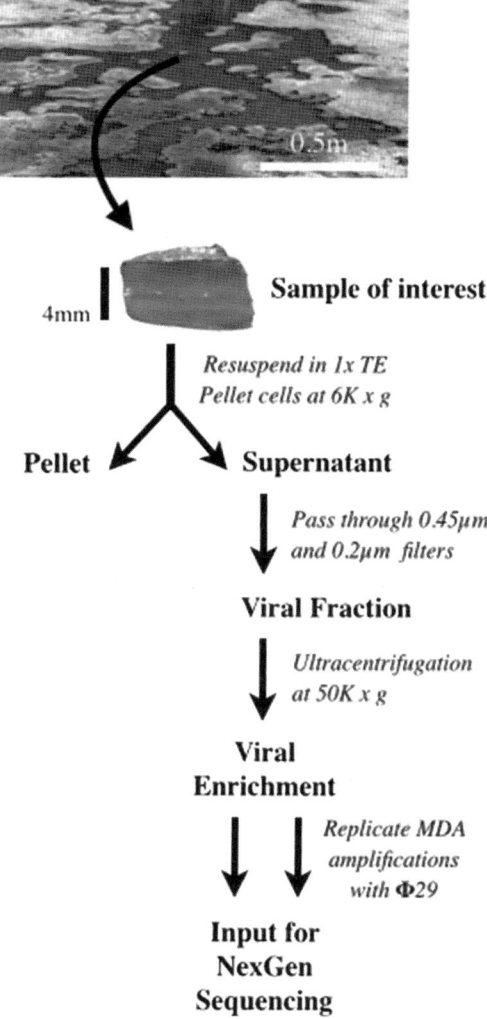

**Fig. 1** Schematic workflow of viral sample preparation for next-generation sequencing

## 3.2 Identification of CRISPR Spacers

### 3.2.1 Identification of CRISPR Spacers Present in Metagenome Reads and Extraction of Novel CRISPR Spacers

To identify reads containing a CRISPR array, pipeline metagenome reads (either from publically available datasets, or optimally a metagenome generated in tandem from the host of interest) through CRISPRFinder to extract both spacers and repeats, then subject repeats to a BLASTN (e-value $<=10^{-5}$) against the nr database to determine the species from which they originated [17–19] (*see* **Note 4**). CRISPR spacer identification can be done with either raw or assembled reads, however, when performed with assembled reads some resolution will be lost as spacers identified will be consensus sequences.

## 3.3 Assembly of Viral Reads

### 3.3.1 Preprocessing of Reads for Viral Assembly

To remove amplification artifacts and low quality/complexity reads from 454 data, use the program prinseq to preprocess viral reads prior to assembly [20]. Reads smaller than 50 bp should be filtered out before further analysis (*see* **Note 5**).

### 3.3.2 Read Filtration (Removing Host/Bacterial DNA)

To find and filter out all hits to known bacterial genomes, and recruit reads to known phage sequences, use NUCmer [21] with the following parameters: --maxmatch –c 30 –l 30 --coords).

### 3.3.3 Genome Assembly

To assemble genomes, firstly, reads confidently classified by FCP (e-value $<=10^{-5}$) as bacterial or archaea should be removed. Then, run an assembly with Newbler version 2.5p1 (params: -minlen 45 -m -cpu 16 -mi 90 -ml 40 -ud -l 1000 -g -e 3 -a 300 -large) (*see* **Note 6**).

### 3.3.4 Celera Assembler Virome Assembly

The Celera Assembler (Celera Genomics, 1999, tuned for metagenomic assembly, utgErrorRate=0.030, utgBubblePopping=1, cgwDistanceSampleSize=10, doToggle=1, toggleNumInstances=0, toggleUnitigLength=1000) should be run on the dataset, for verification of the Newbler output by identifying contigs in disagreement, in addition to improving the virome assembly in terms of contig length and fragmentation.

### 3.3.5 Phage Annotation Pipeline

The phage assembly module, Viritas, exists within the metagenomic assembly pipeline MetAMOS [10] to assemble, bin, annotate and characterize viral sequencing data. The pipeline consists of MetaGeneMark [21] for ORF prediction, PHMMER [22, 28] for homology detection using an annotated phage protein DB (Phantome & EBI), Repeatoire for internal repeat misassembly identification, NUCmer/PROmer for spacer hits to known CRISPR spacers, and tRNAscan for tRNAs [23] frequently found in phages. The specific python command lines for each program are detailed below:

MetaGeneMark
/gmhmmp -o %s.orfs -m MetaGeneMark_v1.mod
tRNAscan
tRNAscan-SE -o trna.out –B
HMMER 3
phmmer --cpu 10 -E 0.1 -o %s.phm.out --tblout %s.phm.tbl --notextw %s.faa ./DBS/allprots.faa"%(prefix,prefix,prefix)
REPEATOIRE
./repeatoire --minreplen=20 --z=11 --extend=0 --allow-redundant=0 --sequence=t1.fna --output=reps.out >& test.out
PROMER
promer --maxmatch -c 4 -l 4 --coords t1.fna ./CRISPR/INPUTCRISPRs.fasta > t1.out 2> t2.err
promer --maxmatch -c 4 -l 4 --coords t1.fna ./CRISPR/INPUTCRISPRhits.fasta > t1.out 2> t2.err
show-coords -I 85 -o -k -L 30 -T -c -r out.delta > out.coords

### 3.3.6 Tetranucleotide Analysis

To group contigs based on tetranucleotide frequency, the following python script was used on the assembled contigs over 1 kb in length:

```python
#count_tetramers.py
#takes an assembly as input, calculates tetramer frequencies stepwise across each contig
#reports frequences in *.tetra output file and provides OBS/EXP values for each tetramer
import os,sys,string,operator,math
#create tetramer hash
def getTetramerDict():
    tdict = {}
    nts = ('A','G','C','T')
    for c1 in nts:
        for c2 in nts:
            for c3 in nts:
                for c4 in nts:
                    tmer = c1+c2+c3+c4
                    ids = [tmer,getRC(tmer)]
                    ids = sorted(sorted(ids), key=str.upper)
                    tdict[ids[0]] = 0
    #print len(tdict.keys())
    return tdict
#get reverse complement
def getRC(tmer):
    rcs = ""
    for nt in tmer:
        if nt == "T":
            rcs += "A"
        elif nt == "A":
            rcs += "T"
        elif nt == "G":
            rcs += "C"
        elif nt == "C":
            rcs += "G"
        elif nt == "N":
            rcs += "N"
    if len(rcs) != 4:
        print "problem in tmer size"
        sys.exit(1)
    rcs = rcs[::-1]
    return rcs
verbose = False
try:
    f = open(sys.argv[1],'r')
except IndexError:
    print "usage: calc_tetramers.py contigs.fa [verbose=0]"
    sys.exit(0)
```

```
        except IOError:
           print "Input file not found!"
           sys.exit(0)
        try:
           verbose = int(sys.argv[2])
        except IndexError:
           pass
        except TypeError:
           pass
        data = f.read()
        data = data.split(">")[1:]
        outf = open(sys.argv[1]+".tetra",'w')
        outf.write("ContigID,")
        cdict = getTetramerDict()
        for tmer in cdict.keys()[:-1]:
           outf.write(tmer+",")
        outf.write(cdict.keys()[-1]+"\n")
        nts = ("A","G","C","T")
        top_mers = {}
        for seq in data:
           hdr,ctg = seq.split("\n",1)
           outf.write(hdr.replace("\n","").split("\t")[0].
        split(" ")[0]+",")
           ctg = ctg.replace("\n","")
           ctg = string.upper(ctg)
           s1 = 0
           s2 = s1+4
           tdict = getTetramerDict()
              while s1+4 < len(ctg):
              tmer = ctg[s1:s2]
              tmer = string.upper(tmer)
              if "N" in tmer:
                 s1 +=1
                 s2 +=1
                 continue
              notok = 0
              for char in tmer:
                 if char not in nts:
                    notok = 1
                    break
              if notok:
                 s1 +=1
                 s2 +=1
                 continue
              try:
              tdict[tmer]
              except KeyError:
                 tmer = getRC(tmer)
```

```python
            tdict[tmer] +=1
            s1 +=1
            s2 +=1
            sum = 0.0
         for tmer in tdict.keys()[:-1]:
            #outf.write("%.5f"%((float(tdict[tmer])*4.0)/
            float(len(ctg)))+",")
            if len(ctg)-4 > 0:
               outf.write("%.5f"%((float(tdict[tmer])))/
               float(len(ctg)-4))+",")
               sum += (float(tdict[tmer]))/float(len(ctg)-4)
         tmer = tdict.keys()[-1]
         #outf.write("%.5f"%((float(tdict[tmer])*4.0)/
         float(len(ctg)))+"\n")
         if len(ctg)-4 > 0:
            outf.write("%.5f"%((float(tdict[tmer])))/
            float(len(ctg)-4))+"\n")
            sum += (float(tdict[tmer]))/float(len(ctg)-4)
         #calc background GC freq
         acnt = ctg.count("A")
         tcnt = ctg.count("T")
         ccnt = ctg.count("C")
         gcnt = ctg.count("G")
         afreq = float(acnt)/float(len(ctg))
         gfreq = float(gcnt)/float(len(ctg))
         cfreq = float(ccnt)/float(len(ctg))
         tfreq = float(tcnt)/float(len(ctg))
         if not verbose:
            continue
         #only run this if verbose
         print hdr
         sorted_x = sorted(tdict.iteritems(), key=operator.itemgetter
         (1))
         sorted_x.reverse()
         print "Tetramer\tCount\tObs/Exp"
         for tup in sorted_x[0:5]:
            if tup[1] > 0:
                  nfreqs = 1
               for nt in tup[0]:
                  if nt == "A":
                     nfreqs *= afreq
                  elif nt == "T":
                     nfreqs *= tfreq
                  elif nt == "G":
                     nfreqs *= gfreq
                  elif nt == "C":
                     nfreqs *= cfreq
               exp = nfreqs*(len(ctg)-4)
```

```
            if exp < 1:
                exp = 1
            #nfreqs = afreq*gfreq*cfreq*tfreq
                printtup[0]+"\t"+str(tup[1])+"\t%f"%(float(tup[1])/
                    (exp)),
            if float(tup[1])/(exp) > 2:
                print "*"
            else:
                print
outf.close()
print "done! tetramer freqs found in: ", sys.argv[1]+".tetra"
```

Contigs less than 1 kb in length often result in "noisy" signatures and should be excluded from further analysis. Use the gplots heatmap.2 R function to visualize frequencies based on a hierarchical clustering of tetranucleotide frequency of the assembled contigs [24].

### 3.4 Mapping of CRISPR Spacers onto Viral Reads

Identified CRISPR spacers are mapped back to viral reads with BLASTN. To be considered a 'putative' match, a spacer had to have at least 85 % identity spanning 70 % of the spacer length [13]. Exclude any reads containing host CRISPR arrays from this BLAST, to prevent artificial inflation of CRISPR hit numbers from spacers matching with 100 % identity to themselves.

### 3.5 Conclusion

We have outlined a method by which viromes can be analyzed, using a multitiered approach with broad applicability. Novel virome sequence can be conservatively assembled with the metAMOS pipeline, similar contigs can then be further clustered based on tetranucleotide signatures, and potential viral-host relationships elucidated by using CRISPR-spacer matching.

## 4 Notes

1. This particular protocol was optimized for use with the microbial mat communities of Yellowstone National Park; however, it can be tailored to suit any sample of interest. Volume of 1× TE should be enough liquid to thoroughly resuspend sample; keep in mind some liquid will be ultimately be lost in the filtering step. For example, we used 50 mL to resuspend an 8 mm circular plug 2 mm deep.

2. A DNase step prior to amplification was intentionally omitted in this case, as it was unknown if the viral particles were intact. Use of DNase on damaged particles can result in complete loss of viral DNA [25].

3. Multiple replicates are highly advised to mediate the effect of $\phi$29 polymerase bias in multiple displacement amplification (MDA) reactions [26].

4. CRISPRFinder: Always manually curate identified spacers to remove any spurious spacer calls, such as repeat-rich sequences, that are not associated with CRISPR loci.

5. All assembly parameters described were optimized for use with Roche 454 Titanium sequencing generated reads.

6. By stringently filtering out all hits to known bacterial and archaeal genomes phage, and prophage genes contained on host genomes will be excluded.

## Acknowledgments

DB acknowledges funding support from the NSF (MCB#1024755) and the Carnegie Institution of Science. This protocol is a modification of the Viritas pipeline described in Davison et al (in prep) which was carried out in collaboration with Mihai Pop, a the University of Maryland and his group. Viritas has been incorporated into the metAMOS pipeline (https://github.com/marbl/metAMOS).

## References

1. Emerson JB, Thomas BC, Andrade K, Heidelberg KB, Banfield JF (2013) New approaches indicate constant viral diversity despite shifts in assemblage structure in an Australian hypersaline lake. Appl Environ Microbiol 79.21(2013):6755–6764

2. Pride DT, Schoenfeld T (2008) Genome signature analysis of thermal virus metagenomes reveals Archaea and thermophilic signatures. BMC Genomics 9:420

3. Fancello L, Raoult D, Desnues C (2012) Computational tools for viral metagenomics and their application in clinical research. Virology 434:162–174

4. Sullivan MB, Huang KH, Ignacio-Espinoza JC, Berlin AM, Kelly L, Weigele PR, DeFrancesco AS, Kern SE, Thompson LR, Young S et al (2010) Genomic analysis of oceanic cyanobacterial myoviruses compared with T4-like myoviruses from diverse hosts and environments. Environ Microbiol 12:3035–3056

5. Chitsaz H, Yee-Greenbaum JL, Tesler G, Lombardo M-J, Dupont CL, Badger JH, Novotny M, Rusch DB, Fraser LJ, Gormley NA et al (2011) Efficient de novo assembly of single-cell bacterial genomes from short-read data sets. Nat Biotech 29:915–921

6. Peng Y, Leung HC, Yiu S, Chin FY (2012) IDBA-UD: a de novo assembler for single-cell and metagenomic sequencing data with highly uneven depth. Bioinformatics 28:1420–1428

7. Bankevich A, Nurk S, Antipov D, Gurevich AA, Dvorkin M, Kulikov AS, Lesin VM, Nikolenko SI, Pham S, Prjibelski AD, Pyshkin AV, Sirotkin AV, Vyahhi N, Tesler G, Alekseyev MA, Pevzner PA (2012) SPAdes: a new genome assembly algorithm and its applications to single-cell sequencing. J Comput Biol 19:455–477

8. Boisvert S, Raymond F, Godzaridis E, Laviolette F, Corbeil J (2012) Ray Meta: scalable de novo metagenome assembly and profiling. Genome Biol 13:R122

9. Namiki T, Hachiya T, Tanaka H, Sakakibara Y (2011) MetaVelvet: an extension of Velvet assembler to de novo metagenome assembly from short sequence reads. In: Proceedings of the 2nd ACM conference on bioinformatics, computational biology and biomedicine. pp. 116–124. Chicago, Illinois: ACM; 2011

10. Peng Y, Leung HCM, Yiu SM, Chin FYL (2011) Meta-IDBA: a de novo assembler for metagenomic data. Bioinformatics 27:i94–i101

11. Treangen T, Koren S, Astrovskaya I, Sommer D, Liu B, Pop M (2011) MetAMOS: a metagenomic assembly and analysis pipeline for AMOS. Genome Biol 12:P25

12. Kultima JR, Sunagawa S, Li J, Chen W, Chen H, Mende DR, Arumugam M, Pan Q,

Liu B, Qin J et al (2012) MOCAT: a metagenomics assembly and gene prediction toolkit. PLoS One 7:e47656
13. Heidelberg JF, Nelson WC, Schoenfeld T, Bhaya D (2009) Germ warfare in a microbial Mat community: CRISPRs provide insights into the Co-evolution of host and viral genomes. PLoS One 4:e4169
14. Deveau H, Barrangou R, Garneau JE, Labonte J, Fremaux C (2008) Phage response to CRISPR-encoded resistance in *Streptococcus thermophilus.* J Bacteriol 190:1390
15. Bhaya D, Davison M, Barrangou R (2011) CRISPR-Cas systems in bacteria and archaea: versatile small RNAs for adaptive defense and regulation. Annu Rev Genet 45:273–297
16. Davison M, Treangen TJ, Koren S, Gosrani S, Pop M, Bhaya, D. Analysis of virome diversity in a polymicrobial community *Manuscript submitted for publication.*
17. Bhaya D, Grossman AR, Steunou A-S, Khuri N, Cohan FM, Hamamura N, Melendrez MC, Bateson MM, Ward DM, Heidelberg JF (2007) Population level functional diversity in a microbial community revealed by comparative genomic and metagenomic analyses. ISME J 1:703–713
18. Grissa I, Vergnaud G, Pourcel C (2007) The CRISPRdb database and tools to display CRISPRs and to generate dictionaries of spacers and repeats. BMC Bioinformat 8:172
19. Altschul SF, Gish W, Miller W, Myers EW, Lipman DJ (1990) Basic local alignment search tool. J Mol Biol 215:403–410
20. Schmieder R, Edwards R (2011) Quality control and preprocessing of metagenomic datasets. Bioinformatics 27:863–864
21. Kurtz S, Phillippy A, Delcher A, Smoot M, Shumway M, Antonescu C, Salzberg S (2004) Versatile and open software for comparing large genomes. Genome Biol 5:R12
22. Zhu W, Lomsadze A, Borodovsky M (2010) Ab initio gene identification in metagenomic sequences. Nucleic Acids Res 38:e132–e132
23. Lowe TM, Eddy SR (1997) TRNAscan-SE: a program for improved detection of transfer RNA genes in genomic sequence. Nucleic Acids Res 25:0955–0964
24. Langfelder P, Horvath S (2008) WGCNA: an R package for weighted correlation network analysis. BMC Bioinformat 9:559
25. Emerson JB, Thomas BC, Andrade K, Allen EE, Heidelberg KB, Banfield JF (2012) Dynamic viral populations in hypersaline systems as revealed by metagenomic assembly. Appl Environ Microbiol 78:6309–6320
26. Ballantyne KN, van Oorschot RAH, Muharam I, van Daal A, John Mitchell R (2007) Decreasing amplification bias associated with multiple displacement amplification and short tandem repeat genotyping. Anal Biochem 368:222–229
27. Sundquist A, Bigdeli S, Jalili R, Druzin M, Waller S, Pullen K, El-Sayed Y, Taslimi MM, Batzoglou S, Ronaghi M (2007) Bacterial flora-typing with targeted, chip-based Pyrosequencing. BMC Microbiol 7:108
28. Eddy SR (2011) Accelerated profile HMM searches. PLoS Comput Biol 7:e1002195

# Chapter 21

## Targeted Mutagenesis in Zebrafish Using CRISPR RNA-Guided Nucleases

Woong Y. Hwang, Yanfang Fu, Deepak Reyon, Andrew P.W. Gonzales, J. Keith Joung, and Jing-Ruey Joanna Yeh

### Abstract

In recent years, the zebrafish has become a critical contributor to various areas of biomedical research, advancing our fundamental understanding of biomedicine and helping discover candidate therapeutics for human diseases. Nevertheless, to further extend the power of this important model organism requires a robust and simple-to-use genome editing platform that will enable targeted gene knockouts and introduction of specific mutations identified in human diseases into the zebrafish genome. We describe here protocols for creating insertion or deletion (indel) mutations or precise sequence modifications in zebrafish genes using customizable CRISPR-Cas9 RNA-guided nucleases (RGNs). These methods can be easily implemented in any lab and may also potentially be extended for use in other organisms.

**Key words** CRISPR, Cas9, Gene-editing, Genome engineering, Zebrafish, T7E1

## 1 Introduction

Over the last few years, the field of genome engineering has generated several breakthrough technologies for constructing site-specific DNA nucleases [1–4]. Customizable site-specific DNA nucleases can be used to create double-stranded DNA breaks (DSBs) at investigator-specified genomic loci. These DSBs can be repaired by an error-prone non-homologous end joining (NHEJ) mechanism, resulting in random-length insertions or deletions (indels) at the site of the break [5–13]. Alternatively, DSBs can also be repaired via homology-directed repair (HDR) using a homologous "donor" template, resulting in a precise modification of the sequence [12–14]. The donor template for HDR can be in the form of either plasmid DNA [15–18] or synthetic single-stranded oligodeoxynucleotides (ssODNs) [19–23]. These plasmid and ssODN template forms respectively contain several hundred base pair (bp) or less than 100 bp homology arms flanking the genomic site of modification. Nuclease-based approaches enable editing of endogenous genes in a wide range of model organisms, including many where targeted genome manipulations were inefficient (e.g., zebrafish, rats, fruit flies, and *C. elegans*) [5–7, 9, 24–33]. Among the various platforms described to date, the CRISPR-Cas9 RNA-guided nucleases are one of the simplest to implement [16, 25, 34–37].

The Type II CRISPR system of *Streptococcus pyogenes* is used by various bacteria as a mechanism to inactivate foreign DNAs [38, 39]. It has been previously shown that the Cas9 endonuclease can function together with a programmable synthetic single guide RNA (sgRNA) to induce targeted site-specific DSBs [38]. The nuclease activity of the Cas9 protein is directed by the first 20 nt of the ~100 nt sgRNA, which are complementary in sequence to one strand of the genomic target DNA site. One additional requirement for the genomic target DNA site is that strand of DNA opposite to the one that interacts with the sgRNA must also contain a proximal protospacer adjacent motif (PAM), which for the *S. pyogenes* Cas9 must be of the form NGG or NAG [38, 40].

We have previously shown that microinjection of sgRNA and Cas9-encoding mRNA into one-cell stage zebrafish embryos results in high rates of somatic indel mutations at 9 of 11 tested endogenous gene target sites [25]. In addition, RGN-mediated indel mutations can be transmitted through the zebrafish germ line efficiently [21]. Furthermore, the RGN system may also be used with single-stranded oligodeoxynucleotides (ssODNs) to introduce precise sequence modifications [21]. Another benefit of this system is the ease with which multiple genes can be targeted at once by introducing several sgRNAs simultaneously ([41] and Hwang, unpublished results).

There are several considerations when designing a gene-targeting experiment in zebrafish using a CRISPR-Cas9 system. Originally, sgRNAs expressed from our vectors begin with a 5′-GG dinucleotide due to the sequence requirements of the T7 promoter we use for in vitro transcription. We have shown that sgRNAs with a 5′-GG overhang that bear either 18 or 20 nucleotides of complementary sequence can still confer efficient target site cleavage in most cases [21]. This simple strategy expands the targeting range of our CRISPR-Cas9 system from 1 in every 128 bp of random DNA sequence to 1 in every 8 bp. Secondly, it is still unclear why some synthesized sgRNAs failed to efficiently induce indels at their genomic targets [21, 25]. One possibility is that certain sgRNA sequences might form secondary structures that either cannot bind to Cas9 or base pair with DNA targets [21]. Thus, we recommend making and testing more than one sgRNA vector per target gene. Finally, several studies have suggested that CRISPR-Cas9 targeted to different sequences may show different ranges of off-target activities [40, 42–44]. Thus, as with any genome editing experiment (regardless of the particular nuclease platform used), one should always perform control experiments to confirm that any phenotype observed is indeed linked to a specific mutation.

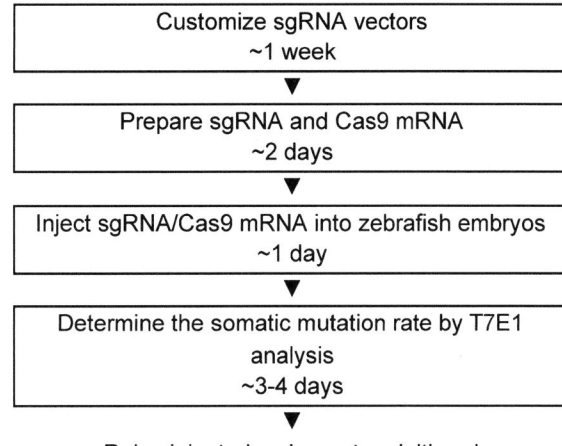

**Fig. 1** The steps for conducting targeted mutagenesis in zebrafish embryos using the customizable CRISPR-Cas9 system are shown in the *boxes*. Protocols for each of these steps are provided in this chapter

In this chapter, we describe our protocols for constructing customized sgRNA vectors. Our system requires the synthesis of only two short complementary oligodeoxynucleotides that encode the 20 nt of complementarity with the target site. Ligation of these annealed oligos into our parental sgRNA vector yields the final plasmid that is ready for in vitro transcription. We also describe a simple method of detecting somatic indel mutations using T7 endonuclease I (T7E1), a mismatch-sensitive enzyme (Fig. 1). Alternatively, somatic indel mutations may be detected by subcloning and sequencing of the target loci, or by restriction digest analysis, both of which have been previously described [7]. In addition, ssODNs containing a predetermined mutation may be co-injected with sgRNA and Cas9-encoding mRNA (*see* Subheading 3.4) [21]. Following the successful demonstration of somatic mutations induced by the CRISPR-Cas9 system, injected embryos may be raised to adulthood and screened for germ line mutations at the target loci. In summary, the CRISPR-Cas9 system is simple to use and enables researchers to rapidly initiate experiments designed to create nearly any desired alteration in zebrafish.

## 2 Materials

### 2.1 Construction of Customized Single-Guide RNA (sgRNA) Vectors

1. The sgRNA vector pDR274 may be obtained through Addgene, a nonprofit plasmid repository (http://www.addgene.org/).
2. Two DNA oligos containing the sequences generated from the ZiFiT Targeter webserver (http://zifit.partners.org).

3. 10× NEBuffer 2 (B7002, New England Biolabs, Ipswich, MA): 50 mM NaCl, 10 mM Tris–HCl, 10 mM $MgCl_2$, 1 mM DTT, pH 7.9.

4. A dry bath with a removable heat block.

5. BsaI (R0535, New England Biolabs) supplied with 10× NEBuffer 4. There is also a high fidelity version BsaI-HF™ (R3535, New England Biolabs) available.

6. An agarose gel electrophoresis apparatus.

7. QIAquick PCR Purification kit (Qiagen, Valencia, CA) supplied with Buffer EB.

8. A spectrometer for measuring DNA concentration.

9. T4 DNA Ligase and 2× Rapid Ligation Buffer in the LigaFast™ System (Promega, Madison, WI).

10. Chemical or electric competent bacterial cells.

11. 6-well LB/kanamycin agar plates: Add 7.5 g agar into 0.5 L of LB. Autoclave. Add 0.5 mL of 30 mg/mL kanamycin stock solution after it cools down but before it solidifies (~60 °C). Mix well and pour 3 mL into each well of 6-well plates. Store at 4 °C for up to 2 months.

12. Sterile bacterial culture tube.

13. LB/kanamycin medium: Add 0.5 mL of 30 mg/mL kanamycin stock solution to 500 mL of autoclaved LB. Store at 4 °C for up to 2 months.

14. QIAprep Spin Miniprep kit (Qiagen).

15. M13F Primer, 5′-ACTGGCCGTCGTTTTAC-3′.

16. QIAGEN Plasmid Mini kit (Qiagen).

## 2.2 Production of sgRNA

1. DraI (R0129, New England Biolabs) supplied with 10× NEBuffer 4.

2. An agarose gel electrophoresis apparatus.

3. QIAquick PCR Purification kit (Qiagen) supplied with Buffer EB.

4. A spectrometer or NanoDrop for measuring DNA and RNA concentration.

5. MAXIscript® T7 kit (Life Technologies, Grand Island, NY).

6. 5 M ammonium acetate, nuclease-free.

7. Ambion® nuclease-free water (Life Technologies), filtered, not DEPC-treated.

8. 100 % ethanol.

9. Nuclease-free 70 % ethanol, prepared using 100 % ethanol and nuclease-free water.

## 2.3 Production of the Cas9 mRNA

1. The Cas9 expression vector JDS246 may be obtained through Addgene (http://www.addgene.org/).
2. QIAGEN Plasmid Purification kit (Qiagen).
3. PmeI (R0560, New England Biolabs) supplied with 10× NEBuffer 4 and 100× BSA.
4. An agarose gel electrophoresis apparatus.
5. QIAquick PCR Purification kit (Qiagen) supplied with Buffer EB.
6. A NanoDrop or spectrometer for measuring DNA and RNA concentration.
7. The mMESSAGE mMACHINE T7 Ultra kit (Life Technologies).
8. 100 % ethanol and 70 % ethanol prepared with nuclease-free water.

## 2.4 Injection of sgRNA and Cas9 mRNA into Zebrafish Embryos

1. Zebrafish: TU (Tübingen) or other wild-type strain may be obtained from the Zebrafish International Resource Center (http://zebrafish.org).
2. Breeding tanks with dividers.
3. A microinjection apparatus and an agarose plate for holding embryos during microinjection.
4. 0.5 % Phenol Red solution (P0290, Sigma, St. Louis, MO).
5. 1× Danieau solution: 58 mM NaCl, 0.7 mM KCl, 0.4 mM $MgSO_4$, 0.6 mM $Ca(NO_3)_2$, 5 mM HEPES, pH 7.6; filter-sterilized. Store at 25 °C for 1 year.
6. Halocarbon oil (Sigma, St. Louis, MO).
7. A micrometer calibration slide.
8. E3 medium: 5 mM NaCl, 0.17 mM KCl, 0.33 mM $CaCl_2$, 0.33 mM $MgSO_4$. Store at 25 °C for 1 month.

## 2.5 Determining the Somatic Mutation Rate in Injected Zebrafish Embryos: T7 Endonuclease I (T7E1) Assay

1. PCR Lysis Buffer: Add 0.5 mL of 1 M Tris at pH 8, 0.2 mL of 0.5 M EDTA, 0.1 mL of Triton X-100, and water to 50 mL. Store at room temperature.
2. Proteinase K, Recombinant, PCR Grade (Roche Applied Science, Indianapolis, IN).
3. Phusion Hot Start II High-Fidelity DNA Polymerase (Thermo Scientific, Pittsburgh, PA).
4. 10 mM dNTPs (Life Technologies).
5. Gene-specific primers: Design primers for each sgRNA target site based on its flanking genomic DNA sequence. Each primer should anneal about 150–250 bp away from the expected double-stranded break location and extend toward it. Ideally, primers should have a melting temperature of approximately

60–65 °C and a length of 23–25 bp. Use free online software such as Primer3 (http://frodo.wi.mit.edu/primer3/) and OligoPerfect™ Designer (http://tools.invitrogen.com/content.cfm?pageid=9716) for designing the primers.

6. An agarose gel electrophoresis apparatus.

7. A thermocycler.

8. Agencourt AMPure XP (Beckman Coulter, Brea, CA) or QIAquick PCR Purification Kit (Qiagen).

9. A NanoDrop machine or a spectrometer for measuring DNA concentration.

10. T7 Endonuclease I (T7E1) (M0302, New England Biolabs).

11. A microfuge that can spin PCR strips or plates.

12. 0.25 M ethylenediaminetetraacetic acid (EDTA).

13. The QIAxcel system (Qiagen). Alternatively, a polyacrylamide gel electrophoresis apparatus, 15 % Tris–borate–EDTA (TBE) acrylamide gels (Bio-Rad, Hercules, CA), 100 bp DNA ladder (New England Biolabs), ethidium bromide, and a gel documentation system.

## 3 Methods

### 3.1 Construction of Customized Single-Guide RNA (sgRNA) Vectors

1. Find 23-basepair (bp) target sites in the form of 5′-GG-(N)18-NGG-3′ in your intended target region. You can use the web-based ZiFiT Targeter program (http://zifit.partners.org) to find sgRNA target sites and to generate the sequences of the DNA oligos required for constructing customized sgRNA vectors (Fig. 2) (*see* **Note 1**). If such a site does not exist in your target gene, it may be fine to allow up to two mismatches at the 5′ end.

2. Obtain complementary DNA oligos containing the sequences generated above from a DNA synthesis company such as Integrated DNA Technologies (Coralville, IA) or Life Technologies (Grand Island, NY). Anneal these two DNA oligos to make a double-stranded DNA fragment using the following protocol. Assemble 45 μL of 100 μM Oligo #1, 45 μL of 100 μM Oligo #2, and 10 μL of 10× NEBuffer 2 in an eppendorf tube. Place the tube in a dry bath with a removable heat block at 90–95 °C for 5 min. Remove the heat block with the eppendorf tube in it from the heating unit and allow it to cool gradually to room temperature (or below 37 °C). Continue to the next step or store the annealed oligonucleotides at −20 °C.

3. Linearize the sgRNA vector pDR274 with BsaI by adding 5 μg of the pDR274 DNA, 10 μL of 10× NEBuffer 4, 1 μL of BsaI

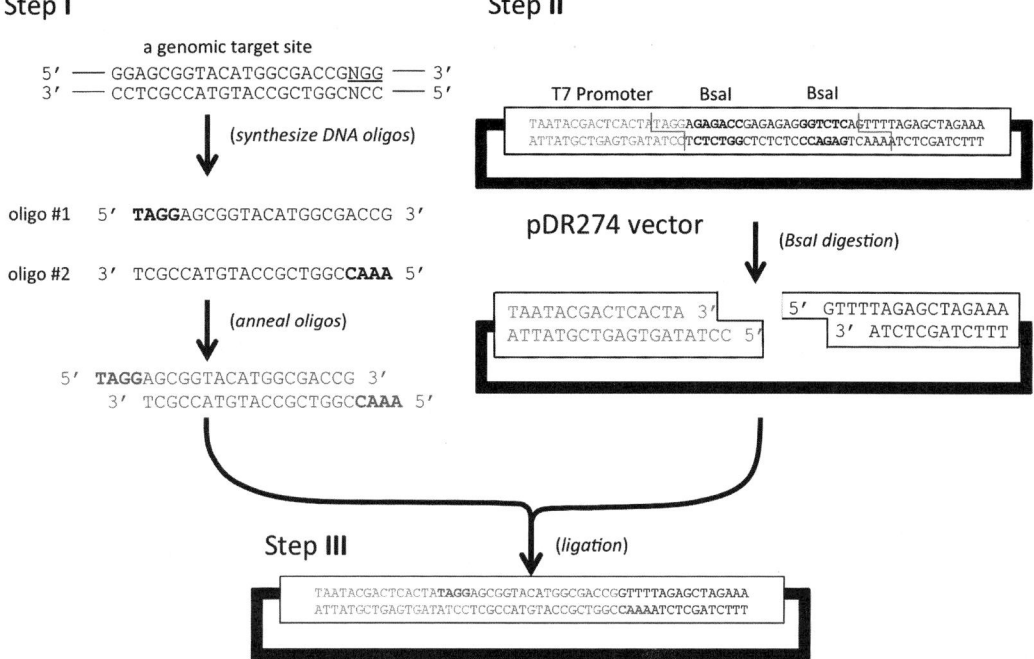

**Fig. 2** Cloning customized target sequences into the sgRNA vector pDR274. Step I shows the DNA oligo sequences for an example target site. Notice that the 3′-NGG (called protospacer adjacent motif or PAM) is required at the target site but is not included in the oligo sequence. Step II shows the cohesive ends generated by BsaI digestion of pDR274. Step III shows the final sequences of the ligated vector

(10 U/μL) and sterile deionized water to a total volume of 100 μL in an eppendorf tube. BsaI should be added last. Incubate the reaction at 37 °C for 2–3 h.

4. Run 50 ng of the uncut vector DNA and 5 μL of the reaction mixture on a 1 % wt/vol agarose gel to verify that the digestion is complete. Purify the digested DNA using the QIAquick PCR Purification kit and elute the DNA with 50 μL of Buffer EB. Measure DNA concentration using a spectrometer and normalize to a final concentration of 5–10 ng/μL with Buffer EB.

5. Ligate the annealed oligos and the linearized vector by mixing 1 μL of the annealed oligos (from **Step 2**), 1 μL of the linearized vector DNA (from **Step 4**), 2.5 μL of 2× Rapid Ligation Buffer, and 0.5 μL of T4 DNA Ligase in an eppendorf tube. Incubate the reaction mixture at either room temperature for 1 h or 4 °C overnight. The ligated product can be used for transformation or stored at −20 °C.

6. Transform competent bacterial cells with 5 μL of the ligation product following the specific instructions for the cells that you are using. At the end of the transformation procedures,

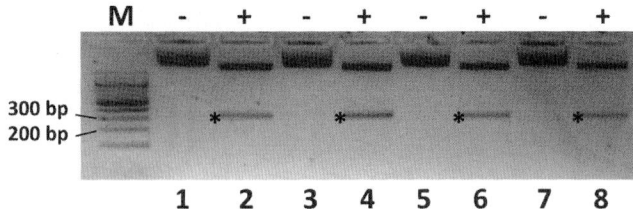

**Fig. 3** A representative gel of DraI-linearized sgRNA vectors. Lane 1–2, 3–4, and 5–6 contain 3 sets of sgRNA vectors. *Minus*, undigested DNA; *plus*, DraI-digested DNA; *asterisk*, the fragments containing both the T7 promoter and the sgRNA sequences; M, 100 bp DNA ladder (New England Biolabs)

spread the cells on a LB/kanamycin plate and incubate the plate at 37 °C overnight.

7. Pick three colonies per transformation and inoculate each colony in a culture tube containing 1.5 mL of LB/kanamycin. Incubate the culture tubes at 37 °C overnight with agitation.

8. On the next day, transfer the bacterial culture to an eppendorf tube for plasmid DNA extraction using the QIAprep Spin Miniprep kit or any plasmid DNA miniprep kit. Verify the sequence of the extracted DNA using M13F primer (*see* **Note 2**). Save the culture tube at 4 °C for re-inoculation.

9. Re-inoculate the bacteria containing sequence-confirmed plasmid DNA by adding 3 mL of LB/Kanamycin to the previous culture tube (**Step 8**). Incubate the culture tube at 37 °C overnight with agitation. Extract plasmid DNA from the culture using the QIAGEN Plasmid Mini kit (*see* **Note 3**). Measure DNA concentration using a spectrometer. This is the customized sgRNA vector.

### 3.2 Production of sgRNA for Microinjection

1. Linearize your sgRNA vector with DraI by mixing 5 μg of the customized sgRNA vector DNA, 10 μL of 10× NEBuffer 4, 1 μL of DraI (10 U/μL) and sterile deionized water to a total volume of 100 μL in an eppendorf tube. DraI should be added last. Incubate the reaction at 37 °C for 2 h.

2. Run 2 μL of the reaction mixture on a 3 % wt/vol agarose gel to verify that the digestion is complete. The digested DNA should contain two fragments at 1.9 kb and 282 bp, respectively (Fig. 3) (*see* **Note 4**).

3. Purify the digested DNA using the QIAquick PCR Purification kit and elute the DNA with 50 μL of Buffer EB (*see* **Note 5**).

4. Measure DNA concentration using a spectrometer. Continue to the next step or store the purified DNA at −20 °C.

5. Produce sgRNA by in vitro transcription using the MAXIscript® T7 kit as follows (*see* **Note 6**). Assemble 1 μg of the linearized

sgRNA vector DNA, 2 μL of 10× transcription buffer, 4 μL of mixed NTPs (1 μL each of the ATP, GTP, CTP, and UTP solutions at 10 mM), 2 μL of T7 Enzyme mix, and nuclease-free water to a total volume of 20 μL in a nuclease-free eppendorf tube (*see* **Note 7**). Incubate the reaction at 37 °C for 2 h to overnight. At the end of incubation, add 2 μL of TURBO DNase and incubate at 37 °C for another 30 min. Add 1 μL of 0.5 M EDTA to stop the reaction.

6. Add 30 μL of nuclease-free water and 5 μL of 5 M ammonium acetate to the reaction mix and mix well (*see* **Note 8**). Subsequently, add 150 μL of 100 % ethanol and mix well. Incubate at −20 °C overnight or on dry ice for 0.5–1 h.

7. Spin the eppendorf tube at >10,000 rpm in a microcentrifuge at 4 °C for 30 min. Discard the supernatant and add 1 mL of nuclease-free 70 % ethanol. Spin the tube again using the same setting for 15 min. Remove the 70 % ethanol as much as you can. Put the tube in a chemical hood and leave the lid open for a few min until the pellet dries completely.

8. Add 11 μL of nuclease-free water to dissolve the pellet (*see* **Note 9**). Take 1 μL of the dissolved RNA solution for measuring RNA concentration using a spectrometer (*see* **Note 10**). Aliquot sgRNA into multiple nuclease-free eppendorf tubes (1 μL/tube) and store at −80 °C. Take 1 μL of the sgRNA, mix it with 5 μL of the Formaldehyde Loading Dye and run these samples on a 3 % wt/vol agarose gel to make sure that the sgRNA is not degraded.

### 3.3 Production of Cas9 mRNA for Microinjection

1. Extract Cas9 vector DNA (JDS246) using a Qiagen Plasmid Purification kit. Linearize Cas9 vector DNA with PmeI by mixing 5 μg of the vector DNA, 10 μL of 10× NEBuffer 4, 1 μL of 100× BSA, 1 μL of PmeI (10 U/μL), and sterile deionized water to a total volume of 100 μL. PmeI should be added last. Incubate the reaction at 37 °C for 2–3 h.

2. Run 50 ng of the uncut vector DNA and 2 μL of the cut DNA on a 1 % wt/vol agarose gel to verify that the vector DNA is completely digested.

3. Purify the digested DNA using the QIAquick PCR Purification kit and elute the DNA with 50 μL of Buffer EB. Measure DNA concentration using a spectrometer. Continue to the next step or store the purified DNA at −20 °C.

4. Synthesize Cas9 mRNA using the mMESSAGE mMACHINE® T7 Ultra kit following the steps below (*see* **Note 11**). Thaw 2× NTP/ARCA and 10× T7 reaction buffer at room temperature. Place 2× NTP/ARCA, but not 10× T7 reaction buffer, on ice immediately once thawed. Keep T7 Enzyme Mix on ice at all times. Vortex the 10× T7 reaction buffer to redissolve any

precipitate in the tube before using. Mix reagents in a nuclease-free eppendorf tube in the following order: 5 μL of T7 2× NTP/ARCA, 1 μL of 10× T7 reaction buffer, 1 μL of T7 enzyme mix, and 3 μL (approximately 1 μg) of linearized Cas9 vector (from **Step 3**). Gently flick the tube and microfuge briefly to collect all of the reaction mixture at the bottom of the tube. Incubate the tube at 37 °C for 2 h. At the end of the incubation, add 1 μL of TURBO DNase. Gently flick the tube to mix and microfuge briefly. Incubate the tube at 37 °C for additional 30 min.

5. During the last incubation period, prepare the poly(A) tailing reaction master mix by combining 21.5 μL of nuclease-free water, 10 μL of 5× E-PAP buffer, 2.5 μL of 25 mM $MnCl_2$, and 5 μL of ATP solution in a nuclease-free eppendorf tube. Add the master mix to the in vitro transcription reaction mixture (**Step 4**). Gently flick and microfuge the tube. Aliquot 2 μL of the reaction mixture to a clean tube and label as "–polyA". Store this tube at –20 °C for a gel analysis later. Add 2 μL of E-PAP enzyme to the reaction mixture and incubate at 37 °C for 1 h. At the end of the reaction, aliquot 2 μL of the reaction mixture to another clean tube and label as "+polyA". Store this tube at –20 °C for a later gel analysis.

6. Add 25 μL of Lithium Chloride Precipitation Solution to the rest of the reaction mixture (*see* **Note 12**). Mix well and incubate the tube at –20 °C overnight or 0.5–1 h on dry ice. We recommend overnight incubation for a better yield.

7. During this incubation, conduct a gel analysis of the "–polyA" and "+polyA" labeled samples stored at –20 °C. Add 5 μL of the Formaldehyde Loading Dye into the 2 μL sample aliquots taken before and after the poly(A) tailing reaction (**Step 5**). Run the samples on a 1 % agarose gel (Fig. 4).

8. After the incubation (from **Step 6**), spin the tube at >10,000 rpm in a microcentrifuge at 4 °C for 30 min. Once the spin is done, check for the pellet at the bottom of the tube and remove the supernatant without disturbing the pellet. Add 1 mL of RNase-free 70 % ethanol and invert the tube several times. Centrifuge the tube at the same setting for 15 min. Check for the pellet once again and remove the solution completely without disturbing the pellet (*see* **Note 13**). Leave the tube with the lid open for 10–15 min until all of the solution evaporated.

9. Dissolve the RNA pellet with 20 μL of nuclease-free water. Once it is dissolved, keep the tube on ice at all time. Take 1 μL of the dissolved RNA solution for measuring RNA concentration using a spectrometer. Aliquot the Cas9 mRNA into multiple nuclease-free eppendorf tubes (1,500 ng/tube) and store at –80 °C (*see* **Note 14**).

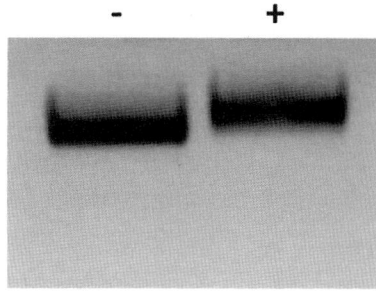

**Fig. 4** A representative gel of in vitro transcribed Cas9 RNA. *Minus*, before the poly(A) tailing reaction; *plus*, after the poly(A) tailing reaction. Notice that the RNA bands are distinct and without smears. Meanwhile, the poly(A) tailing reaction causes an upshift of the RNA band due to an increase in its length

### 3.4 Injection of sgRNA and Cas9 mRNA into Zebrafish Embryos

1. Set up zebrafish for spawning the evening before injection. Place a divider in each breeding tank to separate males and females.

2. On the day of injection, prepare the injection solution by mixing a total of 62.5 ng of sgRNA, 1,500 ng of Cas9 mRNA, 0.5 µL of 0.5 % Phenol Red solution, and 1×Danieau solution to a total volume of 5 µL (*see* **Note 15**). Keep the injection solution on ice to prevent RNA degradation (*see* **Note 16**). In addition, for a predesigned sequence modification, add single-stranded DNA oligos to a final concentration of 1.66 µM (or 25 ng/µL) into the injection solution as the template for DNA repair (*see* **Note 17**). Do not change the final concentrations of sgRNA and Cas9 mRNA.

3. Prepare an agarose plate to hold embryos during microinjection. Calibrate the needle for injection as follows. Add a drop of halocarbon oil on the micrometer calibration slide. Magnify the stereoscope to see the scale. Adjust the injection pressure and/or injection time so that the diameter of each injection droplet is around 0.16 mm, which brings the volume of the droplet to be 2 nL. Once you finish the calibration, pull the divider out from a breeding tank and wait for the fish to spawn.

4. The RNA solution should be injected into the cytoplasm of zebrafish embryos at the one-cell stage. Once you collect the embryos from the breeding tank in E3 medium, orient the embryos in the furrows of the agarose plate so that the cytoplasm is facing toward the needle. Remove most of the E3 medium and leave only enough to cover the embryos (*see* **Note 18**). Carefully inject 2 nL of the RNA solution into the base of the cytoplasm where the nucleus is located. Save some non-injected embryos from each clutch as controls.

5. Pull the divider out from another breeding tank and repeat the process. Incubate both control and injected embryos at 28.5 °C overnight.

6. Examine the injected embryos at 24 h post-fertilization (hpf). Classify the injected embryos as normal, deformed, or dead by visual inspection under a stereoscope. Repeat with the control embryos. Record these numbers (*see* **Note 19**). Meanwhile, the injected embryos may exhibit specific phenotypes resulting from RGN-mediated knockdown of the target gene. In some cases it is necessary to reduce the concentration of sgRNA/Cas9 mRNA for injection if the goal is to obtain founder fish that carry transmittable germ line mutations. After injection and successful identification of somatic mutations (*see* Subheading 3.5), injected embryos that develop normally may be raised to adulthood and screened for founders.

### 3.5 Determining the Somatic Mutation Rate in the Injected Zebrafish Embryos: T7 Endonuclease I (T7E1) Assay

1. Manually dechorionate embryos with forceps. Array 10 injected embryos (for each combination of sgRNA/Cas9 mRNA) and 2 control embryos singly into each well of a 96-well PCR plate. Add 30 μL of PCR lysis buffer with 100 μg/mL of proteinase K. It is important to include 1–2 control embryos from the same clutch of embryos used for microinjection (*see* **Note 20**). Seal and incubate the plate at 50 °C overnight. The next day, gently agitate the plate to make sure that the embryos have been lysed completely. Incubate at 95 °C for 10 min to inactivate Proteinase K (*see* **Note 21**). Spin the plate at 3,000 rpm for 2 min at room temperature to bring down any debris. The supernatant contains genomic DNA and it is now ready to be used for PCR.

2. Set up 50 μL PCR reactions to amplify the genomic DNA encompassing the target site using gene-specific primers (*see* **Note 22**). Assemble 2 μL of the embryo lysate or 100 ng of the genomic DNA (from **Step 1**), 10 μL of 5× Phusion HF Buffer, 1 μL of 10 mM dNTPs, 2.5 μL of the forward primer (10 μM stock), 2.5 μL of the reverse primer (10 μM stock), 0.5 μL of Phusion Hot Start II High-Fidelity DNA polymerase, and sterile deionized water to a total volume of 50 μL. The enzyme should be added last. Use the following cycling condition or a cycling condition that has been optimized for your primer set: Step 1—98 °C, 2 min; Step 2—98 °C, 10 s; Step 3—65 °C, 15 s; Step 4—72 °C, 30 s; Step 5—go back to step 2 for 34 times; Step 6—72 °C, 5 min; Step 7—hold at 4 °C (*see* **Note 23**).

3. Run 2 μL of the PCR products on a 2–3 % agarose gel to verify correct amplification.

4. Purify the products using 90 μL of AMPure XP, elute in 30 μL of 0.1× Buffer EB and recover 25 μL. Alternatively, purify the

products using the QIAquick PCR Purification Kit and elute with 50 μL of Buffer EB. Measure DNA concentration using NanoDrop or a spectrometer. Continue to the next step or store the purified PCR products at −20 °C.

5. Set up the hybridization by mixing 200 ng of the purified PCR product, 2 μL of 10× NEBuffer 2 and sterile deionized water to a total volume of 19 μL in a new PCR plate (see **Note 24**). Put the plate in a thermocycler and run the following cycling condition: Step 1—95 °C, 5 min; Step 2—from 95 to 85 °C, −2 °C/s; Step 3—from 85 to 25 °C, −0.1 °C/s; Step 4—hold at 4 °C (see **Note 25**).

6. At the end of the hybridization cycle, microfuge briefly, add 1 μL of T7E1 to each well. Gently pipette up and down for ~10 times to mix thoroughly. Incubate the plate at 37 °C for exactly 15 min. Stop the reaction by adding 2 μl of 0.25 M EDTA.

7. Analyze the sizes of the DNA fragments using a QIAxcel system or on a 10 % polyacrylamide gel. If the former, the reaction products need to be purified using either AMPure XP or a QIAquick PCR Purification Kit before loading. For the former option, purify the reaction product using 36 μL of AMPure XP, elute in 20 μL of 0.1× Buffer EB and recover in 15 μL.

8. For analysis using QIAxcel, dilute 5 μL of purified T7E1 reaction product with 5 μL of QX Dilution Buffer. Run on QIAxcel using program OM500 and an injection time of 20 s.

    For analysis using a polyacrylamide gel, add 2 μL of 10× gel loading dye to the reaction product and run all of the samples along with the 100 bp DNA ladder on a 15 % TBE acrylamide gel. Stain the gel in water containing 0.5 μg/mL ethidium bromide at room temperature for 30 min. Inspect the gel using a UV light box. Compare the injected samples to the control samples and identify any additional restricted fragments of the expected sizes in the injected samples. The presence of such bands provides evidence of sgRNA/Cas9-induced mutations.

9. To calculate the somatic mutation rate, the sum of the area beneath the injection-specific cleavage peaks (expressed as a percentage of the parent amplicon peak, denoted fraction cleaved) is used to estimate somatic mutation rates using the following equation as previously described: % gene modification = $100 \times (1-(1-\text{fraction cleaved})^{1/2})$ [45]. The fraction cleaved may also be quantified based on the intensities of the parent amplicon band and the T7E1 cleavage bands as determined by a gel documentation system such as the Bio-Rad Gel Doc™ XR+ System with the Quantity One® 1-D Analysis Software.

## 4 Notes

1. On the ZiFiT Targeter website, go to "Examples" → "CRISPR-Cas assembly using T7 promoter" for detailed instructions. In order to be cloned into the pDR274 vector, oligo#1 needs to have TAGG at the 5′ end, which is also a part of the T7 promoter. Oligo #2 needs to have AAAC at the 5′ end, enabling unidirectional cloning into the BsaI-digested pDR274 vector. All sgRNAs transcribed from the T7 promoter will have a GG dinucleotide at the 5′ end.

2. The sequencing results will show the T7 promoter and sgRNA sequences in reverse complement. The full DNA sequence of the pDR274 vector has been previously published [25]. Make sure that the sequenced DNA has a complete T7 promoter, the customized target sequence and the remaining of the guide RNA sequence. Based on our experience, 2 out of 3 colonies usually have correct sequences.

3. Plasmid DNA prep using gravity-flow columns often provides better quality DNA compared to spin columns. Based on our experience, better quality DNA leads to higher yields of sgRNA after in vitro transcription.

4. The smaller DNA fragment, which will be the template for in vitro transcription, contains both the T7 promoter and sgRNA sequences. There is no need to purify this fragment using gel purification.

5. Use a new bottle of Buffer EB to be sure that it is not contaminated with RNase.

6. Alternatively, sgRNA may be produced by using a MEGAshortscript™ T7 kit (Life Technologies, Grand Island, NY).

7. Mixed NTPs are made by mixing an equal volume of 10 mM ATP, CTP, GTP, and UTP from the kit.

8. Alternatively, add 30 µL of nuclease-free water and 25 µL of the lithium chloride precipitation solution (provided in the mMESSAGE mMACHINE® T7 Ultra kit) to precipitate sgRNA. You do not need to add 100 % ethanol subsequently. We recommend incubating the samples at −20 °C overnight for better yield.

9. We use Ambion® nuclease-free water that is not DEPC-treated. We found that injection of DEPC-treated water may induce embryo toxicity.

10. The yield of sgRNA is generally between 80 and 150 ng/µL. We always run 1 µL of sgRNA on a 3 % wt/vol agarose gel to check the integrity of the RNA. It should appear as a distinct band without smearing.

11. The synthesized Cas9 mRNA will contain a 5′ cap and a 3′ poly(A) tail in order to be translated in zebrafish embryos. The mMESSAGE mMACHINE® T7 Ultra Kit has all of the components that you would need. Alternatively, you can use a mMESSAGE mMACHINE® T7 Kit (Life Technologies, USA) and a Poly(A) Tailing Kit (Life Technologies, USA) for these reactions.

12. The volume of the Lithium Chloride Precipitation Solution is half of the volume of the reaction mixture.

13. It is critical to remove the ethanol completely so that the pellet will dry quickly. After removing the solution, one can microfuge the tube briefly and remove the remaining solution once more.

14. The yield of the RNA is generally between 1,000 and 2,000 ng/μL. We generally store RNA in multiple aliquots to prevent freeze-thaw cycles.

15. Multiple sgRNAs that target either the same gene or different genes can be co-injected for simultaneous generation of multiple gene mutations. In a multiplex experiment, the total amount of sgRNAs in the injection solution can be increased up to 100 ng/μL.

16. The injection solution should be freshly prepared. Do not freeze and reuse leftover injection solution.

17. The DNA oligo should contain the desired sequence modification and around 18–25 nt of flanking sequences that are homologous to the genomic target locus. It may be ideal to have the targeted sequence modification near the DSB, which is located at 3 base pairs upstream of the PAM. Using this method, we have successfully identified in injected embryos precise insertions of 3-4 nt, as well as, a precise single nucleotide substitution [21]. Although the efficiency of isolating a precisely modified allele may decrease when the size of the insertion increases, we have successfully inserted up to 40 nt using this method (Gonzales, unpublished data).

18. This is a very important step. Otherwise, embryos will roll around and it will be difficult to inject into the cell.

19. The toxicity of each sgRNA/Cas9 complex varies. Thus, the concentration of the Cas9 mRNA may be reduced to 100 ng/μL if the injection results in less than 50 % healthy embryos.

20. It is important to include non-injected embryos from the same clutches that were used for injections as controls. This will help identify if there is a polymorphism within the PCR amplicon that can result in a cleavage product in this assay.

21. You may use genomic DNA extracted from pooled embryos if you know there is no polymorphism in your target locus. In that case, add 30 μL of PCR lysis buffer with Proteinase K per embryo (e.g. 150 μL for 5 pooled embryos).

22. The primers should anneal to sequences around 150–200 bp upstream and downstream of the target site.
23. This is one PCR condition that generally works for us. It is very important to first optimize the PCR condition and obtain a very specific amplification product. The optimal conditions may differ for different gene targets.
24. It is important to start with the same amount (200 ng) of PCR products for all of the samples.
25. The hybridized DNA will be stable at 4 °C for a couple of hours.

## Acknowledgements

The authors would like to thank Drs. Randall Peterson, Morgan Maeder, and Jeffry Sander for their contributions on various aspects of this study. W.Y.H., A.P.W.G., and J.-R.J.Y. are supported by the National Institutes of Health (R01 GM088040, K01 AG031300, and R01 CA140188) and by the MGH Claflin Distinguished Scholar Award. Y.F., D.R., and J.K.J. are supported by the National Institutes of Health (DP1 GM105378 and R01 GM088040) and by The Jim and Ann Orr MGH Research Scholar Award.

*Conflict of Interest Declaration*: J.K.J. is a consultant for Horizon Discovery. J.K.J. has financial interests in Editas Medicine, Hera Testing Laboratories, Poseida Therapeutics, and Transposagen Biopharmaceuticals. J.K.J.'s interests were reviewed and are managed by Massachusetts General Hospital and Partners HealthCare in accordance with their conflict-of-interest policies.

## References

1. Gaj T, Gersbach CA, Barbas CF 3rd (2013) ZFN, TALEN, and CRISPR/Cas-based methods for genome engineering. Trends Biotechnol 31:397–405
2. Joung JK, Sander JD (2013) TALENs: a widely applicable technology for targeted genome editing. Nat Rev Mol Cell Biol 14:49–55
3. Mali P, Esvelt KM, Church GM (2013) Cas9 as a versatile tool for engineering biology. Nat Methods 10:957–963
4. Urnov FD, Rebar EJ, Holmes MC, Zhang HS, Gregory PD (2010) Genome editing with engineered zinc finger nucleases. Nat Rev Genet 11:636–646
5. Beumer KJ, Trautman JK, Bozas A, Liu JL, Rutter J, Gall JG, Carroll D (2008) Efficient gene targeting in Drosophila by direct embryo injection with zinc-finger nucleases. Proc Natl Acad Sci U S A 105:19821–19826
6. Doyon Y, McCammon JM, Miller JC, Faraji F, Ngo C, Katibah GE, Amora R, Hocking TD, Zhang L, Rebar EJ, Gregory PD, Urnov FD, Amacher SL (2008) Heritable targeted gene disruption in zebrafish using designed zinc-finger nucleases. Nat Biotechnol 26:702–708
7. Foley JE, Yeh JR, Maeder ML, Reyon D, Sander JD, Peterson RT, Joung JK (2009) Rapid mutation of endogenous zebrafish genes using zinc finger nucleases made by Oligomerized Pool ENgineering (OPEN). PLoS One 4:e4348
8. Maeder ML, Thibodeau-Beganny S, Osiak A, Wright DA, Anthony RM, Eichtinger M, Jiang T, Foley JE, Winfrey RJ, Townsend JA, Unger-Wallace E, Sander JD, Muller-Lerch F, Fu F, Pearlberg J, Gobel C, Dassie JP, Pruett-Miller SM, Porteus MH, Sgroi DC, Iafrate AJ, Dobbs D, McCray PB Jr, Cathomen T, Voytas DF,

Joung JK (2008) Rapid "open-source" engineering of customized zinc-finger nucleases for highly efficient gene modification. Mol Cell 31:294–301

9. Meng X, Noyes MB, Zhu LJ, Lawson ND, Wolfe SA (2008) Targeted gene inactivation in zebrafish using engineered zinc-finger nucleases. Nat Biotechnol 26:695–701

10. Perez EE, Wang J, Miller JC, Jouvenot Y, Kim KA, Liu O, Wang N, Lee G, Bartsevich VV, Lee YL, Guschin DY, Rupniewski I, Waite AJ, Carpenito C, Carroll RG, Orange JS, Urnov FD, Rebar EJ, Ando D, Gregory PD, Riley JL, Holmes MC, June CH (2008) Establishment of HIV-1 resistance in CD4+ T cells by genome editing using zinc-finger nucleases. Nat Biotechnol 26:808–816

11. Santiago Y, Chan E, Liu PQ, Orlando S, Zhang L, Urnov FD, Holmes MC, Guschin D, Waite A, Miller JC, Rebar EJ, Gregory PD, Klug A, Collingwood TN (2008) Targeted gene knockout in mammalian cells by using engineered zinc-finger nucleases. Proc Natl Acad Sci U S A 105:5809–5814

12. Moehle EA, Rock JM, Lee YL, Jouvenot Y, DeKelver RC, Gregory PD, Urnov FD, Holmes MC (2007) Targeted gene addition into a specified location in the human genome using designed zinc finger nucleases. Proc Natl Acad Sci U S A 104:3055–3060

13. Lombardo A, Genovese P, Beausejour CM, Colleoni S, Lee YL, Kim KA, Ando D, Urnov FD, Galli C, Gregory PD, Holmes MC, Naldini L (2007) Gene editing in human stem cells using zinc finger nucleases and integrase-defective lentiviral vector delivery. Nat Biotechnol 25:1298–1306

14. Hockemeyer D, Soldner F, Beard C, Gao Q, Mitalipova M, DeKelver RC, Katibah GE, Amora R, Boydston EA, Zeitler B, Meng X, Miller JC, Zhang L, Rebar EJ, Gregory PD, Urnov FD, Jaenisch R (2009) Efficient targeting of expressed and silent genes in human ESCs and iPSCs using zinc-finger nucleases. Nat Biotechnol 27:851–857

15. Hockemeyer D, Wang H, Kiani S, Lai CS, Gao Q, Cassady JP, Cost GJ, Zhang L, Santiago Y, Miller JC, Zeitler B, Cherone JM, Meng X, Hinkley SJ, Rebar EJ, Gregory PD, Urnov FD, Jaenisch R (2011) Genetic engineering of human pluripotent cells using TALE nucleases. Nat Biotechnol 29:731–734

16. Mali P, Yang L, Esvelt KM, Aach J, Guell M, DiCarlo JE, Norville JE, Church GM (2013) RNA-guided human genome engineering via Cas9. Science 339:823–826

17. Zou J, Maeder ML, Mali P, Pruett-Miller SM, Thibodeau-Begganny S, Chou BK, Chen G, Ye Z, Park IH, Daley GQ, Porteus MH, Joung JK, Cheng L (2009) Gene targeting of a disease-related gene in human induced pluripotent stem and embryonic stem cells. Cell Stem Cell 5:97–110

18. Zu Y, Tong X, Wang Z, Liu D, Pan R, Li Z, Hu Y, Luo Z, Huang P, Wu Q, Zhu Z, Zhang B, Lin S (2013) TALEN-mediated precise genome modification by homologous recombination in zebrafish. Nat Methods 10:329–331

19. Bedell VM, Wang Y, Campbell JM, Poshusta TL, Starker CG, Krug RG 2nd, Tan W, Penheiter SG, Ma AC, Leung AY, Fahrenkrug SC, Carlson DF, Voytas DF, Clark KJ, Essner JJ, Ekker SC (2012) In vivo genome editing using a high-efficiency TALEN system. Nature 491:114–118

20. Dicarlo JE, Norville JE, Mali P, Rios X, Aach J, Church GM (2013) Genome engineering in *Saccharomyces cerevisiae* using CRISPR-Cas systems. Nucleic Acids Res 41:4336–4343

21. Hwang WY, Fu Y, Reyon D, Maeder ML, Kaini P, Sander JD, Joung JK, Peterson RT, Yeh JR (2013) Heritable and precise zebrafish genome editing using a CRISPR-Cas system. PLoS One 8:e68708

22. Wang H, Yang H, Shivalila CS, Dawlaty MM, Cheng AW, Zhang F, Jaenisch R (2013) One-step generation of mice carrying mutations in multiple genes by CRISPR/Cas-mediated genome engineering. Cell 153:910–918

23. Chen F, Pruett-Miller SM, Huang Y, Gjoka M, Duda K, Taunton J, Collingwood TN, Frodin M, Davis GD (2011) High-frequency genome editing using ssDNA oligonucleotides with zinc-finger nucleases. Nat Methods 8:753–755

24. Huang P, Xiao A, Zhou M, Zhu Z, Lin S, Zhang B (2011) Heritable gene targeting in zebrafish using customized TALENs. Nat Biotechnol 29:699–700

25. Hwang WY, Fu Y, Reyon D, Maeder ML, Tsai SQ, Sander JD, Peterson RT, Yeh JR, Joung JK (2013) Efficient genome editing in zebrafish using a CRISPR-Cas system. Nat Biotechnol 31:227–229

26. Sander JD, Cade L, Khayter C, Reyon D, Peterson RT, Joung JK, Yeh JR (2011) Targeted gene disruption in somatic zebrafish cells using engineered TALENs. Nat Biotechnol 29:697–698

27. Tesson L, Usal C, Menoret S, Leung E, Niles BJ, Remy S, Santiago Y, Vincent AI, Meng X, Zhang L, Gregory PD, Anegon I, Cost GJ (2011) Knockout rats generated by embryo

microinjection of TALENs. Nat Biotechnol 29:695–696
28. Wood AJ, Lo TW, Zeitler B, Pickle CS, Ralston EJ, Lee AH, Amora R, Miller JC, Leung E, Meng X, Zhang L, Rebar EJ, Gregory PD, Urnov FD, Meyer BJ (2011) Targeted genome editing across species using ZFNs and TALENs. Science 333:307
29. Li W, Teng F, Li T, Zhou Q (2013) Simultaneous generation and germline transmission of multiple gene mutations in rat using CRISPR-Cas systems. Nat Biotechnol 31:684–686
30. Li D, Qiu Z, Shao Y, Chen Y, Guan Y, Liu M, Li Y, Gao N, Wang L, Lu X, Zhao Y, Liu M (2013) Heritable gene targeting in the mouse and rat using a CRISPR-Cas system. Nat Biotechnol 31:681–683
31. Bassett AR, Tibbit C, Ponting CP, Liu JL (2013) Highly efficient targeted mutagenesis of Drosophila with the CRISPR/Cas9 system. Cell Rep 4:220–228
32. Friedland AE, Tzur YB, Esvelt KM, Colaiacovo MP, Church GM, Calarco JA (2013) Heritable genome editing in *C. elegans* via a CRISPR-Cas9 system. Nat Methods 10:741–743
33. Dickinson DJ, Ward JD, Reiner DJ, Goldstein B (2013) Engineering the *Caenorhabditis elegans* genome using Cas9-triggered homologous recombination. Nat Methods 10:1028–1034
34. Cho SW, Kim S, Kim JM, Kim JS (2013) Targeted genome engineering in human cells with the Cas9 RNA-guided endonuclease. Nat Biotechnol 31:230–232
35. Cong L, Ran FA, Cox D, Lin S, Barretto R, Habib N, Hsu PD, Wu X, Jiang W, Marraffini LA, Zhang F (2013) Multiplex genome engineering using CRISPR/Cas systems. Science 339:819–823
36. Jiang W, Bikard D, Cox D, Zhang F, Marraffini LA (2013) RNA-guided editing of bacterial genomes using CRISPR-Cas systems. Nat Biotechnol 31:233–239
37. Jinek M, East A, Cheng A, Lin S, Ma E, Doudna J (2013) RNA-programmed genome editing in human cells. Elife 2:e00471
38. Jinek M, Chylinski K, Fonfara I, Hauer M, Doudna JA, Charpentier E (2012) A programmable dual-RNA-guided DNA endonuclease in adaptive bacterial immunity. Science 337:816–821
39. Wiedenheft B, Sternberg SH, Doudna JA (2012) RNA-guided genetic silencing systems in bacteria and archaea. Nature 482:331–338
40. Hsu PD, Scott DA, Weinstein JA, Ran FA, Konermann S, Agarwala V, Li Y, Fine EJ, Wu X, Shalem O, Cradick TJ, Marraffini LA, Bao G, Zhang F (2013) DNA targeting specificity of RNA-guided Cas9 nucleases. Nat Biotechnol 31:827–832
41. Jao LE, Wente SR, Chen W (2013) Efficient multiplex biallelic zebrafish genome editing using a CRISPR nuclease system. Proc Natl Acad Sci U S A 110:13904–13909
42. Fu Y, Foden JA, Khayter C, Maeder ML, Reyon D, Joung JK, Sander JD (2013) High-frequency off-target mutagenesis induced by CRISPR-Cas nucleases in human cells. Nat Biotechnol 31:822–826
43. Mali P, Aach J, Stranges PB, Esvelt KM, Moosburner M, Kosuri S, Yang L, Church GM (2013) CAS9 transcriptional activators for target specificity screening and paired nickases for cooperative genome engineering. Nat Biotechnol 31:833–838
44. Pattanayak V, Lin S, Guilinger JP, Ma E, Doudna JA, Liu DR (2013) High-throughput profiling of off-target DNA cleavage reveals RNA-programmed Cas9 nuclease specificity. Nat Biotechnol 31:839–843
45. Guschin DY, Waite AJ, Katibah GE, Miller JC, Holmes MC, Rebar EJ (2010) A rapid and general assay for monitoring endogenous gene modification. Methods Mol Biol 649:247–256

# Chapter 22

## Precise Genome Editing of Drosophila with CRISPR RNA-Guided Cas9

Scott J. Gratz, Melissa M. Harrison, Jill Wildonger, and Kate M. O'Connor-Giles

### Abstract

The readily programmable CRISPR-Cas9 system is transforming genome engineering. We and others have adapted the *S. pyogenes* CRISPR-Cas9 system to precisely engineer the Drosophila genome and demonstrated that these modifications are efficiently transmitted through the germline. Here we provide a detailed protocol for engineering small indels, defined deletions, and targeted insertion of exogenous DNA sequences within one month using a rapid DNA injection-based approach.

**Key words** CRISPR, Cas9, Genome engineering, Nonhomologous end joining, Homology-directed repair, Drosophila

### Abbreviations

| | |
|---|---|
| CRISPR | Clustered regularly interspaced short palindromic repeats |
| crRNA | CRISPR RNA |
| DSB | Double-strand break |
| dsDNA | Double-stranded DNA |
| gRNA | Guide RNA |
| HDR | Homology-directed repair |
| Indel | Insertion-deletion |
| NHEJ | Nonhomologous end joining |
| PAM | Protospacer adjacent motif |
| ssDNA | Single-stranded DNA |
| tracrRNA | Trans-activating CRISPR RNA |

## 1 Introduction

The CRISPR-Cas9 system is transforming genome engineering with its simplicity. In Drosophila, the CRISPR-Cas9 system has been employed to interrupt, delete, and replace genes [1–7].

These modifications are efficiently transmitted through the germline for the establishment of stable transgenic lines. The rapid adoption of the CRISPR-Cas9 system illustrates its utility and adaptability, while the variety of complex modifications successfully generated to date offers only a glimpse into the genome manipulations now within reach.

The endogenous *S. pyogenes* CRISPR-Cas9 adaptive immune system has been simplified to two components for use in genome engineering: Cas9 and a single chimeric RNA referred to as a chiRNA or guide RNA (gRNA) [8]. The gRNA comprises CRISPR RNA (crRNA) and trans-activating CRISPR RNA (tracrRNA) sequences that direct sequence-specific cleavage of genomic DNA by Cas9 through a simple base-pairing mechanism. gRNAs are easily programmed to recognize a 20-nt target sequence and direct Cas9-dependent cleavage of both DNA strands at a defined site within the target. The only requirement for a CRISPR target site is the presence of a 3-bp protospacer adjacent motif (PAM) of the form NGG immediately 3′ of the 20-nt recognition sequence. Thus, potential CRISPR-Cas9 target sites are common, occurring on average once in every eight base pairs of genomic sequence.

Cas9-induced double-strand breaks (DSBs) trigger DNA repair via two cellular pathways that can be harnessed for genome editing. Nonhomologous end joining (NHEJ) is an error-prone process that can result in small insertions and deletions (indels) that disrupt function at cleavage sites. Homology-directed repair (HDR) employs homologous DNA sequences as templates for precise repair. By supplying an exogenous donor template, this repair pathway can be exploited to precisely edit genomic sequence or insert exogenous DNA.

In this chapter we describe the method for CRISPR-Cas9-mediated gene targeting in Drosophila developed in Gratz et al. [2] (Fig. 1). Specifically, we detail the use of a DNA injection-based approach to generate gene-disrupting indels and defined deletions via NHEJ and to replace genes with exogenous sequences by HDR employing readily synthesized single-stranded DNA (ssDNA) donors. These modifications can be efficiently generated and transmitted through the germline within a month, making the promise of rapid genome engineering a reality. Throughout the protocol, we highlight key considerations for initiating a CRISPR genome engineering project. We also note alternatives to and extensions of our protocol. Given the accessibility and adaptability of the system, we expect the range of modifications achieved with CRISPR-Cas9 to expand rapidly as this transformative technology is employed by researchers worldwide.

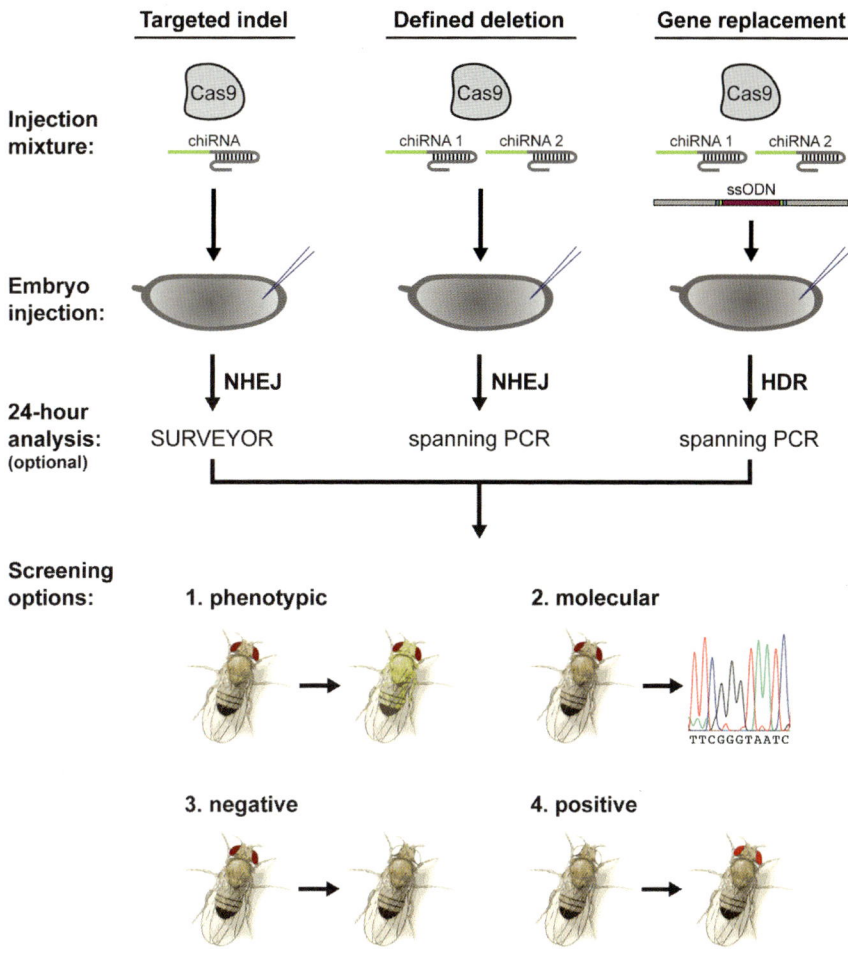

**Fig. 1** Experimental overview. Injection mixtures containing pBS-Hsp70-Cas9, the appropriate pU6-BbsI-chiRNA(s), and, for gene replacement, an ssDNA donor are injected into pre-blastoderm embryos. Twenty-four hours after injection, a subset of embryos can be assayed molecularly for the presence of targeted modifications to assess efficiency. Rear remaining embryos to adulthood and outcross to recover heritable modifications. Methods for identifying flies with the targeted modification include (1) phenotypic screening, if the phenotype of the targeted modifications is known and readily observable, (2) molecular screening by PCR-based analysis, (3) negative screening for the removal of a marked transposable elements in the targeted locus, and (4) positive screening using a dsDNA donor with a visible marker

## 2 Materials

### 2.1 Molecular Biology

1. Cloning and expression vectors: pBS-Hsp70-Cas9 for expression of codon-optimized Cas9 and pU6-BbsI-chiRNA for cloning and expression of specific gRNAs are available from Addgene (plasmid numbers 46294 and 45946).
2. T4 DNA Ligase.
3. T4 Polynucleotide Kinase.

4. *E. coli* DH5α or other suitable strain for general cloning.
5. EndoFree Plasmid Maxi kit (Qiagen).
6. *BbsI* endonuclease.
7. Wizard SV Gel and PCR Clean-up System kit (Promega).
8. Embryo homogenization buffer: 10 mM Tris–HCl (pH 8.2), 25 mM NaCl, 1 mM EDTA, 0.2 % Triton-X100. Immediately before use, add 200 μg/mL proteinase K.
9. Adult fly homogenization buffer: 10 mM Tris–HCl (pH 8.2), 25 mM NaCl. Prior to use, add 200 μg/mL proteinase K.
10. Agarose gel electrophoresis equipment.
11. PCR reagents, including primers designed to amplify a 500–700-bp region flanking your target site.
12. Optional: SURVEYOR Mutation Detection kit (Transgenomic).

## 2.2 Embryo Injections

1. Grape juice agar plates and yeast paste for collecting Drosophila embryos.
2. Population cages for embryo collection.
3. Inverted microscope equipped with a micromanipulator, micropipette holder, and a microinjector.
4. Glass capillary for injection needles.
5. Micropipette puller.
6. Microscope slides and cover slips.
7. Halocarbon oil 700 and 27 (Halocarbon Products Corporation).

## 2.3 Fly Stocks

1. Because CRISPR-Cas9 components are introduced through injection into embryos, any fly stock can be engineered. Lines containing a phenotypically marked element in the targeted locus will allow for negative screening for the genome modifications. In appropriate genetic backgrounds, donor templates carrying visible markers facilitate positive screening. It may also be desirable to carry out modifications in a particular fly strain to control for genetic background, for example, in behavioral, quantitative trait loci, and evolutionary studies. Finally, genome engineering can be carried out in a fly line that transgenically expresses Cas9 in the germline such that only the gRNA vectors and HDR donor vector must be supplied through injection [3–6].

# 3 Methods

## 3.1 Select CRISPR Target Sites

1. Once you have determined the genome modification you wish to make and designed a general strategy (*see* **Note 1**), the first step is to identify best site or sites for targeting Cas9-induced DSBs. For mutagenesis via indel formation, you will need to

identify a single CRISPR target site in a region where a small insertion or deletion has the potential to be disruptive—for example, by inducing a frameshift in the coding sequence of a targeted gene. To generate a defined deletion of, for example, an entire gene, you will need to identify two CRISPR target sites flanking the region to be excised. If your goal is to incorporate a tag, such as FLAG, into the endogenous coding sequences of a gene, you will want to identify a single target site as close as possible to the desired modification.

2. Sequence the target regions in the specific fly line you will be editing. Polymorphisms between fly lines and the reference genome are frequent, especially in intergenic regions, and could significantly decrease cleavage frequency if they occur within your gRNA sequence.

3. For each target region, identify a 20-nt CRISPR target site flanked on the 3′ end by the PAM sequence (NGG) such that the target site is 5′-$(N)_{20}$(NGG). The PAM sequence is not part of the gRNA but is required at the genomic target site for DNA cleavage by Cas9.

4. CRISPR target sites should be selected to minimize potential off-target cleavage. Several studies have shown that the 12-nt "seed" region of the target site immediately adjacent to the PAM is the most critical region for efficient cleavage and should be unique, if possible [9–17]. Furthermore, PAMs of the form NAG permit low-efficiency cleavage in cell lines [12]. Based on current understanding in the field, highly specific gRNAs can be generated by selecting CRISPR target sites with the fewest potential off-target sites in the genome, defined as:

    (a) PAM-adjacent sites with ≥11/12 matches to the target seed sequence.
    
    (b) PAM-adjacent sites with ≥18/20 matches to the full target sequence.
    
    (c) Sites fitting above criteria adjacent to a PAM in the form of NAG as well as NGG.

To facilitate high-quality gRNA design, we have created a web application, CRISPR Optimal Target Finder, that identifies specific target sites in Drosophila and other invertebrate species [3]. The program can be accessed at http://tools.flycrispr.molbio.wisc.edu/targetFinder/.

### 3.2 Generate gRNA Expression Plasmids

1. CRISPR-targeting sequences are synthesized as oligonucleotides, annealed and incorporated into the gRNA expression vector, pU6-BbsI-chiRNA (Fig. 2). Cohesive ends for seamless cloning of targeting sequences into the gRNA backbone are included in the oligonucleotide design. The top strand should be designed in the format of 5′-CTTCG$(N)_{19}$-3′, where

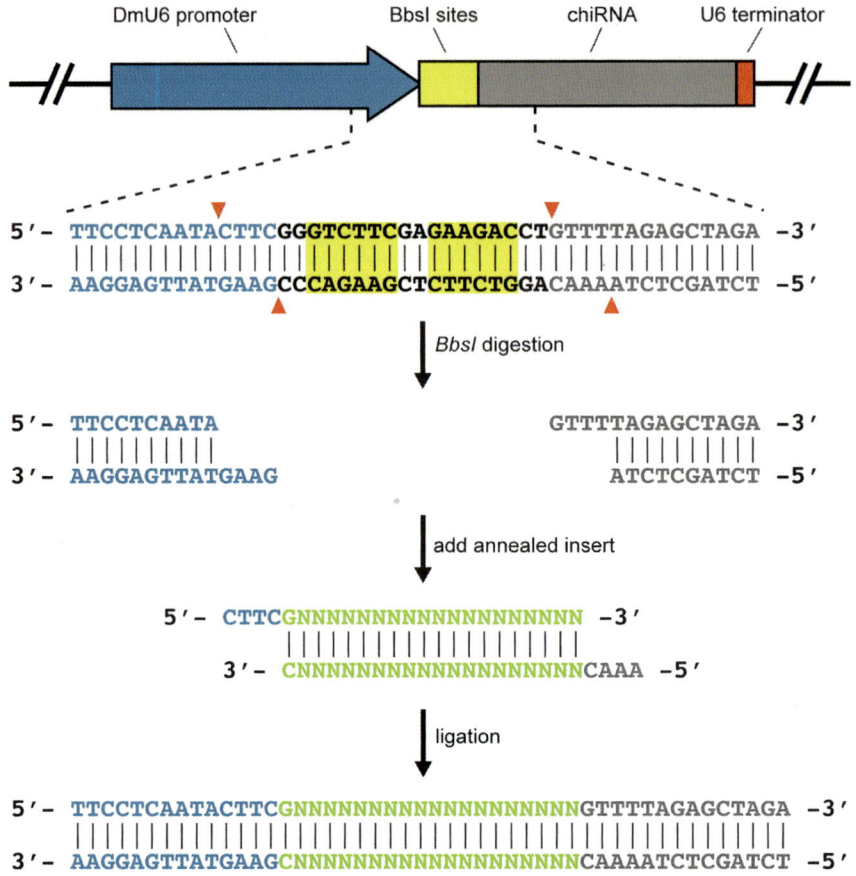

**Fig. 2** gRNA cloning. The pU6-BbsI-chiRNA vector contains two recognitions sites for the type IIs restriction enzyme *BbsI*. Following *BbsI* digestions, unique 4-nt overhangs mediate seamless cloning targeting sequences into the gRNA backbone. Targeting sequence inserts are generated as two 5′ phosphorylated oligonucleotides that are annealed to create complementary overhangs (5′-CTTC, top strand; 5′-AAAC, bottom strand) for ligation into linearized pU6-BbsI-chiRNA

$G(N)_{19}$ corresponds to your unique target-site sequence beginning with a G for efficient transcription from the Drosophila U6 promoter (*see* **Note 2**). The bottom strand is designed in the format of 5′-AAAC$(N)_{19}$C-3′, with $(N)_{19}$C representing the reverse complement of the targeting sequence. You can either order 5′ phosphorylated oligonucleotides or, as described below, use T4 Polynucleotide Kinase (PNK) to add the 5′-phosphates to standard oligonucleotides. Oligonucleotides should be resuspended at a concentration of 100 μM in nuclease-free $H_2O$.

2. Phosphorylate and anneal the target sequence oligonucleotide pairs in one step: Combine 1 μL of the top-strand oligonucleotide (100 μM stock), 1 μL of the bottom-strand oligonucleotide (100 μM stock), 1 μL of T4 DNA Ligase buffer, 6 μL of $H_2O$, and 1 μL of T4 Polynucleotide Kinase. Incubate at 37 °C

for 30 min, and 95 °C for 5 min, and then ramp to 25 °C at a rate of −5 °C/min.

3. Prepare pU6-BbsI-chiRNA for cloning by transforming DH5α cells and selecting colonies on plates containing 100 μg/mL ampicillin. Purify plasmid DNA and resuspend in nuclease-free H$_2$O.

4. Determine the DNA concentration using a spectrophotometer. Digest 1 μg of pU6-BbsI-chiRNA with *BbsI*.

5. Gel purify cut vector to remove any uncut vector. We use the Promega Wizard SV Gel and PCR Clean-up System kit. Determine the DNA concentration using a spectrophotometer.

6. Ligate the annealed insert and purified *BbsI*-digested pU6-BbsI-chiRNA. Combine 1 μL of annealed insert, 50 ng of *BbsI-digested* pU6-BbsI-chiRNA, 1 μL of T4 DNA Ligase buffer, 1 μL of T4 DNA Ligase, and enough H$_2$O to bring the reaction to 10 μL. Incubate at 25 °C for 1 h.

7. Transform the ligation reaction into DH5α cells and select colonies on plates containing 100 μg/mL ampicillin.

8. Isolate plasmids from 2 to 4 individual colonies using a miniprep kit. Screen for plasmids with incorporated oligonucleotides by Sanger sequencing.

9. Prepare DNA for injection from a positive clone using the Qiagen EndoFree Plasmid Maxi kit.

10. Determine the DNA concentration using a spectrophotometer.

11. Aliquot and store at −20 °C.

### 3.3 Prepare Cas9 Expression Plasmid (Unless Injecting into a Cas9-Expressing Line)

1. Transform DH5α cells with pBS-Hsp70-Cas9 and select on ampicillin-containing plates. Purify plasmid DNA using the Qiagen EndoFree Plasmid Maxi kit and resuspend in nuclease-free H$_2$O.

2. Determine the DNA concentration using a spectrophotometer.

3. Aliquot and store at −20 °C (*see* **Note 3**).

### 3.4 Design and Prepare Donor Template for HDR

1. To incorporate exogenous sequences, you will need to generate a donor repair template for HDR. Here we outline the method we used to replace the *yellow* locus with a 50-bp attP ΦC31 phage recombination site using an ssDNA donor template. An analogous strategy can be used to insert other short exogenous sequences or to introduce polymorphisms. The primary advantage of using ssDNA donor templates is that they can be rapidly synthesized, obviating the need for cloning. However, the size of ssDNAs is generally limited to ~200 nt. Larger double-stranded DNA (dsDNA) donor templates can be employed to efficiently incorporate larger exogenous sequences or modify sequences over a large region [3].

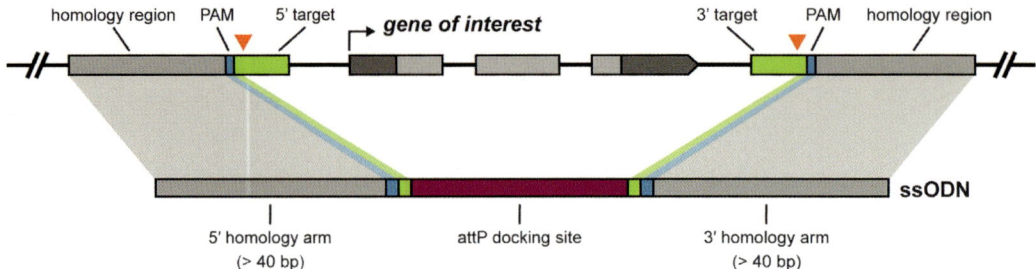

**Fig. 3** ssDNA design. Single-stranded oligonucleotide donors are designed to contain approximately 60 nt of homology to the target locus sequences immediately adjacent to each predicted Cas9-mediated DSB. The homology regions flank a 50-nt attP sequence that provides subsequent access to the targeted locus. *Red arrowheads* mark the predicted Cas9 cleavage sites. Target site sequences (*green*) and PAMs (*blue*) are indicated

Design an ssDNA donor with homology arms corresponding to sequences immediately adjacent to the targeted cleavage sites flanking the attP sequence (Fig. 3). We use the 50-nt attP sequence: GTAGTGCCCCAACTGGGGTAACCTTTGAGTTCTCTCAGTTGGGGGCGTAG. Homology arms of ~40–60 nt have been shown to mediate efficient HDR with ssDNA donors in Drosophila [2, 18, 19]. Cas9-mediated DSBs are generated 3 bp upstream of the PAM, so homology regions in the repair template should be designed to include the portion of the targeting sequence that remains following cleavage. The repair template should also be designed to avoid recutting of the genomic locus following successful HDR. This can be achieved by truncating or mutating the CRISPR target sequence within the donor template. For genome-engineering experiments using two gRNAs, the single-stranded donor can be either the forward or reverse stand. However, when using a sinlge gRNA, the orientation of the single-stranded donor is critical to the success of the experiment. During DNA repair, free 3′ ends created by rescission at the DSB search for and invade homologous DNA. It is therefore essential for incorporation of donor sequences that the ssDNA bears complementarity to a free 3′ end such that synthesis originating from that free end will incorporate the intended modification.

2. Resuspend the ssDNA in nuclease-free $H_2O$ to a final concentration of 1 µg/µL.

## 3.5 Embryo Injections

1. Prepare an injection mixture appropriate for your desired modification. If you are using a single gRNA to generate indels via NHEJ, make an injection mixture containing 500 ng/µL of pBS-Hsp70-Cas9 and 250 ng/µL of the pU6-BbsI-chiRNA vector. To generate defined deletions by NHEJ, make an injection mixture of 500 ng/µL of pBS-Hsp70-Cas9 and 250 ng/µL of each pU6-BbsI-chiRNA vector. For gene replacement by HDR, prepare an injection mixture containing 500 ng/µL of

pBS-Hsp70-Cas9, 250 ng/μL of each pU6-BbsI-chiRNA vector, and 100 ng/μL of the ssDNA donor template (*see* **Note 4**). Prepare injection mixtures in nuclease-free $H_2O$.

2. Establish a small population cage of the fly line to be injected. Provide flies fresh yeast paste on grape-juice plates.

3. Allow flies to lay embryos for 30 min, collect the embryos, and place them in a drop of water on a cover slip. Align the embryos side by side along the edge of a cover slip with the posterior end of the embryos oriented toward the edge for injection through the cuticle (*see* **Note 5**). Allow the embryos to dry until they adhere to the cover slip (1–5 min), and then cover the embryos with a 7:1 mixture of halocarbon oil 700 and halocarbon oil 27. Place the cover slip on a slide for injection. Injections must be completed prior to cellularization of pole cells approximately 1 h after laying at 25 °C. We perform injections at 18 °C to slow development and expand this time frame.

4. Inject embryos with a microinjector and a micromanipulator on an inverted microscope using glass capillary needles pulled on a micropipette puller.

5. After injection, allow the halocarbon oil to drain off the embryos. Transfer the cover slip with attached embryos to a plate of Drosophila agar-based sugar food containing a dab of yeast paste. Upon hatching, larvae will move to the yeast paste. After 3–4 days, transfer yeast paste and larvae to a food vial and rear to adulthood.

### 3.6 Assess Efficiency of Targeting in Injected Embryos (Optional)

1. There is significant variability in targeting efficiency at different genomic loci and, within a single locus, at different target sites [1, 2, 7]. To rapidly estimate the targeting efficiency of our gRNAs before embarking on a full experiment, we collect 10–15 embryos 24 h after injection for molecular analysis (*see* **Note 6**).

2. To isolate embryonic genomic DNA for molecular analysis, place each embryo in a 0.2 mL PCR tube. Using either a P20 or P200 pipette tip, draw up 20 μL of freshly prepared embryo homogenization buffer. Keeping the buffer in the pipette, use the tip to homogenize the embryo against the wall of the tube. Dispense the homogenization buffer.

3. Incubate at 37 °C for 30–60 min followed by a 5-min incubation at 95 °C.

4. Use 2 μL of embryonic genomic DNA in a 50 μL PCR to amplify the region spanning the targeted cleavage site(s). The PCR primers should be designed to generate a product from the mutant locus that can be readily distinguished from the wild-type locus. We generally design our reactions to generate ~500–700-bp products.

5. Defined deletions and gene replacements that result in size differences greater than 25 bp can be readily recognized by the size of the PCR product (*see* **Note 7**), whereas small indels and single base-pair mutations can be efficiently detected using the SURVEYOR Mutation Detection kit (*see* **Note 8**). Briefly, PCR products spanning the targeted site are denatured and reannealed. Any heteroduplexes formed by the annealing of wild-type DNA to indel-containing DNA will be cleaved by the SURVEYOR nuclease.

## 3.7 Screen F1 Progeny for Germline Transmission of Genome Modifications

1. If the phenotype is known and can be used to screen for the desired modification, cross injected flies to an appropriate fly line and screen F1 progeny for the expected phenotype. Chromosomal deletions that uncover the target locus can be used to simplify subsequent molecular characterization of candidates.

2. If you cannot identify your modification by phenotype, you can design your engineering strategy to utilize one of the alternatives outlined below.

   Negative selection: Targeting strategies can be designed to remove a marked element in the targeted locus upon successful editing to enable screening of F1 progeny for loss of the visible marker. For making defined deletions, one consequence of this strategy is an increase in the size of the deletion that must be generated. We have used this strategy to identify defined deletions of approximately 15 kb [3].

   Positive selection: Large dsDNA donors can be designed to include a visible marker, such as mini-*white*, along with the desired modification for positive screening of F1 progeny. The visible marker can be flanked by FRT or LoxP sites for subsequent removal. While not covered in this protocol, we have successfully used CRISPR-Cas9 and dsDNA donors to replace genes that are large as 25 kb with an attP docking site and 3xP3-DsRed marker, to generate conditional alleles and insert in-frame protein tags [3] (O'Connor-Giles, Harrison and Wildonger labs, unpublished data).

   Molecular identification: After outcrossing, individual F1 progeny can be sacrificed for genomic DNA and subjected to PCR-based analysis for identification of targeted events. This approach was successfully applied by Yu et al. [7], who identified Cas9-induced indels in four genes through a combination of direct sequencing and restriction enzyme-based analysis of PCR products (*see* **Note 9**).

## 3.8 Molecularly Confirm Genome Modifications

1. If you have not already done so during your screening process, confirm your modification through PCR and sequencing analysis. Always confirm the entire modification and surrounding DNA because cellular DNA repair processes can result in unexpected modifications or rearrangements, such as repair events that are correct on one end but not the other.

### 3.9 Evaluate Potential Sites of Off-Target Cleavage

1. Once fly lines with the correct targeted modification have been recovered, an assessment of off-target mutations can be conducted. Using the criteria outlined in Subheading 3.1 or CRISPR Optimal Target Finder, identify potential off-target cleavage sites for evaluation by PCR and sequence analysis (*see* **Note 10**).

2. To isolate genomic DNA for molecular analysis, anesthetize a single fly and place it in a 0.2 mL PCR tube. Using a P200 pipette tip draw up 50 µL of freshly prepared adult fly homogenization buffer. Keeping the buffer in the pipette, use the tip to homogenize the fly. Once the fly is homogenized, dispense the remaining buffer.

3. Incubate at 37 °C for 30–60 min followed by a 5-min incubation at 95 °C.

4. Use 1 µL of adult genomic DNA in a 50 µL PCR to amplify 500–700-bp regions spanning any potential off-target cleavage sites.

5. Sequence the PCR products to identify any small indels induced by Cas9 at potential off-target cleavage sites.

6. If you identify any off-target mutations, standard recombination methods can be used to separate them from the intended genome modification. *See* **Note 11** for additional suggestions for reducing off-target cleavage.

## 4 Notes

1. A key consideration in designing your strategy is determining how you will identify flies in which the desired modification was induced, and it is essential that you consider this critical step early in your planning process. We discuss several screening options in Subheading 3.7.

2. A study in zebrafish and our experiences indicate that G nucleotides can be added as an additional base pair on the 5′ end of the 20-nt target sequence without significantly affecting efficiency [20]. This can be useful when a suitable target beginning with G is not available.

3. Injection of Cas9 and a targeting gRNA as mRNA and RNA, respectively, has also been shown to efficiently generate indels in Drosophila [1, 7].

4. The concentrations of the CRISPR-Cas9 components can be titrated to balance cleavage efficiency and the potential for off-target cleavage [12, 14]. As high concentrations induce more off-target cleavage and may be toxic, it might be necessary to try multiple concentrations if viability or fertility is low. Based on our molecular analysis of targeting in embryos, concentrations

as low as 50 ng/μL of pBS-Hsp70-Cas9 and 25 ng/μL of pU6-BbsI-chiRNA can catalyze cleavage in somatic cells.

5. Dechorionated embryos can also be injected using standard protocols.

6. The factors that determine either CRISPR target-specific or locus-specific differences in efficiency are not understood. Thus, a rapid method for assessing whether a particular targeting experiment is efficiently generating the intended modification is desirable. Molecular characterization of injected embryos can be used to assess somatic transformation rates. However, these numbers may not correlate directly with germline transmission rates.

7. In gene-replacement experiments, some cells will repair the Cas9-induced DSB by HDR and incorporate the donor sequence, while others will employ NHEJ and generate a defined deletion lacking the exogenous sequence. In our gene replacement experiments, we generally observe doublets that reflect repair by both pathways in somatic cells of injected embryos.

8. High-resolution melting analysis offers a highly efficient alternative for the detection of indels [1].

9. The percentage of analyzed progeny with induced mutations ranged from 2 to 99 % [7].

10. Whole-genome analyses of the off-target effects of Cas9-induced DSBs have not yet been conducted in Drosophila. Because off-target cleavage is expected to be sequence based, individual inspection of the most likely off-target cleavage sites should be effective. However, this approach may not detect larger rearrangements that can occur during DSB repair.

11. As noted earlier, careful CRISPR target selection and low concentrations of CRISPR components can reduce off-target cleavage. Alternatively, a version of Cas9 that has been modified into a nickase by disrupting one of its nuclease domains can be utilized in HDR experiments [8]. Because DNA nicks are less prone to NHEJ but still catalyze HDR, use of the nickase version of Cas9 may substantially reduce off-target effects. Fly lines expressing the nickase version of Cas9 were generated by Kate Koles and Avi Rodal's lab and are available at http://www.crisprflydesign.org/flies. Finally, gRNA length has been correlated with specificity, with longer gRNAs showing less specificity but higher activity [14].

## Acknowledgements

We are grateful to members of the Harrison, O'Connor-Giles, and Wildonger labs for their help in establishing CRISPR-Cas9 protocols in Drosophila, and to Dustin Rubinstein for comments on this chapter. Our work has been funded by start-up funds from the University of Wisconsin to M.M.H., J.W., and K.O.C.G. and grants from the National Institutes of Health to J.W. (R00 NS072252) and K.O.C.G. (R00 NS060985 and R01 NS078179). Plasmids and transgenic fly lines described here are available through the nonprofit distributor Addgene and the Bloomington Drosophila Stock Center, respectively. Detailed reagent information is available at http://flycrispr.molbio.wisc.edu.

## References

1. Bassett AR, Tibbit C, Ponting CP, Liu JL (2013) Highly efficient targeted mutagenesis of drosophila with the CRISPR/Cas9 system. Cell Rep 4(1):220–228. doi:10.1016/j.celrep. 2013.06.020, S2211-1247(13)00312-4 [pii]

2. Gratz SJ, Cummings AM, Nguyen JN, Hamm DC, Donohue LK, Harrison MM, Wildonger J, O'Connor-Giles KM (2013) Genome engineering of Drosophila with the CRISPR RNA-guided Cas9 nuclease. Genetics 194(4): 1029–1035. doi:10.1534/genetics.113.152710

3. Gratz SJ, Ukken FP, Rubinstein CD, Thiede G, Donohue LK, Cummings AM, O'Connor-Giles KM (2014) Highly specific and efficient CRISPR/Cas9-catalyzed homology-directed repair in Drosophila. Genetics 196(4):961–971. doi:10.1534/genetics.113.160713

4. Kondo S, Ueda R (2013) Highly improved gene targeting by germline-specific Cas9 expression in Drosophila. Genetics 195(3):715–721. doi:10.1534/genetics.113.156737

5. Ren X, Sun J, Housden BE, Hu Y, Roesel C, Lin S, Liu LP, Yang Z, Mao D, Sun L, Wu Q, Ji JY, Xi J, Mohr SE, Xu J, Perrimon N, Ni JQ (2013) Optimized gene editing technology for Drosophila melanogaster using germ line-specific Cas9. Proc Natl Acad Sci U S A 110(47):19012–19017. doi:10.1073/pnas. 1318481110

6. Sebo ZL, Lee HB, Peng Y, Guo Y (2013) A simplified and efficient germline-specific CRISPR/Cas9 system for Drosophila genomic engineering. Fly (Austin) 8(1)

7. Yu Z, Ren M, Wang Z, Zhang B, Rong YS, Jiao R, Gao G (2013) Highly efficient genome modifications mediated by CRISPR/Cas9 in Drosophila. Genetics 195(1):289–291. doi: 10.1534/genetics.113.153825

8. Jinek M, Chylinski K, Fonfara I, Hauer M, Doudna JA, Charpentier E (2012) A programmable dual-RNA-guided DNA endonuclease in adaptive bacterial immunity. Science 337 (6096):816–821. doi:10.1126/science.1225829, science.1225829 [pii]

9. Cho SW, Kim S, Kim Y, Kweon J, Kim HS, Bae S, Kim JS (2014) Analysis of off-target effects of CRISPR/Cas-derived RNA-guided endonucleases and nickases. Genome Res 24(1): 132–141. doi:10.1101/gr.162339.113

10. Cradick TJ, Fine EJ, Antico CJ, Bao G (2013) CRISPR/Cas9 systems targeting beta-globin and CCR5 genes have substantial off-target activity. Nucleic Acids Res 41(20):9584–9592. doi:10.1093/nar/gkt714

11. Fu Y, Foden JA, Khayter C, Maeder ML, Reyon D, Joung JK, Sander JD (2013) High-frequency off-target mutagenesis induced by CRISPR-Cas nucleases in human cells. Nat Biotechnol 31(9):822–826. doi:10.1038/nbt.2623

12. Hsu PD, Scott DA, Weinstein JA, Ran FA, Konermann S, Agarwala V, Li Y, Fine EJ, Wu X, Shalem O, Cradick TJ, Marraffini LA, Bao G, Zhang F (2013) DNA targeting specificity of RNA-guided Cas9 nucleases. Nat Biotechnol 31(9):827–832. doi:10.1038/nbt.2647

13. Mali P, Yang L, Esvelt KM, Aach J, Guell M, DiCarlo JE, Norville JE, Church GM (2013) RNA-guided human genome engineering via Cas9. Science 339(6121):823–826. doi: 10.1126/science.1232033

14. Pattanayak V, Lin S, Guilinger JP, Ma E, Doudna JA, Liu DR (2013) High-throughput profiling of off-target DNA cleavage reveals RNA-programmed Cas9 nuclease specificity. Nat Biotechnol 31(9):839–843. doi:10.1038/ nbt.2673

15. Semenova E, Jore MM, Datsenko KA, Semenova A, Westra ER, Wanner B, van der Oost J, Brouns SJ, Severinov K (2011) Interference by clustered regularly interspaced short palindromic repeat (CRISPR) RNA is governed by a seed sequence. Proc Natl Acad Sci U S A 108(25):10098–10103. doi:10.1073/pnas.1104144108

16. Wiedenheft B, van Duijn E, Bultema JB, Waghmare SP, Zhou K, Barendregt A, Westphal W, Heck AJ, Boekema EJ, Dickman MJ, Doudna JA (2011) RNA-guided complex from

a bacterial immune system enhances target recognition through seed sequence interactions. Proc Natl Acad Sci U S A 108(25):10092–10097. doi:10.1073/pnas.1102716108

17. Yang H, Wang H, Shivalila CS, Cheng AW, Shi L, Jaenisch R (2013) One-step generation of mice carrying reporter and conditional alleles by CRISPR/Cas-mediated genome engineering. Cell 154(6):1370–1379. doi:10.1016/j.cell.2013.08.022

18. Banga SS, Boyd JB (1992) Oligonucleotide-directed site-specific mutagenesis in Drosophila melanogaster. Proc Natl Acad Sci U S A 89(5):1735–1739

19. Beumer KJ, Trautman JK, Mukherjee K, Carroll D (2013) Donor DNA utilization during gene targeting with zinc-finger nucleases. G3 (Bethesda). doi:10.1534/g3.112.005439

20. Hwang WY, Fu Y, Reyon D, Maeder ML, Kaini P, Sander JD, Joung JK, Peterson RT, Yeh JR (2013) Heritable and precise zebrafish genome editing using a CRISPR-Cas system. PLoS One 8(7), e68708. doi:10.1371/journal.pone.0068708, PONE-D-13-13968 [pii]

# Chapter 23

# Targeted Transcriptional Repression in Bacteria Using CRISPR Interference (CRISPRi)

John S. Hawkins, Spencer Wong, Jason M. Peters, Ricardo Almeida, and Lei S. Qi

## Abstract

Clustered regularly interspersed short palindromic repeats (CRISPR) interference (CRISPRi) is a powerful technology for sequence-specifically repressing gene expression in bacterial cells. CRISPRi requires only a single protein and a custom-designed guide RNA for specific gene targeting. In *Escherichia coli*, CRISPRi repression efficiency is high (~300-fold), and there are no observable off-target effects. The method can be scaled up as a general strategy for the repression of many genes simultaneously using multiple designed guide RNAs. Here we provide a protocol for efficient guide RNA design, cloning, and assay of the CRISPRi system in *E. coli*. In principle, this protocol can be used to construct CRISPRi systems for gene repression in other species of bacteria.

**Key words** CRISPRi, dCas9, sgRNA, *Escherichia coli*

## 1 Introduction

About 40 % of bacteria and 90 % of archaea possess an endogenous clustered regularly interspersed short palindromic repeats (CRISPR) system for defense against foreign DNA elements, such as viruses, bacteriophages, and plasmids [1, 2]. CRISPR utilizes CRISPR-associated (Cas) proteins and noncoding CRISPR (cr) RNA elements to confer genetic immunity [1–4]. In type II CRISPR systems, a single endonuclease protein Cas9, in complex with the mature form of crRNA-transacting (tracr) RNA complex, binds specifically to the DNA target by sequence complementarity and induces a double-stranded break. Jinek et al. have shown that a designed chimeric single-guide (sg) RNA derived from *Streptococcus pyogenes* could solely direct the DNA targeting and cleavage activities of Cas9 [5]. Thus the system allows easy programming of DNA target specificity, offering a modular, efficient, and multiplexable tool for genome editing [6–8]. In addition to using CRISPR for gene editing, we recently developed a new

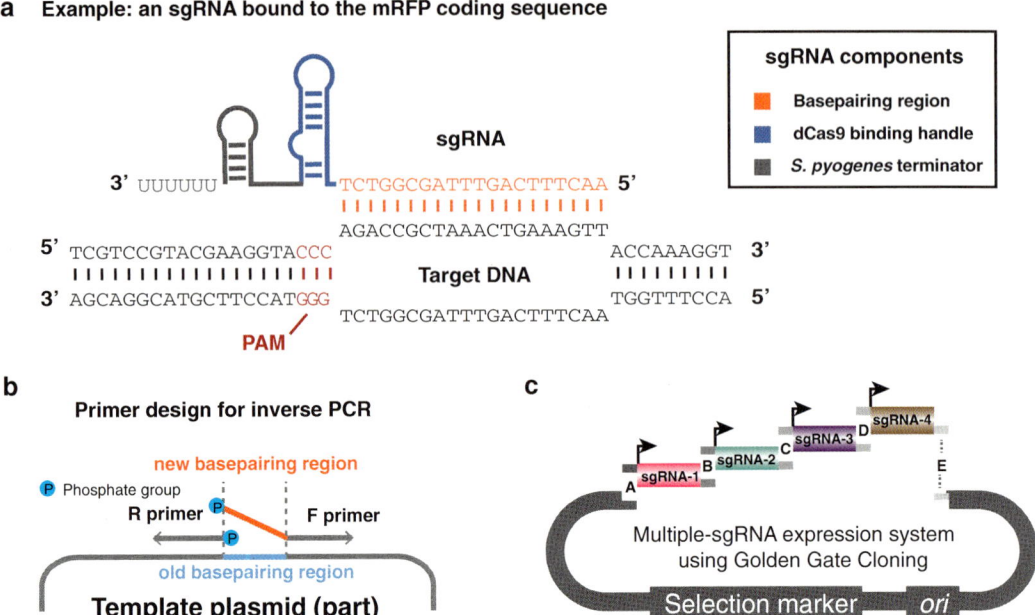

**Fig. 1** Design and cloning of sgRNAs for targeted gene repression. (**a**) Schematic of sgRNA binding to target DNA. The sgRNA consists of a 20-nt base-pairing region (*orange*), a 42-nt dCas9 binding handle (*blue*), and a 40-nt *S. pyogenes* transcription terminator (*grey*). The PAM is shown in *red*. Depicted is DNA from the coding sequence of RFP. (**b**) Schematic of generating vectors with new sgRNA sequences using inverse PCR. 5′ phosphorylation of oligos is indicated by *blue circles*. (**c**) Schematic of cloning multiple sgRNAs into a single vector using Golden Gate cloning

technology, CRISPR interference (CRISPRi) [9], that repurposes the CRISPR system for use in transcription regulation. In CRISPRi, a catalytically dead variant of Cas9 (dCas9) without endonucleolytic activity is coupled with designed small guide RNAs (sgRNAs) complementary to a desired DNA target. The dCas9-sgRNA complex acts to sterically hinder transcription of the targeted DNA, causing gene repression. Because repression by CRISPRi depends on base pairing between a short segment of the sgRNA and the DNA target, new DNA targets can be specified simply by altering the sgRNA sequence. Furthermore, repression by CRISPR is titratable and reversible through use of inducible promoters for both Cas9 and sgRNA expression. The current sgRNA design is a 102-nt chimeric noncoding RNA consisting of a 20-nt base pairing region, followed by a 42-nt Cas9-binding "handle," and a 40-nt transcription terminator (Fig. 1a). The 20-nt base pairing region can be freely modified, allowing broad flexibility in targeting genes for repression. Here, we provide guidance on sgRNA design and plasmid construction for repression of specific genes of interest

[10, 11]. We also provide a protocol for the modular assembly of multiple sgRNA expression cassettes into a single vector using Golden Gate Cloning [12]. Finally, we provide methods for quantifying CRISPRi repression.

## 2 Materials

### 2.1 Materials for sgRNA Design

1. The genome sequence of the target bacterial strain (e.g., a FASTA reference genome).
2. The 102-nt sgRNA reference sequence (see below).
3. (Optional) An RNA secondary structure prediction algorithm such as the Vienna software suite for RNA folding analysis.

### 2.2 Materials for Single-sgRNA Cloning

- *E. coli* sgRNA expression plasmid with ampicillin resistance (Addgene ID no. 44251). The sgRNA is expressed from a strong constitutive promoter J23119 (http://parts.igem.org/Part:BBa_J23119).
- Designed forward sgRNA primers (e.g., 5′-$N_{20}$GTTTTAGAGCTAGAAATAGCAAGTTAAAATAAGGC-3′ for Addgene ID no. 44251, $N_{20}$ is the sgRNA base pairing region).
- Universal reverse primer (e.g., 5′-ACTAGTATTATACCTAGGACTGAGCTAGC-3′ for Addgene ID no. 44251).
- 10 mM ATP (NEB).
- T4 Polynucleotide Kinase Kit (NEB).
- 2× Phusion PCR Master Mix (NEB).
- QiaQuick PCR Purification Kit (Qiagen).
- DpnI Kit (NEB).
- QiaQuick Gel Extraction Kit (Qiagen).
- Quick Ligase Kit (NEB).
- One Shot TOP10 chemically competent *E. coli* cells (Life Technologies).
- LB plates with 100 µg/mL ampicillin.
- Ampicillin (Sigma).
- TYGPN broth (can be purchased from Amresco) (*see* **Note 1**).
- Breatheable Sealing Films (e.g., AeraSeal Breatheable Sealing Films; Excel Scientific).
- QiaPrep 96 Turbo Miniprep Kit (Qiagen).
- QiaVac 96 vacuum manifold (Qiagen).

| | |
|---|---|
| **2.3 Cloning Multi-sgRNA Constructs for Multiplex CRISPRi** | - T4 DNA Ligase Kit, 400,000 units/mL (NEB).<br>- BsaI (NEB).<br>- 2× Taq Master Mix (NEB). |
| **2.4 Materials for CRISPRi Repression Assays** | - *E. coli* test strain (e.g., K12-strain MG1655).<br>- *E. coli* dCas9 expression plasmid with chloramphenicol resistance (Addgene ID no. 44249). The *dcas9* gene is under the control of an anhydrotetracycline-inducible promoter $P_L TetO$-1.<br>- LB plates with 100 μg/mL ampicillin and 20 μg/mL chloramphenicol.<br>- Chloramphenicol (Sigma).<br>- Anhydrotetracycline (Clontech).<br>- RNeasy RNA Purification Kit (Qiagen).<br>- DNA-free Kit (Ambion).<br>- Superscript III First Strand Synthesis System (Life Technologies).<br>- Random Hexamers (Life Technologies).<br>- qPCR primers.<br>- Brilliant II SYBR Green Master Mix (Agilent). |

## 3 Methods

### 3.1 sgRNA Design

The CRISPRi system requires two components, the dCas9 protein and a sequence-specific sgRNA. The key to successful gene repression is to custom design an appropriate sgRNA for the gene target of interest. The main body of the sgRNA, the Cas9 handle and transcriptional terminator, remains constant for different DNA targets. The region that must be designed specifically for each DNA target is a 20-nt base pairing region of the sgRNA (*see* **Note 2**).

*Example: Design an sgRNA target to repress mRFP [13] expressed in E. coli.*

There are four primary constraints to consider in designing the sgRNA base pairing region:

1. Choice of target site: For better repression efficiency, the base pairing region should bind to the non-template DNA strand of the coding region for the gene to be repressed. Targeting the template DNA strand of the coding sequence is generally ineffective, at most leading to mild repression (~50 %) [9]. Choosing a target closer to the 5′ end of the gene generally results in greater repression efficiency.

```
  Non-template DNA strand
5'ATGGCGAGTAGCGAAGACGTTATCAAAGAGTTCATGCGTTTCAAAGTTCGTATGGAAGGTTCCGTTAA
3'TACCGCTCATCGCTTCTGCAATAGTTTCTCAAGTACGCAAAGTTTCAAGCATACCTTCCAAGGCAATT
  Template DNA strand
```
**PAM     Basepairing**
```
CGGTCACGAGTTCGAAATCGAAGGTGAAGGTGAAGGTCGTCCGTACGAAGGTACCCAGACCGCTAAACTG
GCCAGTGCTCAAGCTTTAGCTTCCACTTCCACTTCCAGCAGGCATGCTTCCATGGGTCTGGCGATTTGAC
```

***region***
```
AAAGTTACCAAAGGTGGTCCGCTGCCGTTCGCTTGGGACATCCTGTCCCGCAGTTCCAGTACGGTTCCA
TTTCAATGGTTTCCACCAGGCGACGGCAAGCGAACCCTGTAGGACAGGGGCGTCAAGGTCATGCCAAGGT


AAGCTTACGTTAAACACCCGGC 3'
TTCGAATGCAATTTGTGGGCCG 5'
```

**Fig. 2** An example for designing an sgRNA that targets the mRFP gene. The PAM sequence (CCC) is *underlined* and *bold*. The base pairing sequence is in *bold italics*. The *bottom* DNA strand is the DNA transcription template strand

*Example: See Fig. 2 for initial sequence of* mRFP.

PAM adjacency: In addition to the 20-bp base pairing region, the Cas9-sgRNA complex requires a juxtaposed *p*roto-spacer *a*djacent *m*otif (PAM) to bind DNA [5, 14, 15], which is specific to the Cas9 being used [16, 17]. In the case of *S. pyogenes* Cas9, the motif is NGG or NAG (N as any nucleotide) [6]. Thus, the targetable sites are restricted to 20-nt regions 5' to NGG in the genome (Fig. 1a).

*Example:* To target a particular site on the template DNA strand, search for $N_{20}$-N(G/A)G, and $N_{20}$ is the sequence of the base pairing region of sgRNA; to target a site on the non-template strand, search for C(C/T)N-$N_{20}$, and the reverse complementary sequence of $N_{20}$ is the sequence of the sgRNA base pairing region. For example, to target the non-template DNA strand of the *mRFP* sequence listed above, one target site is AGACCGCTAACTGAAAGTT (in bold italic) on the non-template strand, with a PAM (CCC, in bold underline). The sgRNA base pairing sequence is thus the reverse complementary, i.e., AACTTTCAGTTTAGCGGTCT.

2. Genomic specificity: The dCas9-sgRNA complex will bind to any sufficiently similar DNA sequences that are adjacent to PAM sites. To identify such sequences, run BLAST (blastn; default settings) [18] searches with each designed sgRNA base pairing region against the complete genome of the organism in which the system will be used. Potential off-target DNA-binding sites with partial complementarity to the sgRNA should be evaluated for degree of homology; nucleotides closer to the PAM have a greater impact on relative binding affinity.

One or two mismatches, particularly in the PAM-adjacent 12-nt portion of the target (the "seed" region), are sufficient to reduce binding efficiency by an order of magnitude or more [9].

*Example: BLAST [19] the above sgRNA base pairing region (AACTTTCAGTTTAGCGGTCT) against the E. coli genome. There should be no exact 20-nt matches with adjacent PAM sites.*

3. Folding quality (optional): Generate the full sgRNA sequence by appending the target DNA sequence (replacing T with U) to the 5′ end of the sgRNA sequence (5′ 20-nt base pairing region + GUUUUAGAGCUAGAAAUAGCAAGUUAAA AUAAGGCUAGUCCGUUAUCAACUUGAAAAAGUGG CACCGAGUCGGUGCUUUUUUU -3′). Using a folding algorithm such as the Vienna suite [20], compare the predicted structure of the sgRNA containing the new 20-nt base pairing region to the structure of a functionally validated sgRNA. If the dCas9 binding handle is disrupted the sgRNA will bind poorly to dCas9, and repression efficiency will be reduced.

## 3.2 Single-sgRNA Cloning Using Inverse PCR

Primers for the target-specific inverse PCR (iPCR; described in [21] and Fig. 1b) are paired with a universal reverse primer to generate new sgRNA expression vectors. These vectors are transformed into a cloning strain of *E. coli* and then grown to a sufficient quantity for plasmid purification. The forward and reverse primers are homologous to the template vector, with the forward primer containing a 20-nt base pairing region at the 5′ end unique to each sgRNA (Fig. 1b). The forward primer anneals to the part of the vector that encodes the dcas9-binding handle, immediately downstream of the 20-nt base pairing region. The following protocol addresses parallel or pooled cloning of many vectors, each of which expresses a single sgRNA (*see* **Note 3**).

1. iPCR can be performed using individual primer pairs (i.e., one sgRNA product per reaction) or by pooling sets of forward primers that contain different sgRNA sequences with a universal reverse primer (i.e., numerous sgRNA products per reaction). Pooling can reduce labor and reagent consumption, but may cause bias (e.g., differential amplification of specific sgRNA plasmids) or other problems (e.g., mispriming) in the PCR reaction. To form pools, combine multiple forward primers in equal portion so that the *total* concentration of the primer pool is the concentration listed below. (The concentration of any individual forward primer will be 1/N times the total concentration, where N is the number of primers in the pool. The reverse primer should be used at the same concentration, as listed.) For construction of large sgRNA libraries (>96), we typically start with multiple iterations of a pooled approach followed by individual cloning of sgRNA plasmids to

complete the library. For small libraries (<96), individual cloning of the entire library may be more efficient.

2. Inverse PCR requires primer phosphorylation (required for the subsequent blunt-end ligation step) prior to PCR, which can be done as a pool or as multiple sub-pools (which may reduce PCR bias). To prepare a 100-µM primer pool, mix 1 µL of individual 100 µM primers. To phosphorylate primers:

| | |
|---|---|
| 100 µM Forward primer (or primer pool) | 1 µL |
| 100 µM Universal reverse primer | 1 µL |
| 10× PNK buffer | 1 µL |
| 10 mM ATP | 1 µL |
| T4 PNK | 0.5 µL |
| ddH$_2$O | 6.5 µL |
| Total | 10 µL |

Incubate at 37 °C for 30 min, and then heat-inactivate the reaction at 65 °C for 20 min.

3. Amplify new sgRNA vectors using iPCR:

| | |
|---|---|
| Phosphorylation reaction | 10 µL |
| 100 ng/µl *E. coli* sgRNA plasmid template | 0.5 µL |
| 2× Phusion PCR Master Mix | 12.5 µL |
| ddH$_2$O | 2 µL |
| Total | 25 µL |

4. Perform iPCR using the following thermal cycler program:

| Step | Temperature (°C) | Duration |
|---|---|---|
| Initial denaturation | 98 | 30 s |
| 25 cycles | 98 | 10 s |
| | 62 | 30 s |
| | 72 | 1 min |
| Final extension | 72 | 5 min |

5. If more than one pool is used for iPCR, combine pools for the remaining steps.

6. Purify PCR products using the QiaQuick PCR Purification Kit (Qiagen) following the manufacturer's instructions. Elute using 30 µL of EB.

7. Treat the purified PCR products with DpnI to degrade the PCR template plasmid.

| Purified PCR products | 10 μL |
|---|---|
| DpnI restriction endonuclease | 1.25 μL |
| 10× CutSmart Buffer | 2.5 μL |
| ddH$_2$O | 11.25 μL |
| Total | 25 μL |

Incubate at 37 °C for 1 h

8. Run 2 μL of the DpnI-digested PCR product on a 1 % agarose gel, and then stain with 1 μg/mL ethidium bromide for 20 min. If using the *E. coli* sgRNA plasmid as a template (Addgene ID no. 44251), the iPCR product should be ~2,500 bp in size. Wrong PCR products may interfere with subsequent ligation or may form a circularized plasmid that can transform *E. coli*, which greatly reduces the cloning efficiency. To remove wrong PCR products, purify the ~2,500 bp band from a 1 % agarose gel using the QiaQuick Gel Extraction Kit (Qiagen) following the manufacturer's instructions. Elute using 30 μL of EB.

9. Circularize the PCR products by ligating the ends:

| DpnI-treated PCR product | 9 μL |
|---|---|
| 2× Quick Ligase Buffer | 10 μL |
| Quick Ligase | 1 μL |
| Total | 20 μL |

Incubate at 25 °C for 5 min.

10. Transform 10 μL of the ligation reaction into One Shot TOP10 chemically competent *E. coli* cells following the manufacturer's instructions.

11. Spread onto LB agar plates containing ampicillin (100 μg/mL) and incubate overnight at 37 °C.

12. Prepare 96-well 2 mL deep-well blocks with 1.7 mL per well of TYGPN liquid medium + 100 μg/mL ampicillin.

13. Use pipette tips to inoculate individual wells of a 96-well deep-well block with single colonies.

14. Cover the plate with an AeraSeal breatheable film. Incubate overnight at 37 °C without shaking.

15. Pellet cells by centrifugation at 2,000 × $g$ on a tabletop centrifuge for 10 min.

16. Aspirate medium off of the cell pellets.

17. Prep plasmid DNA using a QIAprep 96 Turbo miniprep kit and a QIAVac 96-well vacuum manifold following the manufacturer's instructions. Elute in 100 μL EB. Alternatively, minipreps can be performed on individual cell pellets using the QIAprep Spin Miniprep kit or an equivalent kit from another manufacturer.

## 3.3 Cloning Multi-sgRNA Constructs for Multiplex CRISPRi

This protocol describes construction of a single vector that contains multiple sgRNAs for targeting two or more genes in *E. coli* using Golden Gate cloning [12], which facilitates ligation of up to ten DNA fragments in one reaction. Golden Gate cloning employs BsaI, a type IIS restriction enzyme that recognizes a sequence 5′ of the actual cut site, allowing creation of arbitrary 4-nt overhangs (Fig. 3a). Ligation of non-palindromic 4-nt overhangs allows efficient assembly of sgRNAs in a defined order (Fig. 1c). We use primers with added BsaI sites and overhangs to amplify sgRNA expression cassettes using PCR (Fig. 3b). BsaI recognition sites are removed from the sgRNA insert during digestion and thus are not

**a**
```
5' - GGTCTCNNNNNN   - 3'              5' - GGTCTCN          - 3'
3' - CCAGAGNNNNNN   - 5'              3' - GGTCTCNNNNN - 5'
                          + BsaI                    Overhang
                          enzyme
```

**b**
Primer design: gccgcgGGTCTCaNNNNNNNNNNNNNNNNN
                      BsaI site    Homologous to vector
                              Overhang

**c** An orthogonal set of overhangs for Golden Gate Cloning

| Overhang | Sequence |
|----------|----------|
| A | AGGT |
| B | CCAC |
| C | GGTA |
| D | CCGA |
| E | TACG |
| F | TGAA |
| G | ATCG |
| H | GAAC |
| I | GCTT |

**Fig. 3** Primer design and overhang sequences for Golden Gate cloning. (**a**) BsaI is a type IIS restriction enzyme that cuts outside of the recognition site. (**b**) The primers (both forward and reverse) contain a BsaI site, a unique 4-nt overhang, and homologous sequence binding to the PCR vector. (**c**) An orthogonal set of overhangs for Golden Gate cloning. These orthogonal overhangs (A–I) prevent erroneous ligation during the ligation step

present in the final plasmid. Each sgRNA expression fragment contains a constitutive promoter (promoter J23119; http://parts.igem.org/Part:BBa_J23119) that drives expression of a single sgRNA. The Golden Gate cloning procedure described below streamlines creation of multi-sgRNA vectors for multiplex CRISPRi (Fig. 1c) (*see* **Note 4**):

1. Design primers to amplify the sgRNA expression fragments with added BsaI sites and custom overhang sequences (Fig. 3c).
2. PCR amplify fragments using NEB Phusion Polymerase Master Mix, with 0.5 µL of 100 µM primers:

| sgRNA expression vector | 0.5 µL |
| Forward primer (100 µM) | 0.5 µL |
| Reverse primer (100 µM) | 0.5 µL |
| 2× Phusion PCR Master Mix | 25 µL |
| ddH$_2$O | 23 µL |
| Total | 50 µL |

3. Run the PCR products on 1 % agarose gel to verify that the PCR is successful and specific (no secondary bands).
4. PCR purification using standard protocol using, e.g., QiaQuick PCR Purification Kit.
5. Quantify the concentration of each purified PCR product using a Nanodrop.
6. Set up the Golden Gate reaction in a PCR reaction tube:

| DNA fragments | 25 ng each |
| Vector DNA | 75 ng |
| T4 DNA Ligase | 15 units |
| BsaI | 10 units |
| 10× T4 DNA Ligase Buffer | 1.5 µL |
| ddH$_2$O | Up to 15 µL |

7. In a thermocycler, perform the program:

| Step | Temperature (°C) | Duration (min) |
| --- | --- | --- |
| 50 cycles of digestion and ligation | 37 | 10 |
|  | 16 | 10 |
| Final digestion | 50 | 10 |
| Heat inactivation | 80 | 20 |

8. Add 10 units of DpnI and incubate at 37 °C for 1 h to remove the template vector.

9. Directly transform ligation reactions into One Shot TOP10 chemically competent *E. coli* cells following the manufacturer's instructions.

10. Spread the cells on LB plates with 100 μg/mL ampicillin. Grow cells overnight at 37 °C.

11. Inoculate and miniprep cells as described in Subheading 3.2, **steps 13–17**.

12. While incubating the cells, perform colony PCR to validate clones prior to miniprep. Use Primer3 software (http://biotools.umassmed.edu/bioapps/primer3_www.cgi) to design PCR primers; the two primers are designed to flank the sgRNA expression cassette and generate approximately 1-kb PCR product. Primers should have annealing temperatures of 60 °C. Set up the following PCR reaction:

| | |
|---|---|
| Cell culture | 1 μL |
| Forward colony PCR primer, 10 μM | 0.5 μL |
| Reverse colony PCR primer, 10 μM | 0.5 μL |
| Taq Master Mix, 2× | 12.5 μL |
| ddH2O | 10.5 μL |
| Total | 25 μL |

13. Perform colony PCR using the following program:

| Step | Temperature (°C) | Duration |
|---|---|---|
| Initial denaturation | 95 | 30 s |
| 30 cycles | 95 | 20 s |
| | 55 | 30 s |
| | 68 | 1 min per kb |
| Final extension | 68 | 5 min |
| Hold | 10 | |

14. Select positive clones for DNA miniprep.

## 3.4 CRISPRi Gene Repression Assay Using Quantitative PCR

If the target gene encodes a fluorescent protein, transcriptional repression using CRISPRi can be assayed using methods such as flow cytometry [9]. More generally, transcriptional repression of endogenous genes can be assayed using quantitative PCR (qPCR) (*see* **Note 5**).

1. Co-transform the cloned sgRNA vector with an inducible dCas9-expressing vector (e.g., Addgene cat. no. 44249) into

desired *E. coli* strain (e.g., MG1655) for targeted gene expression.

2. Grow overnight without inducers at 37 °C.

3. Inoculate single colonies into a 2-mL deep-well 96-well plate. Each well should contain 200 μL of LB with ampicillin (100 μg/mL) and chloramphenicol (20 μg/mL) but without inducers.

4. Grow overnight at 37 °C at 900 RPM shaking speed in a deep-well plate shaker.

5. Dilute overnight cultures 1:1,000 into 3 mL LB in 18 mL glass culture tubes, supplemented with 1 μM anhydrotetracycline to induce dCas9 expression. Grow cultures for 6 h at 37 °C in a culture tube rotator.

6. When cells reach late-exponential phase, mix 600 μL of culture with 600 μL of −20 °C methanol in a microcentrifuge tube, and then pellet cells at maximum speed (e.g., $18,000 \times g$) for 1 min.

7. Pipette off the supernatant and then extract the RNA using Qiagen RNeasy RNA purification kit following the manufacturer's instructions.

8. Following RNA isolation, remove genomic DNA using Ambion DNA-Free kit following the manufacturer's instructions.

9. Arbitrarily choose six DNA-free RNA samples and add 1 μL of 100 mg/mL RNase A. These will be used to ensure that there is no contaminating DNA in the samples when the qPCR is run.

10. Synthesize cDNA using Invitrogen SuperScript III First-Strand Synthesis System with random hexamers following the manufacturer's instructions. Make working stocks by diluting the cDNA 1:10.

11. Using Primer3 software [22], design primers with annealing temperatures of approximately 60 °C flanking a 200-bp region in the middle of the target gene. Design primers to amplify a nontargeted control gene in the genome (a gene unlikely to be perturbed by the knockdown, such as the *rpoD* gene that encodes the primary σ factor in *E. coli*) as control for qPCR reactions.

12. Set up the qPCR reaction:

| Agilent Brilliant II SYBR Green qPCR 2× Master Mix | 12.5 μL |
| --- | --- |
| Each qPCR primer, 5 μM | 3 μL |
| Diluted cDNA | 1 μL |
| Nuclease-free water | Up to 25 μL |

Include three biological replicates for each gene target.

13. Using a real-time thermal cycler (e.g., Stratagene Mx3005P qPCR System), perform the qPCR reaction.

14. Calculate the relative gene expression using the cycle threshold values (Ct). First, subtract the experimental strain's control gene's average Ct from the experimental strain's target gene's average Ct (difference called A for simplicity). Then subtract the control strain's control gene's average Ct from the control strain's target gene's average Ct (difference called B for simplicity). Finally, subtract B from A and use this difference as an exponent for 2, i.e., $2^{(A-B)}$ to calculate gene expression. Errors are calculated with the formula for propagation of error [23].

## 4 Notes

1. *TYGPN medium:*
   For 1 L: 750 mL ddH$_2$O, 5 g Na$_2$HPO$_4$, 10 g KNO$_3$, 20 g tryptone, 10 g yeast extract, 10.7 mL 75 % glycerol
   Be sure to dissolve the salts thoroughly first and then mix in the remainder of the reagents. Autoclave for 45 min.

2. The final repression effect of a given sgRNA may depend on multiple factors, including sequence composition of the base-pairing region, off-target binding, target location relative to the transcription start site, and possibly even local chromosome architecture [9].

3. Protocol 3.2 should be preferentially performed in a single day, or samples should be stored at −20 °C and used soon after to avoid loss of phosphate groups prior to the ligation step.

4. For protocol 3.3, we found that DNA containing eight or more tandem sgRNA cassettes does not efficiently amplify by PCR. If desired constructs consist of eight or more sgRNA cassettes, it may be necessary to use two compatible plasmids for co-expression of multiple sgRNAs.

5. Repressing the first gene in an operon likely has a similar repressive effect on all downstream genes in the same operon.

## Acknowledgements

We thank the Lei Qi lab, Carol Gross lab, and Wendell Lim lab for their support. J.S.H. acknowledges the support from Biophysics Graduate Program at UCSF. Spencer Wong acknowledges the support from Summer Research Training Program (SRTP) at UCSF. This work was supported by NIH P50 (grant GM081879, L.S.Q.), NIH Director's Early Independence Award (grant

OD017887, L.S.Q.), and a Ruth L. Kirschstein National Research Service Award (F32GM108222-01, J.M.P.).

**References**

1. Wiedenheft B, Sternberg SH, Doudna JA (2012) RNA-guided genetic silencing systems in bacteria and archaea. Nature 482:331–338
2. Barrangou R, Fremaux C, Deveau H, Richards M, Boyaval P, Moineau S, Romero DA, Horvath P (2007) CRISPR provides acquired resistance against viruses in prokaryotes. Science 315:1709–1712
3. Wiedenheft B, Lander GC, Zhou K, Jore MM, Brouns SJ, van der Oost J, Doudna JA, Nogales E (2011) Structures of the RNA-guided surveillance complex from a bacterial immune system. Nature 477:486–489
4. Brouns SJJ, Jore MM, Lundgren M, Westra ER, Slijkhuis RJH, Snijders APL, Dickman MJ, Makarova KS, Koonin EV, van der Oost J (2008) Small CRISPR RNAs guide antiviral defense in prokaryotes. Science 321:960–964
5. Jinek M, Chylinski K, Fonfara I, Hauer M, Doudna JA, Charpentier E (2012) A programmable dual-RNA–guided DNA endonuclease in adaptive bacterial immunity. Science 337:816–821
6. Jiang W, Bikard D, Cox D, Zhang F, Marraffini LA (2013) RNA-guided editing of bacterial genomes using CRISPR-Cas systems. Nat Biotechnol 31:233–239
7. Mali P, Yang L, Esvelt KM, Aach J, Guell M, DiCarlo JE, Norville JE, Church GM (2013) RNA-guided human genome engineering via Cas9. Science 339:823–826
8. Cong L, Ran FA, Cox D, Lin S, Barretto R, Habib N, Hsu PD, Wu X, Jiang W, Marraffini LA, Zhang F (2013) Multiplex genome engineering using CRISPR/Cas systems. Science 339:819–823
9. Qi LS, Larson MH, Gilbert LA, Doudna JA, Weissman JS, Arkin AP, Lim WA (2013) Repurposing CRISPR as an RNA-guided platform for sequence-specific control of gene expression. Cell 152:1173–1183
10. Huang S-H (1994) Inverse polymerase chain reaction. Mol Biotechnol 2:15–22
11. Quan J, Tian J (2011) Circular polymerase extension cloning for high-throughput cloning of complex and combinatorial DNA libraries. Nat Protoc 6:242–251
12. Engler C, Gruetzner R, Kandzia R, Marillonnet S (2009) Golden Gate shuffling: a one-pot DNA Shuffling method based on type IIs restriction enzymes. PLoS One 4:e5553
13. Campbell RE, Tour O, Palmer AE, Steinbach PA, Baird GS, Zacharias DA, Tsien RY (2002) A monomeric red fluorescent protein. Proc Natl Acad Sci U S A 99:7877–7882
14. Deveau H, Barrangou R, Garneau JE, Labonté J, Fremaux C, Boyaval P, Romero DA, Horvath P, Moineau S (2008) Phage Response to CRISPR-Encoded Resistance in Streptococcus thermophilus. J Bacteriol 190:1390–1400
15. Garneau JE, Dupuis MÈ, Villion M, Romero DA, Barrangou R, Boyaval P, Fremaux C, Horvath P, Magadán AH, Moineau S (2010) The CRISPR/Cas bacterial immune system cleaves bacteriophage and plasmid DNA. Nature 468:67–71
16. Mojica FJ, Díez-Villaseñor C, García-Martínez J, Almendros C (2009) Short motif sequences determine the targets of the prokaryotic CRISPR defence system. Microbiology 155:733–740
17. Shah SA, Erdmann S, Mojica FJ, Garrett RA (2013) Protospacer recognition motifs: mixed identities and functional diversity. RNA Biol 10:891–899
18. Altschul SF, Gish W, Miller W, Myers EW, Lipman DJ (1990) Basic local alignment search tool. J Mol Biol 215:403–410
19. BLAST: Basic Local Alignment Search Tool, http://blast.ncbi.nlm.nih.gov/Blast.cgi.
20. Gruber AR, Lorenz R, Bernhart SH, Neuböck R, Hofacker IL (2008) The Vienna RNA websuite. Nucleic Acids Res 36:W70–W74
21. Ochman H, Gerber AS, Hartl DL (1988) Genetic applications of an inverse polymerase chain reaction. Genetics 120:621–623
22. Rozen S, Skaletsky H (2000) Primer3 on the WWW for general users and for biologist programmers. Methods Mol Biol 132:365–386
23. Clifford AA (1973) Multivariate error analysis: a handbook of error propagation and calculation in many-parameter systems. Wiley, New York. ISBN 0470160551

# INDEX

## A

Adaptation .................... 35, 48, 92, 149, 151–152, 155–158, 202, 251, 265
Affinity protein purification .................... 186, 191
Agarose EMSA .................... 173–177, 183
Archaeal viruses .................... 223–231
ATPase assay .................... 280–281, 283–285

## B

*Bacillus halodurans* .................... 63, 266, 270, 273, 274
Bacteriophage insensitive mutants (BIMs) .................... 197–203, 206–210, 212, 216–218
 production .................... 206–209, 216–218
Base hydrolysis mapping .................... 39, 41
Binding assay .................... 278
BioNumerics .................... 123
Biotin labelling .................... 92, 114, 116, 121, 124
Branched DNA preparation .................... 259

## C

Cas5a, Cas3′, Cas3″ and Cas8a2 .................... 24, 28–30
Cascade .................... 23, 24, 27–30, 48, 50, 51, 54–56, 63–65, 134, 136, 137, 149, 171, 174–179, 182, 278, 279, 281–283, 285, 288–290, 293–304
Cas5d .................... 36–38, 42, 43
Cas6e .................... 7, 35, 36, 38, 42, 43, 66, 134, 141, 144, 278
Cas genes .................... 16, 35, 49–61, 70, 71, 94, 147–150, 196, 202, 234, 247, 265, 278
Cas protein complexes .................... 23–32
Cas1 purification .................... 252, 257–258
cDNA library .................... 2, 10–14, 17, 18
Chemical DNA footprint .................... 294
Chemically-competent *E. coli* .................... 153, 351, 356, 359
Cmr6 .................... 189, 191, 193
CMR complex .................... 5, 12, 16, 17, 55, 63, 66, 185–194
Competent *sulfolobus* cells .................... 189–190
Complementation .................... 190, 235
Conjugative plasmid .................... 224, 230
Constructing single guide RNA (sgRNA) vectors .................... 319–320, 322–324
*Corynebacterium diphtheriae* .................... 92, 113

CRISPR
 array direction .................... 77, 80
 identification .................... 4, 52, 77, 309
 RNA biogenesis .................... 1–18
 screening .................... 155–156
 sequence analysis .................... 211–212
 target selection .................... 346
CRISPR associated proteins
 annotation .................... 47–71
 bioinformatics .................... 52
 Cas1 .................... 48–54, 56, 58–63, 69, 71, 202, 247, 251–266
 Cas2 .................... 48–54, 56, 58–61, 63, 69, 202, 208, 247, 251, 265, 266, 268–275, 278
 Cas3 .................... 48–50, 52–54, 56, 68, 147, 148, 150, 151, 247, 277–291
 Cas6 .................... 7, 35, 36, 38, 42, 50, 52, 54, 56, 60, 65–67, 251
 Cas9 .................... 15, 48, 49, 51, 55, 56, 58, 59, 63, 65, 171, 202, 277, 318, 319, 321, 325–329, 331, 335–346, 349, 350, 352, 353
 classification .................... 47–71
 evolution .................... 61
 nomenclature .................... 49, 52–61, 70
 phylogenomics .................... 49–70
CRISPRDetect .................... 77, 78, 80
CRISPRDirection .................... 77, 80, 88
CRISPRi .................... 349–361
CRISPRTarget .................... 77–89, 202
crRNA
 identification .................... 4, 17
 purification .................... 136, 137
Cruciform DNA .................... 260, 261
Csa5 and Cas7 .................... 24, 28–30

## D

dCas9 .................... 349–361
Deep sequencing .................... 1, 3, 5, 16, 17, 214
Denaturing PAGE .................... 261
Differential RNA-seq (dRNA-seq) .................... 3, 6, 7, 9–15, 17
DNA cleavage mapping .................... 283
DNase
 assay .................... 274, 298
 treatment .................... 9, 10
Donor template preparation .................... 341–342

Double stranded breaks (DSBs) .............................. 317, 318, 321, 336, 338, 342, 346, 349
dRNA-seq. *See* Differential RNA-seq (dRNA-seq)
*Drosophila* ................................................................ 347, 348
DSBs. *See* Double stranded breaks (DSBs)

### E

Electron microscopy ......................... 224, 226–228, 230, 231
Electrophoretic mobility shift
  assay (EMSA)......................................... 171–184, 298
Electroporation................................. 164, 166, 169, 190, 198, 203, 209, 238, 241, 244
Electrospray ionization mass spectrometry.............. 133–145
Embryo injection............................................. 338, 342–343
EMSA. *See* Electrophoretic mobility shift assay (EMSA)
Endonuclease assay........................................... 321, 328–329
Endoribonuclease ............................... 7, 12, 35–38, 40, 42–44, 54, 65, 133, 144, 266
Enzymatic DNA footprint ............................................. 294
*Escherichia coli* (*E.coli*) .............................. 2, 9, 14, 23, 27, 29, 31, 35, 54, 61, 63, 68, 77, 81, 86, 94, 134, 141–144, 147–159, 161–169, 185, 195, 199, 200, 207, 213, 216, 229, 234–238, 243, 246, 251, 252, 259–261, 263, 266–268, 278, 302, 338, 351, 352, 354–357, 359, 360
  transformation ........................................... 153, 161–169
Exonuclease III................................................ 295–301, 303

### G

Gel-shift.......................................................................... 290
Gene repression assay .............................................. 359–361
Genome
  editing............................. 15, 59, 202, 318, 335–346, 349
  engineering ................................................ 317, 335–346
Genotyping ............................................................. 111–129
Germline screening .................................................. 338, 344

### H

HDR. *See* Homology-directed repair (HDR)
Helicase
  activity assay ..................................................... 277–291
  polarity assay................................................ 286–287, 291
His tag ....................................................... 16, 31, 194
Holliday junctions ......................................................... 259
Homology-directed repair (HDR) ........................ 317, 336, 338, 341–342, 346
HtpG.......................................................... 243–245, 248
Hyperthermophiles........................................................ 185

### I

*Iap*.................................................................................... 94
Inclusion bodies............................................ 23–25, 27–29

Interference
  assay........................................................... 352, 359
  efficiency.................................................... 161
In vitro transcription .................. 42, 318, 319, 324, 326, 330
Isothermal titration calorimetry (ITC)................... 266, 268, 272–273, 275

### L

Large subunits ............................................. 50, 51, 54–57, 60, 61, 64–68
Liquid chromatography ......................................... 133–145

### M

Macroarrays............................................................ 111–129
Membrane
  hybridization......................................................... 92, 122
  stripping ............................................................ 123, 128
Metagenomics ................................... 78, 196, 307, 308, 310
Microbead hybridization ................................................ 106
Microinjection.......................................... 318, 321, 324–328
Mongo Oligo Mass Calculator................................ 138, 140
Mutagenesis..................................... 188, 234, 236, 239–240, 248, 317–332, 338
*Mycobacterium tuberculosis* ......................... 91–108, 111, 114

### N

Next generation sequencing............................. 2, 3, 308, 309
NHEJ. *See* Non-homologous end joining (NHEJ)
Non-homologous end joining (NHEJ) .................. 317, 336, 342, 346
Nuclease activity assays............................ 269–272, 274, 288
Nuclease assays ....................................... 253–254, 258–260, 266–268, 270, 275, 279–282
Nuclease P1 ...................................................... 183, 294–304

### O

Oligonucleotide radiolabelling ......................................... 179

### P

PCR amplification of CRISPR loci ................................ 211
Periodate oxidation/base elimination..................... 39–42, 44
Phage
  annotation............................................................... 310
  resistance .......................................... 196, 206, 216–218
  resistance assays ...................................................... 217
  sensitivity test (assay)........................................ 156–158
PIM production ...................................... 197–201, 203–204, 209–210, 213
Plasmid interfering mutants (PIMs) .............. 198–203, 210, 212, 216
Plasmid transformation .................... 161–169, 203, 210, 220
PNK treatment................................................. 4, 7, 9, 11–12

Polyacrylamide gel electrophoresis (PAGE)
  EMSA .......................................... 173–174, 176–180
Positive selection.................................... 233–248, 344
Potassium permanganate .................................. 295
Pre-crRNA cleavage ........................................ 7, 35–45, 66
Preparation of Cas9 mRNA ........................ 321, 326–327
Preparation of phage genomic DNA ............................ 154
Primed adaptation assay ........................ 151–152, 155–158
Primed (escape) protospacer ................... 149, 151, 154–155
Priming................................................ 2, 149–154, 156, 262
Protein co-refolding .................................................... 30
Protein–DNA interaction........................................ 180, 183
Protein purification .................................... 187, 252, 257–258
Protospacer adjacent motif (PAM)...................... 55, 58, 63,
  65, 78, 83–86, 149, 150, 152, 153, 157, 158, 161, 163,
  168, 202, 278, 280, 282, 288, 290, 295, 300, 302, 318,
  323, 331, 336, 339, 342, 350, 353, 354
  identification................................................ 278
PyrEF selection ................................................ 190

## R

RAMPs ............................ 50, 51, 53–55, 60, 61, 64–67, 341
Reverse hybridization ............................................ 122
Ribonucleoprotein complexes.............................. 15–17, 23,
  133, 171–184, 279
R-loop ................................ 63, 65, 182, 278, 282, 293–304
RNA
  isolation .................................................. 9, 16, 360
  processing ........................................................ 15
RNA-guided nucleases (RGNs)............................... 317–332
RNase mapping........................................................ 134, 138
RNA-sequencing (RNA-seq)........................ 1–18, 255, 264
RNase T1 mapping ................................................ 39, 41, 42

## S

*Salmonella enterica*............................................ 7, 11, 91–108
Sequence-specific gene repression .................................... 352
Sequencing gel............................................ 256, 262, 297
sgRNA
  cloning.......................................... 351, 354–357
  design ........................................................ 350–354
  expression ........................................ 350, 351, 354, 357–359
  production ........................................................ 354
Signature Cas proteins............................................ 49, 56, 277
Single-stranded oligodeoxynucleotides
  (ssODNs) .................................................. 317–319
SITVIT2 .................................................. 113
SITVITWEB.................................................. 111, 123
Small subunits .............................................. 50, 51, 55–57,
  60, 61, 64, 65, 67
Solubilisation of inclusion bodies .................... 24–25, 27–28
Spacer
  acquisition........................ 61, 69, 148, 195–220,
  223–231, 234, 266
  identification.................................................. 309
  mapping.................................................. 42, 314
  targets .................................................. 149
Spoligotyping ........................ 91, 93, 97–99, 102–106,
  108, 111–115, 117, 118, 120, 121, 123–129
Spoltools.................................................. 123, 129
*Streptococcus pyogenes* ........................ 5, 6, 14, 318, 336,
  349, 350, 353
Streptavidin tag .................................................. 186
*Streptococcus thermophilus*........................ 78, 79, 92,
  196–199, 202–213, 215–219, 278, 279
  phages .................................................. 204–205, 208
Structural probing .................................................. 296
Subtypes .................................. 7, 24, 35, 49–62, 64–67,
  71, 94, 196, 233, 235, 265
Sulfolobales .................................................. 223–231
*Sulfolobus*
  *S. islandicus*.................................................. 224, 230, 231
  *S. solfataricus* .................................................. 5, 12, 16, 63, 67,
  185–194, 224, 227, 229–231, 252, 266
Surveillance/effector complexes............................ 48, 50, 51,
  53, 56, 57, 60, 62–68

## T

Tandem mass spectrometry .................................. 17, 143
Tandem tags .................................................. 186, 188, 194
Target interference motif (TIM) ............................ 161–169
T7 endonuclease I (T7E1) assay ............. 321–322, 328–329
Terminator exonuclease (TEX) .................. 3–11, 13, 15, 17
Tetranucleotide analysis.................................................. 311–314
*Thermoproteus tenax* .................................................. 24, 28, 29
*Thermus thermophilus* .................................. 4, 5, 12, 36, 43,
  63, 66, 266, 274, 288
Tobacco acid pyrophosphatase (TAP)
  treatment .................................................. 9, 12–13, 17
T4 phage .................................................. 69, 151, 157
Transformation efficiency.............................. 165, 168, 169,
  235, 238–239, 244
Transposon insertion site PCR............................ 236–237,
  242–243, 246
Transposon mutagenesis........................ 234, 236, 239–240
Type
  I.................................................. 4, 7, 35–38, 42, 48, 53–56,
  60–68, 85, 134, 171, 202, 212, 223, 277
  I-A.................................................. 62, 67
  I-C.................................................. 38, 56, 67
  I CRISPR-Cas .................. 53–56, 60, 67, 85, 134, 278
  I-E.................................................. 7, 65, 77, 86, 235, 300, 303
  II.................................................. 48, 52, 56, 58–60, 62,
  65, 171, 202, 277
  II CRISPR-Cas.................................. 6, 15, 17, 55–59, 63, 212
  III-B .................................................. 60, 185–194
  III CRISPR-Cas .................................. 59–60, 65, 67, 69
  IV CRISPR-Cas .................................................. 60–61

## V

Viral
- diversity ... 307
- genome assembly ... 307
- shuttle vector ... 186

Virome ... 307–315
Virus-like particles ... 225–228, 230
Virus purification ... 231

## Y

YgbT ... 259–261, 263
*ygcF* gene ... 94

## Z

Zebrafish ... 317–332, 345
Zebrafish embryo injection ... 321, 327–329

Printed by Printforce, the Netherlands